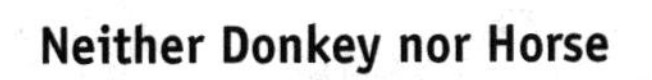

Neither Donkey nor Horse

Studies of the Weatherhead East Asian Institute, Columbia University

The Studies of the Weatherhead East Asian Institute of Columbia University were inaugurated in 1962 to bring to a wider public the results of significant new research on modern and contemporary East Asia.

Neither Donkey nor Horse: Medicine in the Struggle over China's Modernity

Sean Hsiang-lin Lei

The University of Chicago Press :: Chicago and London

The University of Chicago Press, Chicago 60637
The University of Chicago Press, Ltd., London

Paperback edition 2016
Printed in the United States of America

25 24 23 22 21 20 19 18 17 16   2 3 4 5 6

ISBN-13: 978-0-226-16988-0 (cloth)
ISBN-13: 978-0-226-37940-1 (paper)
ISBN-13: 978-0-226-16991-0 (e-book)
DOI: 10.7208/chicago/9780226169910.001.0001

Published with the support of the Chiang Ching-kuo Foundation for International Scholarly Exchange.

Library of Congress Cataloging-in-Publication Data

Lei, Xianglin, author.
Neither donkey nor horse : medicine in the struggle over China's modernity / Sean Hsiang-lin Lei.
pages cm.—(Studies of the Weatherhead East Asian Institute, Columbia University)
Includes bibliographical references and index.
ISBN 978-0-226-16988-0 (cloth : alk. paper)—ISBN 978-0-226-16991-0 (e-book) 1. Medicine—China—History—20th century. I. Title. II. Series: Studies of the Weatherhead East Asian Institute, Columbia University.
R601.L45 2014
610.951—dc23

2014001559

∞ This paper meets the requirements of ANSI/NISO Z39.48-1992 (Permanence of Paper).

To my mother, Ni Mei-an, and in memory of my father, Lei Bin-sheng

Contents

1 Introduction

During the last months of his life, Sun Yat-sen (1866–1925), the founding father of the Republic of China and a practitioner of Western medicine, was forced to take a personal stand on the issue of traditional Chinese medicine. Faced with the possibility of losing his life to liver cancer, his deliberation on and ultimate acceptance of treatment with Chinese medicine became a highly symbolic event, followed with great interest by countless observers from early January to his death on March 12, 1925. If Sun had refused to accept Chinese medicine even in this most vulnerable and hopeless situation, his decision would have been interpreted as a testimony to his steadfast belief in modernity, which he personified. According to the account provided by Lu Xun (1881–1936), one of the fathers of modern Chinese literature and once a student of biomedicine himself, the dying Sun in the end reasoned this way: "While some Chinese drugs might be effective, knowledge of a [biomedical] diagnosis is lacking [in Chinese medicine]. How can one take drugs without a trust-worthy diagnosis? There is no need to take them."[1] After hearing that Sun had consistently refused to take Chinese medicine (which turned out not to be true in the end),[2] Lu could not contain his feelings and stated that Sun's decision "moved me so much that it is no less [important] than his life-long commitment to revolution."[3] As revealed by Lu's comment, to many of Sun's comrades and progressive intellectuals of the times, the very act of taking Chinese drugs amounted to a public betrayal of the notion of modernity.

An equally symbolic event, albeit with totally different meanings, took place at the same Hospital of Peking Union Medical College nearly half a century later. In 1971, the *New York Times* journalist James Reston (1909–95), who had traveled to China as part of an advance team before President Nixon's historic visit, underwent an emergency appendectomy there. As ordered by Premier Zhou Enlai (1898–1976), a team of leading medical specialists cooperated in the management of his case. The surgery went well, but Reston suffered serious postsurgical pain. After that pain was alleviated with acupuncture, he published a report in the *Times* on both his personal experience and his observations of the effectiveness of acupuncture on other patients. For many Westerners, this was the first time they had heard about acupuncture. Reston's groundbreaking report was later credited for "help[ing to] open the doors in this country [United States] to an exploration of alternative medicine."[4] Reston was keenly aware of the historical irony of the event he had helped to create. The hospital where he had received surgery had been established by the Rockefeller Foundation in 1916 to serve as an elite model to instill "the scientific spirit" in Chinese minds,[5] but Reston's report noted that "like everything else in China these days, it is on its way toward some different combination of the very old and very new."[6] While Chinese medicine had been seen as the antithesis of modernity at Sun's deathbed, fifty years later, in the same hospital, it had become a message to the world that China had developed a very different kind of medicine and, by implication, a very different kind of modernity.

Progressive intellectuals in the early twentieth century would undoubtedly have found it appalling that the view of Chinese medicine had been transformed in these fifty years from a burdensome tradition into an inspiration for the acceptance of alternative medicine around the world. They would have been even more puzzled had they known that during the same period China had dramatically increased access to modern health-care services for its citizens. By the time that Reston published his report, China had developed a primary health-care system that, in "achieving 90% coverage of a vast population, was the envy of the world," according to Dr. Margaret Chan, director-general of the World Health Organization.[7] Thus it was precisely in the process of the global expansion of modern (i.e., "Western") medicine into China that traditional Chinese medicine went through its historic transformation. Against the background of this surprising and puzzling history, I attempt in this book to answer an apparently simple question: How was Chinese medicine transformed from the antithesis of modernity to one of the most potent symbols for China's exploration of its own modernity?

When Chinese Medicine Encountered the State

It is a great irony that the turning point in the modern history of Chinese medicine was an event that was meant to put an end to it. In 1928, amid civil war, social unrest, and foreign occupation, the Nationalist Party (*guomindang*, also known as KMT) finally ended the political chaos of the Warlord period (1916–28) and formed a new government for China. Even though the Nationalist Party controlled only certain regions of the country, the regime nevertheless dedicated itself to the project of state building, establishing the Ministry of Health in the new capital of Nanjing. With the exception of the Song dynasty (960–1279),[8] this was the first time that China had a national administrative center to take charge of all issues related to health care. In the next year, the first National Conference on Public Health, which was dominated by practitioners of Western medicine, unanimously passed a proposal to abolish the practice of Chinese medicine. To the great surprise of almost everyone concerned, this resolution mobilized the previously unorganized practitioners of Chinese medicine into a massive National Medicine Movement (*guoyi yundong*), formally instigating a decade-long collective struggle between the two factions of medical practitioners.

Half a century later, the widely respected pioneer of public health in China, Chen Zhiqian (also known as C. C. Chen, 1903–2000), reflected on this historical confrontation as follows:

> In the 1920s, modern physicians, including Chinese nationals, inadvertently delayed the diffusion of scientific medicine, probably by many decades, through their demands for the abolition of traditional medicine. Fear generated by their actions caused a powerful coterie of traditional scholar-physicians in the cities to organize for collective action and to seek the intervention of high officials on their behalf. Respected by officials and the public alike, the scholar-physicians were able not only to defend

> what they already had but even to expand their influence. More than fifty years later, the two systems of medicine stood on equal footing in China, each with its own schools, treatment facilities, and highly placed friends in the bureaucracy.[9]

In short, from the viewpoint of biomedical practitioners such as C. C. Chen, an unforgivable miscalculation was made in proposing to abolish Chinese medicine in the spring of 1929. This miscalculation not only delayed the "diffusion" of Western medicine by decades, but it also gave birth to what we now know as a bifurcated medical field in China.

Presupposing the global diffusion of scientific medicine, Chen assumed that what he had witnessed in the 1920s was just a "delay," a local suspension of the necessary triumph of biomedicine and of the unavoidable extinction of an indigenous medical tradition. Although Chen made the comment on this event quoted above in the late 1980s, like so many others among China's modernizers, he still regarded the ultimate replacement of local medical traditions by scientific biomedicine as merely a matter of time. Nevertheless, the practice of traditional Chinese medicine is today legalized by the governments on both sides of the Taiwan Strait and has gradually spread into virtually all other parts of the world.[10] The World Health Organization considers Chinese medicine a seminal field in the area of alternative and complementary medicine, and it is increasingly accepted into mainstream health-care services in nations around the world.[11] Instead of treating this history as simply an unfortunate delay or transition stage in the inevitable progression toward scientific medicine, I intend to examine the modern history of Chinese medicine as an essential part of the Chinese exploration of modernity. For this purpose, the following pages explore the events through which traditional Chinese medicine crossed the threshold to modernity—politically, institutionally, and epistemologically. As I explain below, this was also the point in time (as noted in the heading for this section) when Chinese medicine encountered the state.

First of all, my subject matter is traditionally considered to have been shaped by a cultural confrontation between two knowledge systems: scientific Western medicine and premodern Chinese medicine. This confrontation is generally taken as an unavoidable local event in the global spread of modern science and technology around the world. Nevertheless, by the 1920s practitioners of the two styles of medicine had already coexisted for decades in China without directly competing against each other. I argue, therefore, that the struggle between these two medical factions would never have taken place, or at least would have taken a very different form, if the state had not intervened to abolish the practice of Chinese medicine in the late 1920s. In this sense, the historic confrontation did not take place directly between the two styles of medicine but between Chinese medicine and the modernizing Chinese state.

Second, the two historic events that took place in 1929—the declaration of the government's intention to abolish Chinese medicine and the rise of the National Medicine Movement—fundamentally transformed the logic of competition between Chinese and Western medicine in China. Instead of competing for individual patients, as they might have done before these events, practitioners of the two styles of medicine now occupied themselves with competing for alliance with the state. More importantly, these historic events provoked traditional practitioners to pursue actively and collectively the series of professional interests, institutional infrastructure, and governmental recognition that the state had just started to grant to Western medicine. In order to pursue the interests created and offered by the state, practitioners and advocates of Chinese medicine dedicated themselves to reforming Chinese medicine, thereby adapting it to the modernizing agenda of the state. To highlight their new vision, practitioners of Chinese medicine decided to call their style of medicine "national medicine" (*guoyi*). This name illustrates how the proponents of Chinese medicine, when forced to cope with the threat from the state, responded by striving to link the future of their profession closely to that of the state.

Third, I demonstrate that the 1929 confrontation also constituted an epistemological event for the remaking of traditional Chinese medicine. In addition to coping with the political challenges posed by the arrival of biomedical knowledge, Chinese medicine was confronted with epistemic violence when its leaders started to embrace what I characterize as discourses of modernity and committed themselves to reforming Chinese medicine on the basis of these discourses. Because these globally circulated discourses were designed to demarcate a divide between the modern and the premodern, reform-minded traditional practitioners encountered critical challenges in their efforts to reform Chinese medicine on the basis of these new conditions of knowledge. As their endeavors radically transformed the theories, practices, pedagogy, and social network of Chinese medicine, they paved the way for the full-scale creation of the standardized, textbook-based system of Traditional Chinese Medicine (TCM) that emerged in Communist China in the mid-1950s.[12] In this sense, the creation of modern Chinese medicine began at the moment when Chinese medicine of the 1920s encountered the state.

Beyond the Dual History of Tradition and Modernity

With a few notable exceptions, the majority of scholarly works on the history of medicine during the Republican period (1911–49) can be sharply divided into two largely independent categories:[13] histories of biomedicine in China and histories of traditional Chinese medicine.[14] Although no scholar has ever stated this explicitly, and most scholars might not be fully conscious of this fact, the rigorous divide between these two types of histories implies that Chinese medicine was of little relevance to the development of modern health care in China. It is certainly true that historians have been keen to document how modern institutions, values, and knowledge influenced the trajectory of Chinese medicine on many fronts. Much less, however, is known about how the advocates and practitioners of traditional medicine influenced either the introduction of Western medicine into China or the construction of public health and medical administration. Clearly, the historiography of medicine has reproduced a binary opposition between tradition and modernity, thereby preventing us from viewing Chinese medicine as a constitutive part of China's medical modernity.

This division of intellectual labor is a great pity, because what has made Chinese medicine unique is precisely its contested and ambiguous relationship with China's modernity. As Benjamin Elman concludes in his monumental study on science in China, among all the scientific fields of study in Imperial China, Chinese medicine was the only traditional discipline that "survive[d] the impact of modern science between 1850 and 1920."[15] The word *survive* serves as a reminder that the arrival of modern science and biomedicine in China, as in almost every other part of the non-Western world, often meant the mass marginalization—if not extinction—of traditional knowledge. No matter how hard historians of modern science have striven to go beyond the linear, teleological framework of the Enlightenment, they almost inevitably end up with the familiar story that traditional practices were pushed to the periphery, if not replaced completely by modern science. Against the unfortunate fate of either extinction or marginalization that has been common to almost all traditional practices around the globe, Chinese medicine stands out as a unique case that not only survived the attack of science and modernity but also flourished and became accepted into state-sanctioned public knowledge.[16]

Unfortunately, many people continue to view Chinese medicine as a "survivor" or remnant of premodern China. Although scholars like Elman have been careful to emphasize that the entity that survived the attack of science is "a modernized version of Chinese medicine,"[17] people often assume without much reflection that Chinese medicine has survived as an anachronism in the fragmentary enclaves of a modernized China. If we view the modern history of Chinese medicine merely as a history of the endeavor to preserve and even "modernize" Chinese medicine—done only for the sake of Chinese medicine itself and relevant only in such enclaves—this history would not have much to do with the development of modern mainstream health care in China. Conversely, if the contested history of Chinese medicine is purged from the history of medicine proper, the resultant "history of modern medicine in China" looks conveniently close to what medical historian Warwick Anderson has criticized as "local variations of a master narrative called 'the development of modern medicine.'"[18] As Anderson points out, such a historiography of medicine merely reflects and reinscribes the general idea that the modern history of China, just like that of many other non-Western nations, "tends to become variations on a master narrative that could be called 'the history of Europe.'"[19] That is, it suggests that every major aspect of modernity developed first in Europe and then everywhere else.

If we recognize that the modern history of Chinese medicine is an integral part of China's history of modern health care, this history will have the potential to challenge the above-mentioned Eurocentrism and historicism that Dipesh Chakrabarty criticizes in his influential book *Provincializing Europe: Postcolonial Thought and Historical Difference*. Although I emphasize this critical potential of the history of Chinese medicine, I certainly do not mean to romanticize the "surviving tradition" of Chinese medicine as a radical alternative to biomedicine. The romanticization of Chinese medicine presupposes, and thereby reinforces, a clear-cut separation between the modern history of traditional Chinese medicine and that of biomedicine in China—the former being what survives of a traditional discipline and the latter being an importation to China from modern Europe. I argue that neither of these two characterizations is historically true. As scholars, we must therefore go beyond the conventional framework that treats this complex and interwoven history as two separate historical processes: the survival of traditional medicine on the one hand and the development of biomedicine on the other.

Toward a Coevolutionary History

While it is important to recognize the unique status of Chinese medicine as a survivor of science and modernity, the concept of speciation is more appropriate for writing the modern history of Chinese medicine. First of all, what the advocates of Chinese medicine strove for during this period was neither the preservation of Chinese medicine nor even the so-called modernization of Chinese medicine.[20] As many leading figures of the National Medicine Movement explicitly pointed out, they strove to create a "new Chinese medicine" (*xinzhongyi*), namely, a recognizably new species of Chinese medicine. The metaphor of biological evolution can help us to appreciate the crucial difference between the survival of Chinese medicine and the speciation of what they called National Medicine. While the modernization of Chinese medicine can be likened to the transformation of a group of organisms (i.e., Chinese medicine), the vision of a National Medicine actually involved a dual transformation of both this group of organisms and of its new niche, that is, the emerging nation-state of China. The term *survivor* fails to capture the speciation process of a new Chinese medicine that coevolved with its new niche in the environment of the modern Chinese state.

In order to go beyond the dual history of the "survival of traditional medicine" and the "development of modern medicine," I have foregrounded the reciprocal interactions among Chinese medicine, Western medicine, and the state. To highlight these otherwise obscure interactions, I have synthesized into a single historical narrative what has previously been separated into three independent histories: the history of Western medicine in China, the history of Chinese medicine, and the political history of the state. This unconventional research design keeps me from doing full justice to any one of these three histories; nevertheless, it does reveal the surprising alliances among historical entities that were considered to be geographically separated, categorically distinct, and therefore practically remote from each other.

For one concrete but simplified example, consider the control of the Manchurian plague in 1911, whose history I present in chapter 2. It was widely considered to be a landmark event in the history of Western medicine in China, because the control of the Manchurian plague demonstrated the political and indispensable role of Western medicine in state building. In the process of confining the plague, practitioners of Western medicine successfully defended China's territorial sovereignty against encroachment from Japan and Russia. This landmark success of Western medicine depended on the combined forces of a new geopolitical power (resulting from the sovereignty crisis) and a new knowledge (the germ theory of the plague). To understand why Chinese medicine revealed itself as irrevocably inferior to Western medicine in this event, we have to take into account the unprecedented alliance between sovereignty and the microscope (see chapter 2).

In addition to tracing the reciprocal interactions and surprising alliances among these three histories, there are practical reasons to include them all in this study. In terms of narrating the modern history of Chinese medicine, the emergence of the nation-state and of Western medicine in Republican China were simply too radical, too profound, and too directly related to our topic to be treated as mere historical background. After the suppression of the Manchurian plague, the advocates of Western medicine came to the ironic conclusion that the scientific truth of biomedicine could not diffuse in China on its own but demanded as its vehicle the modern state and its accompanying institutions. In order to implement this newly developed strategy of "popularizing Western medicine by means of the state" (see chapter 3), they endeavored to turn Western medicine into a tool for state building. In response, the struggling practitioners of Chinese medicine soon adopted a similar goal as their crucial challenge. Striving to link Chinese medicine to the modern nation-state, they started the historic process of politicizing Chinese medicine. In other words, the rise of National Medicine as a collective vision for Chinese medicine was a direct response to a recently developed local feature of Western medicine in China.

As a result of the politicization of both styles of medicine, the state became the undisputed driving force behind the history of medicine in modern China. In practice, the relationship between the state and medicine went through a profound transformation during this period. Although the late Qing state had shown little concern for medical and public health matters until the outbreak of the Manchurian plague in 1910, by 1928 the Nationalist government had committed itself to building an independent Ministry of Health—only one decade after Britain had established the world's first Ministry of Health in 1919. In 1947, the Nationalist government included in its first constitution a national policy of State Medicine (*gongyi*), an unusual form of national health system staffed, sponsored, and controlled completely by the government. Within less than forty years, China thus went from being regarded as "one of the last of the world's communities to undertake the utilization of scientific medicine"[21] to having a government with a firm commitment to providing modern health-care services to all of its citizens, foreshadowing the further development of the world-famous primary health-care system under the Communist government. This dramatic transformation of the state radically altered the social space for both styles of medicine; simultaneously, though, the interaction of the government with the two styles of medicine also resulted in some important features for the Chinese state. In this sense, the present book is intended both as a political history of medicine and as a medical history of the Chinese state.

To summarize, it is important to assimilate the history of Chinese medicine into the general history of modern medicine in China because the crucial features of modernity were not produced in any one of three insulated, categorically pure historical processes, but in the conjunction of three interrelated histories. By considering these three histories as moving targets, rapidly influencing and being influenced by each other, this book intends to capture the reciprocal relationships and coevolutionary processes among these three histories. From this viewpoint, the unique "survival" of so-called traditional Chinese medicine has been influenced greatly by the knowledge, discourses, and institutions that originated in Europe and Japan; conversely, what makes modern health care in China a unique contribution to global medicine is precisely its incorporation of these newly transformed indigenous health practices. Instead of reproducing the dichotomy between tradition and modernity, this coevolutionary history reveals the intensive interplay between them in a concrete historical process, thereby elaborating the features of manifested modernity in China.

China's Modernity

So far, my discussion has focused on the policy aspect of the struggle between these two styles of medicine. Nevertheless, as Ralph Croizier cogently pointed out in his pioneering work published almost four decades ago, the debate over Chinese medicine concerned much more than medical policy.[22] Contextualizing this struggle within the cultural milieu of the May Fourth Movement, Croizier convincingly explained why this controversy over medical policy drew in so many passionate participants from outside the medical circle, including the most prominent intellectuals and political leaders of that time. In his view, the controversy over Chinese medicine was a manifestation of a larger conflict, and a crucial paradox, between two modernist ideological forces in China, namely, cultural nationalism and scientism. Building upon his ideological analysis of this medical struggle, I argue that there were actually two separate but related struggles—a policy struggle over the role of Chinese medicine in the national health-care system and an ideological struggle over the nature of China's modernity.

Personified as "Mr. Science" (*sai xiansheng*), scientism became a dominant force in modern Chinese thinking in the context of the May Fourth Movement in 1918. Its influence peaked in the debate between science and metaphysics in 1923, which addressed the question of whether one's "philosophy of life" could be determined in a scientific fashion.[23] As this controversy reveals, progressive intellectuals strove to turn science into the representative of modernity so that the directions of various rampant changes of China could all be determined with confidence by the so-called scientific method. Because of this historical context, science was not considered merely a specialized way of knowing about and controlling the natural world; it was also used as an ideological authority to excoriate the Confucian religion and to direct social and cultural changes. Paradoxically, as intellectual historian Wang Hui has pointed out, the leaders of the May Fourth Movement did not advocate science as the antithesis of Confucian religion. Instead, they strove to promote a "belief" (*xinyang*) in science and advocated the "replacement of religion by science."[24]

They thus endeavored to introduce to China a monotheistic framework that was to be implemented in various aspects of culture and daily life, including political ideology, historical development, love affairs, and the doctor-patient relationship.[25] Instead of attacking the religious concept of belief, science was supposed to serve as the defining example of what constituted a monotheistic belief.

This ideological function of science was by no means a Chinese phenomenon. On the basis of his study of British colonialism in India, political historian David Arnold has emphasized the crucial role that science played in giving shape and authority to the understanding of modernity.[26] In my opinion, his conception of "science as modernity" is more appropriate than the concept of scientism for understanding why Chinese intellectuals treated the debate over Chinese medicine as a life-or-death struggle over China's modernity. In the context of Chinese medicine, the modernist debate centered not on scientism in the sense of whether the scientific method could be applied to realms apparently *outside* of science, such as the "philosophy of life," but on whether modern science could claim a monopoly on cultural authority *within* the realms presumably belonging to the natural sciences, such as medicine. While both of these debates were concerned with the cultural authority of science, they represented two extreme poles of a spectrum. When viewed against this background—progressive intellectuals fighting aggressively to maximize the authority of science in a variety of cultural realms—it becomes understandable why they appeared so angry and anxious when they were forced to defend the authority of science in the struggle over Chinese medicine. Fu Sinian (1896–1950), one of the leaders of the May Fourth Movement, made the following strong statement in 1934:

"The most shameful, the most resentful, and the most saddening aspect of contemporary China is the so-called struggle between Chinese and Western medicine."[27] Fu was right to feel that way, because what was at stake was much more consequential than the scientism debate over "philosophy of life." As the struggle over Chinese medicine called into question the very foundation of science as modernity in China, it also constituted the most dangerous and unacceptable challenge to the universalist conception of modernity for the world beyond.

In the eyes of these participants, the central question of this struggle was whether there was any room at all for presumably premodern, nonscientific Chinese medicine in the project of defining China's modernity. The answer to this question would determine whether China's modernity would have to be a faithful copy of the universalist modernity. Nowadays, some scholars of modern China tend to use the term *modernity* in a rather loose fashion, identifying as modernity almost any aspect of the historical processes that took place during the nineteenth and twentieth centuries. The unfortunate result of such historiography is that we end up losing sight of both its epistemic violence and the painful confinement that historical actors faced in their efforts to cope with the universalist conception of modernity. As Arnold has summarized insightfully in the Indian context, "acceptance of modernity as partisanly presented by the colonial officials, missionaries, educationalists and scientists would always confine Indians to a state of tutelage and subordination, always leave them one step behind, second-best and imperfect copies of a Western ideal."[28] In order to understand this difficult position, scholars need to take seriously the normative as well as confining effect of modernity when conceived in terms of science.

To begin with, science has been considered the most objective and universal knowledge that humans could ever possess. Other aspects of modernity may have been open to negotiation and local modification, but it was much more difficult to consider meddling with science, whose universality was guaranteed by "nature," which presumably stood outside history and culture. Once conflated with nature, science thus conceived was able to serve as the most solid foundation for a culture-free, universalist conception of modernity.[29] To resist this modernity, or to propose its local modification, amounted to embracing obscurantism, if not irrationality and superstition (*mixin*).[30]

The ideological struggle over Chinese medicine was deeply embedded within this framework of "science as modernity." As revealed in the comment by Lu Xun with which I opened this introduction, in the early twentieth century, Chinese medicine was considered one of the most salient opposites of science. It was in this ideological context that the act of summoning a traditional physician to Dr. Sun Yat-sen's deathbed became such a high-profile controversy.[31] If Dr. Sun, the most admired persona of modernity in China, openly embraced Chinese medicine, the perceived opposite of science, it would have spelled a "split personality" for Mr. Science and thereby the bankruptcy of science as modernity.

While the struggle over Chinese medicine indeed took place within the larger context of ideological conflict, and the two struggles thus became entangled, the struggle over medical policy should not be equated with, or reduced to, the ideological struggle over the concept of modernity. I emphasize the crucial importance of differentiating these two conflicts and yet keeping track of both within one study, because the ideological struggle over China's modernity did not take place in a purified discursive or conceptual space. Equally relevant were the practical context of resolving China's health-care problem and the sociotechnical processes of transforming, or "scientizing," traditional Chinese medicine. If one reduces the struggle over Chinese medicine to a purely ideological struggle over China's modernity, the multilayered attempts to transform Chinese medicine—institutional, epistemological, and technical—fall by the wayside. While Croizier brilliantly reconstructed the larger ideological context for the struggle over Chinese medicine, this book is devoted to contextualization in the reverse fashion—that is, to reconstructing the medical and sociotechnical contexts for the purpose of understanding the ideological struggle over China's modernity.

It is important to situate the ideological struggle over China's modernity within these medical and practical contexts. In a nutshell, modernity is not "a thing in itself,"[32] let alone a fixed entity insulated from real-life concerns and sociotechnical developments in specific local settings. In order to investigate the nature of China's modernity with regard to the issue of Chinese medicine, we must suspend the abstract concept of a universalist modernity. Instead, we need to consider the possibility that a radically transformed modern Chinese medicine might have contributed to and even given shape to the modernity that came to be historically realized in China. For the sake of exploring this possibility, I examine the specific processes through which advocates of Chinese medicine embraced and negotiated the discourses of modernity and reassembled Chinese medicine. In this sense, this book is devoted to tracing the coproduction of modern Chinese medicine and China's modernity.[33]

The Discourse of Modernity

Throughout the first half of the twentieth century, Chinese medicine encountered numerous new ideas, artifacts, people, and institutions that had not existed in China just decades earlier: the microscope, the steam engine, *Gray's Anatomy*, the germ theories of disease, the modern hospital, social surveying, the Ministry of Health, the Rockefeller Foundation, and professionalism, to cite just a few examples that are discussed in the following chapters. Although practitioners of Chinese medicine felt shocked, threatened, and sometimes enthralled by these various aspects of the modern world, none of these innovations—with the exception of the modern state—was as crucial as what I characterize as the "discourse of modernity" in shaping the modern history of Chinese medicine. Because the discourse of modernity was instrumental in turning Chinese medicine and science into incompatible opposites, it constituted the most daunting challenge for the advocates and reformers of Chinese medicine.

Central to the discourse of modernity was what philosopher of science Ian Hacking has characterized as the representationist conception of reality.[34] As a philosophy of science, it suggests that the objective of science is to create representations of the natural world, which is assumed to be unique, eternal, and transcultural—that is, natural and thus universal. Moreover, the act of making representations should imitate the act of unmediated seeing, as if scientists were doing nothing but holding up a mirror in front of nature, without disturbing in any way the reality to be represented. From this point of view, the main function of theoretical concepts is to represent the material entities that already exist in the real and eternal world. Because this conception of reality makes it possible to conflate science with nature, it provides the unquestionable basis for science to serve as the guardian of modernity. Although this philosophy would now have few supporters among researchers in science and technology studies, it persists as the commonsensical view of reality for both the general public and most practicing scientists. More importantly, historians have yet to understand the historical influence of this philosophy in shaping the modern history of China, not to mention the full consequences of abandoning it.

The embrace of this representationist conception of reality fundamentally transformed the relationship between the two competing styles of medicine. First of all, although none of the historical participants in the struggle over Chinese medicine ever mentioned this commonsensical philosophy of science explicitly, their arguments put into practice this theory of scientific knowledge, which was rather novel to many of their contemporaries. As the following chapters show, when critics of Chinese medicine accused it of being antiscientific—its cardinal sin and the ultimate rationale for its abolition—they meant to say that its theories failed to represent the known natural world as it was infallibly represented by science and biomedicine. As the result of this accusation, ontology thereby came to be valorized as the key to deciding the fate of Chinese medicine and China's health-care policies. Moreover, by conflating science with nature and reality, this theory of science turned the competition between "scientific medicine" and Chinese medicine into a zero-sum game. From now on, Chinese medicine had formally entered the strange ontological space of modernity, a space that was deemed too crowded to allow Chinese medicine to coexist with biomedicine.

It is remarkable how little we know about the historical processes through which Chinese people came to embrace this fundamental framework about knowledge and reality. Scholars of modern history have been so focused on tracing the eye-catching changes in the intellectual landscape that it has become easy to miss the restructuring of the conditions of knowledge underlying it. Because the modern history of Chinese medicine involved an intensive public debate about the project of "scientizing Chinese medicine," a debate concerned with both the nature of science and the controversial idea of "scientizing" its salient Other, this history offers a rare perspective from which to examine how the Chinese people came to embrace and later to negotiate these new conditions of knowledge.

Neither Donkey nor Horse

The key to answering the question posed in the beginning of this introduction lies in the book's title, *Neither Donkey nor Horse* (*feilü feima*). To summarize the answer, the key development in this historic transformation was the rise of a new "species" of Chinese medicine that was seen by its detractors as "neither donkey nor horse," a new medicine that demonstrated in concrete terms that the relationship between Chinese medicine and modernity was not to be antithetical. Although I feel slightly apprehensive about choosing a phrase that was meant to be derogatory as the title of this book on the modern history of Chinese medicine, I believe that this Chinese idiom correctly captures the historic challenge that the reformers of Chinese medicine took on—the challenge to realize a project that was assumed to be unimaginable and doomed to fail.

It may be useful to first spell out the origin and connotations of this expression, because there is no direct counterpart for this Chinese idiom in English.[35] Its earliest use is found in the Book of the Han Dynasty (*Hanshu*) from about two thousand years ago, which concerns the King of Qiuci, a Central Asian country in contemporary Xinjiang. This king admired the Han culture so much that he ordered his court to construct a Han-style palace, to wear Han-style clothes, and to adopt Han-style rituals and institutions. The non-Han people in his region bitterly commented: "It looks like a donkey but is not a real donkey. It looks like a horse but is not a real horse. The King of Qiuci is nothing but a mule."[36] Criticizing the king's efforts as a double betrayal of both cultural traditions, this idiom thus became an emotionally charged expression intended to show one's strong disapproval of boundary-crossing cultural integration.

This expression is illuminating in understanding the modern history of Chinese medicine for several reasons. First of all, it registered the rise of a new kind of medical syncretism at a specific historical juncture. Instead of being a tool of analysis, like the valuable postcolonial concept of hybridity,[37] the idiom "neither donkey nor horse" was used by the actors at the time (and is intended in this book) as a reference to a new medicine in formation. It was not until the late 1920s that this idiom, which had existed in Chinese literature for roughly two thousand years, was used to criticize efforts to integrate Chinese and Western medicine. When Tang Zonghai (1851–1908), the widely acclaimed founder of the "school of converging Chinese and Western medicine" (*zhongxiyi huitong pai*), launched his medical syncretism in the 1890s, almost no one attacked him with such a strongly worded expression. In fact, Tang's positive and optimistic attitude toward cultural integration was representative of the reformist group in the late Qing period. Liang Qichao (1873–1929), a leading reformist and the most influential scholar-journalist in the late Qing, took pride in his devotion to creating "a new school of learning which would be 'neither Chinese nor Western but in fact both Chinese and Western' (*buzhong buxi jizhong jixi*)."[38] There was no sign then that such an effort of integration was either ill-conceived or doomed to fail.

As I argue in detail in chapter 6, it was only after the 1929 confrontation, especially after the establishment of the Institute of National Medicine by the Nationalist government in 1931 in order to "scientize" (*kexuehua*) Chinese medicine, that people from both sides of this medical struggle started to recognize the rise of a "mongrel medicine" (*zazhongyi*) that was "neither donkey nor horse." In other words, both the popularity of a new kind of medical syncretism (among reformers of Chinese medicine) and its derogatory characterization as a "mongrel medicine" were closely associated with the notion of science and the related discourse of modernity. In order to transform it into a state-sanctioned modern profession and a legitimate member of world medicine, reformers of Chinese medicine accepted the project of scientizing Chinese medicine and committed themselves to a path of integrating Chinese medicine with biomedicine. Risking the accusations and derision implied by the stigmatizing expression "neither donkey nor horse," many leaders of the National Medicine Movement nevertheless devoted themselves wholeheartedly to the creation of this new Chinese medicine.

In addition, this expression highlights the fact that this medical syncretism was doomed to fail because it was an artificial creation against nature. As the story about the King of Qiuci reveals, this Chinese idiom conceptualizes a cultural phenomenon in terms of a biological metaphor, comparing cross-cultural integration to cross-species reproduction. The reference to the King of Qiuci as a mule clearly draws on the biological knowledge that cross-species breeding between a female horse and a male donkey always leads to an infertile creature. Just like mules, which are incapable of reproducing, the "neither donkey nor horse" medicine, no matter how vigorous it appeared at the present moment, was seen as nothing but a monstrous creature that was destined to have no future.

During the 1930s, even reformers of Chinese medicine found it hard to imagine a future for this new Chinese medicine that they had struggled to reassemble. Citing as an example the successful integration of Confucianism and Buddhism that had taken place during the Song dynasty (960–1278), some reformers of Chinese medicine clearly aimed at a similar kind of audacious and creative integration between Chinese and foreign cultures.[39] Nevertheless, because the foreign culture involved in this case was "science," which supposedly held a monopoly on the truth about the natural world, it became impossible to imagine how one could integrate Chinese medicine with science without corrupting the latter's "truth contents." As these reformers were not able to articulate their vision for this hybrid medicine and to defend its values, they either silently endured the stigmatization or were forced to denounce the concept of a "mongrel medicine."[40] And the medical reformers who aimed at the kind of cross-cultural integration comparable to the one between Confucianism and Buddhism had to live with the fact that their enterprise had been successfully defined, not by themselves but by their critics, with an explicitly derogatory term. Because the vision for this medicine appeared utterly unthinkable and unspeakable, and also because the advocates of Chinese medicine indeed often appealed to the rhetoric of cultural nationalism, it was just too easy to fail to properly appreciate this crucial development in the modern history of Chinese medicine—that is, the rise of a new Chinese medicine that was self-consciously "neither donkey nor horse."

Even though such innovative efforts have now been going on for more than eight decades, some people still find it hard to imagine a future for this kind of "neither donkey nor horse" medicine. Even within China, there have been widely supported efforts to say "farewell to Chinese medicine and Chinese herbs" as recently as 2006.[41] Unlike other histories of modern science, the modern history of Chinese medicine remains a highly contested and therefore open-ended history. Precisely because the future of this "neither donkey nor horse" medicine continues to be hotly debated today, it is helpful to understand the process by which reformers of Chinese medicine have carved out a future for it since they began to commit themselves to the project of "scientizing Chinese medicine" in the 1930s.

Finally, I want to highlight the fact that the use of this idiom was an emotional tactic. Instead of arguing against the specific content of this medical syncretism, critics often used this idiom to remind everyone that the only appropriate way of dealing with such a self-contradictory endeavor was to laugh it off as a joke. Nowadays, when I mention the idiom "neither donkey nor horse" in public lectures, the audience invariably responds with friendly laughter. This response reveals the huge emotional distance between us and the actors in the 1930s. Since one of my objectives is to give the readers a sense of the emotional and epistemological violence that the reformers of Chinese medicine had to endure during that time, I have decided to follow my actors' use of this idiom and to render it as *mongrel medicine* in English. Only when readers have acquired a sense of this taken-for-granted disdain for mongrel medicine, a disdain perceived as so natural that there was no need to spell out its rationale, can they begin to understand the historic challenges associated with supporting such an endeavor in 1930s China, and perhaps also in many other parts of the world. The title of my book is thus intended to highlight this enterprise as a historical challenge—a challenge to realize a project that its critics had successfully defined as impossible, pathological, and self-contradictory.

Conventions

It is essential to clarify from the outset what I mean by "practitioner of Western medicine" (*xiyi*) and "practitioner of Chinese medicine" (*zhongyi*). For the sake of argument in this book, *practitioners of Western medicine* refers to Chinese nationals educated in Western-style medical schools, either in China or overseas, roughly starting from the last two decades of the nineteenth century. Those whom I call "practitioners of Western medicine" therefore do not include foreign doctors and medical missionaries. An important detail in this context is that the majority of these "practitioners of medicine," in fact, acquired their medical training either in Japan or in Japanese-influenced medical colleges in China. On the eve of the historical events of 1929, the number of practitioners of Western medicine was about two thousand.[42]

It is more difficult to define what I mean by "practitioners of Chinese medicine." By the term *Chinese medicine* I mean primarily the medicine of the scholarly elites during the Imperial period as well as the subsequent development of this medicine in the modern period. These scholarly elites were particularly important because they led the way in organizing the National Medicine Movement and subsequently transforming Chinese medicine into a modern profession. During the Imperial period, however, as Nathan Sivin has pointed out, the scholar-physicians "were not organized, did not think of themselves as a group, and could not set or enforce common standards of medical education, skill, or compensation."[43] As the result, before the government stepped in to regulate practitioners of indigenous medicine in the 1920s, the barrier to professional entry into the medical field amounted to nothing. Therefore, *practitioners of Chinese medicine* refers broadly to everyone who practiced traditional Chinese medicine and lacked formal medical training in biomedicine. Since membership identification is precisely one of the key issues at stake in this study, it would be counterproductive to provide more precise definitions of either kind of medical practitioner at this point in our investigation. With the help of a highly informative chart entitled "Medical Environment in Shanghai," drawn by a vocal critic of Chinese medicine in 1933, I analyze the complexity and heterogeneity of both Chinese medicine and Western medicine in 1930s Shanghai in detail in chapter 6. Thanks to the history documented in this book, what we now called Chinese medicine and Western medicine (in China) gradually took shape as they struggled against each other in the field of the state.

2 Sovereignty and the Microscope: The Containment of the Manchurian Plague, 1910–11

Not Believing That "This Plague Could Be Infectious"

In terms of accelerating China's acceptance of modern medicine, no event has been celebrated more than the containment of the Manchurian plague at the end of the Qing dynasty. Medical historians have described this achievement as a "watershed event in the history of modern medicine [in China]," "a lesson in the importance of epidemic control and preventive medicine," an "acknowledgement of the superiority of modern medicine [over Chinese medicine]," and a benchmark that "establish[ed] the importance of public health as a national responsibility."[1] In view of such accolades from medical historians, one thing becomes puzzling. Why did the four relatively independent trends of development suggested in these quotes—increasing exposure to Western medicine, greater control of epidemic diseases, the state's growing responsibility for public health, and the gradual acknowledgement of Chinese medicine's inferiority—have to wait for this plague to take off all at once?

At first the answer appears to be very straightforward: modern Western medicine achieved monumental success in containing this devastating plague, whereas Chinese medicine failed miserably. This explanation, however, unavoidably raises the question of why similar results did not take place earlier in other parts of China, a country at that time infamously known as the "fountainhead" of epidemics throughout the world.[2] To put the puzzle in the form of a more specific and useful question: What was so special about the history of the Manchurian plague that it could bring about the concurrence of multiple breakthroughs in these four aspects of medical history?

To the officials in charge of dealing with this tragic event, the Manchurian plague was a very special kind of epidemic. In the preface of the two-volume official report submitted to the Qing court after the event, Viceroy Xi Liang (1853–1917)— the highest-ranking official in Manchuria—reflected on the major difficulties encountered in controlling the plague. Calling the foremost obstacle "the difficulties caused by the lack of learning among the officials and physicians," Xi elaborated as follows:

> In the beginning, [we] did not believe that this plague could *chuanran* [spread by contagion or infection]; all the protective and therapeutic measures were based on the conventional ways that China used to cope with "febrile epidemics" (*wenyi*). Because practitioners of traditional Chinese medicine did not take advantage of the microscope, they had no way of sorting out the real cases of plague from those that merely resembled the plague in terms of their symptoms. As a result, whenever practitioners of Chinese medicine succeeded in curing a patient who actually suffered from common cold and fever, the stubborn and conservative society would immediately grab onto this cure and claim that epidemics could be treated easily.[3]

If one suspends the baffling assertion that "In the beginning, [we] did not believe that this plague could *chuanran* (be infectious)"—a cryptic but crucial statement that I decode below—this recollection clearly suggests that the Manchurian plague fundamentally challenged the traditional definition of febrile epidemics—from the conceptualization of *chuanran*, to antiplague measures and the clinical procedures for identifying genuine plague cases. With this in mind, the historic containment of the Manchurian plague ought to be studied as the beginning of the process of constructing, instituting, and thereby coping with a new category of disease—*chuanranbing* (infectious disease)—in China.

This radical paradigm shift from febrile epidemics to modern *chuanranbing* took place, in the opinion of many of the participants, because the handling of the Manchurian plague was a successful, closely watched "public experiment" that validated the new conception of *chuanranbing* in practice. The success of this public experiment was by no means simply a result of the new conceptualization; it had everything to do with the specific features of this plague. As the medical historian Charles Rosenberg points out in his introduction to *Framing Disease*, disease "serves as a structuring factor in social situations, as a social actor and mediator." Therefore, "each disease is invested with a unique configuration of social characteristics, and thus triggers disease-specific responses."[4] In light of this understanding, scholars have to ask how the special characteristics of the Manchurian plague, a pneumonic plague that was "the deadliest of all diseases" according to the historian Fabian Hirst,[5] contributed to the historic defeat of Chinese medicine. In order to articulate the specific features of this plague and the social responses that it triggered, I maintain throughout this chapter a comparative perspective between the Hong Kong plague (1894) and the Manchurian plague (1910).

Finally, let us turn to the most revealing point about Xi Liang's statement—his emphasis on the role of the microscope. It is surprising how little emphasis scholars have placed on the role of the germ theory in containing the Manchurian plague. This dearth of information is unfortunate. because that aspect of the story is actually part of a history whose significance goes beyond medicine in China and into the heart of what Andrew Cunningham and Perry Williams call "the laboratory revolution in medicine."[6] Between the outbreaks of the Hong Kong plague (1894) and the Manchurian plague, the scientific community transformed the identity of the plague with the aid of bacteriological research, by discovering the bacterium *Yersinia pestis* to be the "causative micro-organism of plague."[7] Dr. Wu Liande (also spelled Wu Lienteh, 1879–1960)—a Cambridge-trained physician, Malaysian Chinese, and hero of the Manchurian plague—was a key figure in the history of the creation of this new knowledge. To highlight the global significance of this local history, this chapter provides an analysis that integrates the sociopolitical context in Manchuria, the scientific controversy over new medical knowledge, and the implementation of laboratory-based antiplague measures. This integrative approach sheds some light on the reasons why the four trends of development had to wait for the Manchurian plague to take off all at once.

Pneumonic Plague versus Bubonic Plague

In October 1910, a mysterious plague broke out in Manchuria, causing great terror due to its exceptional communicability. As migrant coolies returned home by train for Chinese New Year, this epidemic spread with a speed that was unprecedented in premodern China. Originating in Manzhouli on the Chinese-Russian border, the plague traveled more than two thousand miles to the province of Fengtian within two months, threatening both the capital, Beijing, and the Chinese heartland. The unintended consequences of advancements in industrial technology soon made the three railway centers of Harbin, Changchun, and Shenyang (also known as Mukden) epicenters of the plague throughout the Manchurian hinterland. By any standard, the plague was a calamity.

From its very beginning, however, the Manchurian plague was a threat to the Qing government's already shaky sovereignty. The government did not take this plague seriously until the diplomatic corps, frightened by the plague's steady progress toward Beijing, began to exert pressure on the central government.[8] According to Carl Nathan,[9] what really provoked the Qing state to take unusual action was the concern that Japan and Russia would use plague containment as an excuse to expand their influence in Manchuria. As a result of the Russo-Japanese War (1904–5), which had ended just five years previously in Manchuria, Russia controlled the Chinese Eastern Railway, which ran through the north of Manchuria, while Japan dominated the Southern Manchuria Railway.[10] Because the Manchurian railway system was controlled by three countries with conflicting geopolitical interests, as William Summers points out, "The complexity of the railway administration, their paramount political importance, and their crucial role in transporting people infected with plague ensured that the railroads in Manchuria became the focus of plague control efforts as well as main locus of contention."[11]

The only way to resolve this sovereignty crisis was, as Nathan put it, to organize a Chinese plague service that would "preserve, as far as possible, formal Chinese control of its administration" and simultaneously "be as 'Westernized' as possible in medical practice."[12] This was not the first time that the Qing government had been forced to adopt Western public health measures for the sake of protecting its sovereignty. In order to regain sovereignty over Tianjin, the government established China's first municipal department of health in 1902, analogous to the one created by the foreign occupiers after the Boxer Uprising.[13] Perhaps having such an uneasy precedent in mind, China's vice minister of foreign affairs, Shi Zhaoji (Alfred Sao-ke Sze, 1877–1958), urgently called upon Dr. Wu Liande, his old friend, to rush to Manchuria.

In the first paragraph of Dr. Wu's autobiography, *The Plague Fighter*, the young physician recalls the chilly afternoon of December 24, when he arrived in Harbin—the center of the then-rampant plague. Even before introducing himself to the reader, the author zooms in on his own hand, in which he was carrying "a compact, medium-sized British-made Beck microscope fitted with all necessaries for bacteriological work."[14] Opening the chapter on the "black death" in this way, Dr. Wu skillfully hinted that, more than anything else, these materials were to play a pivotal role in containing this horrific plague. Dr. Wu's microscope remained untouched for three days, until he finally had an opportunity to perform a postmortem on a Japanese woman in guarded secrecy. At once he detected some organisms in this corpse that appeared to be identical with the *Bacillus pestis* that the Japanese scientist Kitasato Shibasaburo (1853–1931) and French scientist Alexandre Yersin (1863–1943) had codiscovered during the Hong Kong plague in 1894. Under his microscope, Dr. Wu further observed the novel fact that these bacilli were found exclusively in the victim's lungs. This important discovery suggested that the Manchurian plague was very different in nature from the Hong Kong plague.

The Manchurian plague, as Dr. Wu reasoned, was in fact an airborne disease; its bacilli were transmitted directly through person-to-person contact instead of only through rat fleas, as had been the case in Hong Kong's bubonic plague.[15] As a result, the pneumonic plague was considered more virulent and infectious than the insect-borne bubonic plague. Moreover, different strategies were required to control the spread of these two distinct epidemics. In order to block the transmission of pneumonic plague via human contact, public health authorities had to exercise rigorous control over human movement, while the containment of the bubonic plague had focused on controlling rats, which bore the fleas that were the primary plague carriers.

Dr. Wu immediately informed both the local officials and the officials in Beijing of his discovery. The local magistrate and chief of police were invited to look into the microscope and convince themselves of the true cause of the plague. Their collective skepticism, however, made Dr. Wu conclude that "it was not always easy to convince persons who lack the foundations of modern knowledge and the science."[16] The microscope could make visible both the plague-causing organisms and their channel of transmission, but it could have only very limited influence on people's beliefs and prior knowledge, not to mention their entrenched ways of coping with epidemic disease.

The problem of learning new knowledge was not restricted to the Chinese people; foreign doctors trained in modern methods also learned the hard way about the pneumonic nature of the plague. Most of them felt confident that they were equipped with the most advanced knowledge developed since the Hong Kong plague. Four years after Kitasato and Yersin had identified the plague bacillus in 1894, the French scientist Paul-Louis Simond further discovered that the plague was transmitted by rat fleas. The pneumonic plague was considered to be merely a derivative phenomenon, and Dr. Wu was not the first to distinguish it from the bubonic form or to describe its clinical characteristics. Pneumonic plague was initially thought to arise when a victim was infected by a bite; instead of a bubo developing, the disease spread through the bloodstream to the lungs, causing pneumonia. In previous centuries, however, pneumonic plague had been very rare, and outbreaks had remained remarkably circumscribed. According to Hirst, "It was the great pneumonic epidemic of 1910–1911 in north Manchuria which first aroused universal interests in this unfamiliar type of plague."[17]

Nothing better symbolized foreign doctors' resistance to the idea of a distinct form of pneumonic plague than the controversy over gauze masks. To protect people from direct infection via the respiratory passages, Dr. Wu designed gauze masks and demanded that both sanitary staff members and the general public wear them properly. Confident of their updated knowledge about the bubonic plague, however, many foreign doctors—including staff from Japan, Russia, and France—doubted Dr. Wu's analysis and therefore refused to wear masks, even when they were in close contact with terminally ill plague patients. According to Dr. Wu's recollection, when the French doctor Gérald Mesny, his senior colleague on the Chinese antiplague team and the head professor of Beiyang Medical College, expressed strong resistance to his discovery, Dr. Wu felt so indignant that he submitted his resignation to the Qing court. A few days later, however, the news arrived that Dr. Mesny had become infected with the plague while visiting the Russian epidemic hospital without wearing a mask. Six days later, this leading figure of the antiplague team passed away, and panic erupted in Manchuria. From this point on, people started accepting the dreadful nature of the plague, and "almost everyone in the street was seen to wear one form of mask or another."[18] The death of Dr. Mesny was only the first in a series of incidents by which certain responses to the plague unintentionally aided Dr. Wu by taking the lives of those who disagreed with him.

In contrast to the local officials, the officials in Beijing, especially Dr. Wu's old friends in the Ministry of Foreign Affairs who had sought his assistance in the first place, did not need the microscope to be convinced that it was a pneumonic plague. As the tragic death of Dr. Mesny further confirmed their trust in Dr. Wu, they were eager to turn Dr. Wu's discovery into effective antiplague measures in order to resolve the sovereignty crisis. In a telegram to Vice Minister Shi Zhaoji, Dr. Wu summarized his plan for containing the plague: "It spread almost entirely from man to man, and the question of rat infection may, for the time being, be left out, so that all efforts at suppression of the present epidemic may be concentrated upon the movements and habits of man."[19] It was very daring for a young, inexperienced physician such as Dr. Wu to suggest a suspension of all the recently established measures against rat infection.[20] Just a few weeks after Dr. Wu sent his telegram, a Japanese-owned newspaper in Manchuria published an editorial titled "Anti-Plague Administration Should Absolutely Pay Attention to Catching Rats." The editorial emphasized that Taiwan, where the plague had once been endemic, had become virtually plague-free as a result of the Japanese colonial government's rigorous implementation of regulations against rats.[21] In order to reproduce their success in Dalian, the Japanese city in Manchuria, the Japanese devoted much of their antiplague efforts to the control of rats, catching more than twenty thousand rats and mice by the end of February.[22]

Even after it became clear that almost all plague victims were infected with the pneumonic form of plague, and that no plague bacilli were found in dissected rats, Dr. Kitasato, the eminent plague authority, still insisted that the most urgent need was rodent extermination when he visited Mukden in late February. He reasoned that once the weather warmed up and the rats came out of hibernation, they would become infected by contact with plague victims, and a whole new wave of bubonic plague would join the pneumonic strain and wreak havoc across the country.[23] In contrast, he denounced the practice of wearing masks, at that time adopted by many Japanese in Dalian, as "unnecessary and exaggerated."[24] Most importantly, when Dr. Kitasato gave a lecture to the assembled staff of the Japanese consulate, he emphasized that pneumonic plague, relatively easy to keep from spreading overseas, might soon turn into bubonic plague, which might be carried all over the world by shipboard rats.[25] In short, the nature and potential transformation of the plague—which was contested at that time—influenced the Qing court's struggle to protect its sovereignty in Manchuria.

"The Most Brutal Policies Seen in Four Thousand Years"

Knowing no cure for the present plague, however, Dr. Wu focused his energies on identifying plague cases, separating patients from non-coughing suspects, devising plans for the proper detention of contacts, and teaching people to wear gauze masks properly.[26] In order to "police the sick,"[27] Wu recruited six hundred policemen to be trained in antiplague work, to replace the previously established untrained police.[28] As the plague spread south along the railroad, Wu also suggested that "all railway traffic between Manzhouli on the Siberian border and Harbin be strictly controlled."[29] As a result, the Chinese government mobilized its troops to restrict traffic on trains and to deter foot travelers from crossing the Great Wall.

Once the antiplague campaign began to focus on controlling the movement of microbe-carrying patients, the microscope became an indispensable tool because it provided the ultimate criterion for the diagnosis of the plague. First of all, when conducting house-to-house visits in search of plague patients, inspectors were not allowed to fill in the space for "diagnosis of illness" on the registration form unless they had first secured a report of a microscopic test from a medical doctor.[30] This form had lines for the result and date of the test. In this sense, the identity of the infectious disease was inseparable from the use of both the microscope and the germ theory. Nevertheless, it is hard to believe that the microscope was involved in most of the diagnostic procedures. There were not enough instruments or trained personnel for such a comprehensive effort, and, more importantly, a microscopic test was far from the most cost-effective method for identifying plague patients.[31]

During the 1910 outbreak, confirmed plague patients were placed in a plague hospital, while their contacts were sent to emergency detention camps that had been constructed from 120 railway wagons lent by the Russian Railway. These plague contacts had their pulses and temperatures taken every morning and evening, and anyone showing fever was at once isolated in a separate wagon. Once bacteriological tests had determined that someone had been infected with the plague, the person was transferred immediately to the actual plague hospital, where he or she usually died within a couple of days after admission.[32] Since the mortality rate in plague hospitals was 100 percent, the microscope's function of determining one's disease status gave the instrument the power to judge over life and death.

The Qing state empowered the microscope by basing its antiplague infrastructure on the capabilities of this tool; in return, the microscope legitimated the antiplague measures, which, even in the eyes of Xi Liang, were "the most extreme and brutal policies seen in four thousand years."[33] It quickly becomes apparent why the containment policies of the Qing government were viewed as brutal. As the historian

Carol Benedict has pointed out, in many ways the Chinese response to febrile epidemics was very similar to the European response to the plague: both peoples sought spiritual assistance, found fault with the larger environment, and tried to shelter themselves from the poisonous atmosphere. In contrast to the situation in Europe, however, "where government-imposed quarantines directly affected people's lives, the imperial Chinese state did not impose forceful public health measures."[34] In other words, what differentiated the Chinese and Western responses to the plague was exactly the strategy Dr. Wu suggested: rigorous control over people's movements and habits.

The Chinese people strongly resisted the measures implemented during both the Hong Kong and Manchurian plagues. When the British public health authority imposed house-to-house inspections in Hong Kong and removed patients for quarantine to the ship *Hygeia*, several riots broke out among the enraged and fearful local Chinese. There was so much tension that Western doctors felt the need to carry revolvers as they reached the agitated neighborhood around Donghua (also known as Tung Wah) Hospital.[35] Furthermore, to escape from the dreadful measures imposed on them, one-third to one-half of the Chinese population fled from Hong Kong to Guangzhou, where the plague was known to be no less rampant.[36] The Chinese residents of Hong Kong apparently feared the antiplague measures more than the plague itself.

In Manchuria, Wu Liande implemented antiplague measures that were much more rigorous and intrusive than those carried out during the earlier Hong Kong plague. With nearly twelve hundred soldiers outside and six hundred policemen on duty inside Harbin, no citizen was allowed to enter or leave his or her designated section, let alone the city limits. In addition to controlling the population, Wu had to contend with the more difficult problem of two thousand unburied corpses already piled up on the ground. Stretching over one mile, the corpses could not be buried because the soil had frozen in temperatures of −30°C. Given the Chinese reverence for ancestors, Wu did not dare to carry out a mass cremation with anything less than an imperial edict. Throughout the process of plague containment, Wu's greatest concern appeared to be whether the emperor would sanction a mass cremation. Afterward, he often cited the imperial edict granting this as a milestone in the introduction of Western medicine into China.[37]

The majority of the Chinese population—who had no knowledge of the existence of *Yersinia pestis* and little concern for sovereignty—were horrified to learn that no one taken into detention ever returned alive. In their eyes, Wu's medically directed police action and sanitary measures seemed arbitrary, despotic, and destructive. To fight the plague, Wu allowed the police force to restrict people's movements, interfere with their normal business, burn their residences and belongings, and take away their relatives—all without saving the life of anyone who was infected. Heartbreaking stories testified to the perceived cruelty: to avoid involving family members in despotic police protocols, many plague patients crawled out of their houses in the middle of the night, to die in the street.[38] In addition, rumors circulated widely throughout the whole ordeal. Some Chinese believed that the plague had been created by the Japanese poisoning of wells;[39] others claimed that plague patients were being buried alive by the administration. Often it was claimed that traditional Chinese medicine did succeed in curing certain patients.[40] The authority and legitimacy of the Plague Prevention Service was challenged from many directions.

Since Western medicine offered no cure for the plague, Wu and his colleagues had no immediate means through therapeutics to win the trust of the Chinese people. Instead, they had to justify their apparently brutal measures on the basis of absolute necessity. From their point of view, as long as confirmed plague patients were doomed to die, the only viable option was to segregate the sick from the healthy and to put the former group in a plague hospital to await their final destiny. The measures undertaken were unable to cure a single patient, but that was never their intended goal. Rather, by isolating diagnosed carriers of the pneumonic plague from the general public, Wu and his colleagues could save many lives by limiting the further spread of the disease. While no one can say with certainty what finally stopped the plague, the total number of plague deaths per month in Harbin dropped from 3,413 to zero in just thirty days after the enforcement of antiplague measures.[41] The harsh measures were thus the best available solution, even if no one was able to cure those patients infected with plague.

Challenges from Chinese Medicine: Hong Kong versus Manchuria

Practitioners of traditional Chinese medicine took offense at the conclusion that the plague could not be cured, both in Hong Kong and Manchuria. But they encountered variant forms of the plague and therefore experienced very different results. By claiming success in treating victims of both plagues, some practitioners of Chinese medicine stood up to challenge the necessity of antiplague measures, which they also perceived as brutal and cruel. On the basis of Chinese medicine, the Chinese-run Donghua Hospital competed with the colonial government in solving the crisis of the Hong Kong plague. The hospital's interference so irritated the colonial governor that he once ordered the gunboat *Tweed* to be anchored opposite Donghua.[42] The challenge was more daunting for the Qing state in Manchuria, where even its local officials at first agreed with lay people that native medicine could be effective in treating plague patients.[43] If that had indeed been the case, forcing patients to enter plague hospitals would have deprived the patients of their only chance of being cured. In this sense, Chinese medicine constituted a serious challenge to the legitimacy of antiplague measures.

Two interesting cases reveal how the microscope played a crucial role in meeting the challenge from Chinese medicine. Told by a biomedical doctor, the first story is about a famous practitioner of Chinese medicine who was hired by Donghua Hospital because he claimed to be able to treat the plague during the Hong Kong epidemic. After he failed to cure even a single plague patient over the course of a month, he resigned from the post with the explanation that the "plague in Hong Kong was very different from that of other places in China." The traditional practitioner was actually correct; the plague cases he treated were indeed different. The biomedical doctor went on to remind the reader that "before sending patients to Donghua Hospital, the British health authority always used a microscopic test to make sure they were genuine cases of plague. As a result, this time the traditional practitioner had to deal with the genuine cases and he surely failed in curing anyone."[44] It seems that if the famous traditional practitioner had ever enjoyed success in treating "plague" patients in the past, it was because they were not actually plague cases. As the practitioner of Western medicine emphasized, it was crucial for the public health authorities to control the definition of plague with microscopic examinations; otherwise, it would face serious challenges from traditional practitioners and their self-serving and misguided claims of success.

The second case was documented in the official plague report to the Qing court. Early on, when the plague was rampant in Harbin but had not yet reached Mukden, the antiplague authority rushed to send all the microscopes in Mukden north to Harbin. As the plague spread south, the resulting microscope shortage required public health officers to identify plague patients by means of manifested symptoms, even when there were "cases of doubt." Since this method unavoidably misdiagnosed some plague patients, lay people got the wrong message that native treatments could cure plague patients. Very soon, local officials started boasting about successful cures and dismissed the modern measures of disinfection and isolation as ineffective.[45] The situation got so out of control that the viceroy had to personally order local officials to refrain from circulating these groundless success stories. In short, only when public health officers had microscopes to help identify true plague patients—or, to put it more bluntly, only when they made the microscopic test the effective criterion for identifying borderline plague cases—could they convince local officials and the general public of the truth that pneumonic plague was universally fatal.

It cannot be overemphasized that the two plagues were very different in terms of communicability, virulence, and transmission. More than twenty-five hundred people died during the Hong Kong plague; sixty thousand suffered the same fate in Manchuria. Being an airborne disease, the Manchurian plague was transmitted through person-to-person interaction; the bubonic plague in Hong Kong was transmitted indirectly through rat fleas. When biomedical doctors first learned about the germ theory of the plague in Hong Kong, they did not know that the plague was an insect-borne disease. Nevertheless, they had already recognized that the plague was not a very contagious disease, and direct body-to-body contagion seemed to be an unlikely way for it to spread.[46] Having such different characteristics, the two plagues posed very different challenges for the struggle between biomedicine and traditional Chinese medicine.

Perhaps due to the differences between the bubonic and pneumonic strains of plague, practitioners of Chinese medicine were spared a fatal lesson in Hong Kong but were forced to observe firsthand their ineffectiveness in Manchuria. Throughout the Hong Kong plague, Donghua Hospital continued to demand that the British government send Chinese patients to a glassworks factory, where traditional practitioners had set up a temporary plague hospital. Until the very end of the outbreak, some local Chinese still preferred medical assistance from traditional practitioners.[47] Neither historical actors nor historians of the Hong Kong plague claim that the involvement of traditional medicine resulted in higher mortality rates. In fact, because citizens repeatedly debated the efficacy of Chinese medicine in the decade-long plague outbreaks, Henry Blake (1840–1918), the governor of Hong Kong from 1898 to 1904, decided in 1903 to conduct a controlled experiment to compare patients treated with Western medicine and those treated with Chinese medicine. To his surprise, the difference between the two medicines in mortality rate was only 1.83 percent. The governor also mentioned a successful treatment of a plague patient by Chinese medicine, a case which the principal medical officer approved.[48] The governor concluded that "the prescriptions given by the Chinese doctors are good, so far as they go."[49] By contrast, during the Manchurian plague, Dr. Dugald Christie asserted that "the fatal venture of the merchants [with Chinese medicine] was the turning point" at which the public resistance toward the antiplague measures eased.[50]

Dr. Christie, a Scottish Presbyterian medical missionary and later the founder of Mukden Medical College in 1912, personally witnessed those "fatal ventures." According to him, the most serious resistance to modern intervention came from local merchants, who found that the public health measures severely interfered with their commerce. To alleviate the impact on their business, the local Chamber of Commerce decided to establish its own plague hospital and invited Dr. Christie to take charge of its operation. After he turned down their invitation, the merchants invited two famous practitioners of Chinese medicine to run their hospital. Clearly these merchants did not prefer Chinese medicine per se; their concern was mainly with business. Although the two traditional practitioners treated plague patients with acupuncture and herbs, their plague hospital did try to observe the modern principle of isolation by separating its compounds into two sections, one for plague patients and one for their contacts. Nevertheless, the traditional practitioners did not bother to wear masks and therefore became infected and carried the disease to the non-plague ward. Within twelve days, the two traditional practitioners and their two hundred and fifty patients and contacts had all succumbed to the plague. Overall, there was a 50 percent mortality rate among practitioners of Chinese medicine in contrast to a 2 percent rate for practitioners of Western medicine.[51] "It was a costly experiment, but it taught [Fengtian] a lesson," Dr. Christie concluded.[52]

The tragic sacrifice of human lives was more than just the "cost" of the so-called public experiment; it was perhaps the necessary condition for delivering a valuable lesson.[53] This point becomes evident when we compare the two plagues. Until recently, for instance, historians and practitioners of Chinese medicine still suggested that "Chinese medicine was the savior during the Hong Kong plague" and celebrated the names of the three famous practitioners of Chinese medicine who courageously served native citizens during the plague.[54] By contrast, the pneumonic and virulent nature of the Manchurian plague made similar conclusions impossible in Manchuria. In the final analysis, the pneumonic plague—"the deadliest of all diseases," in Hirst's words—played a crucial role in the historic defeat of Chinese medicine. Precisely because neither type of medicine could offer any effective cure to individual patients, Western medicine could demonstrate its relative, albeit nontherapeutic, strength in diagnosing, preventing, and containing the plague, thereby defending China's sovereignty.[55]

Chuanran: *Extending a Network of Infected Individuals*

From the evidence presented thus far, it may seem tempting to dismiss traditional Chinese medicine as irredeemably foolhardy. After all, Xi Liang's assertion seemingly displayed a stunning ignorance of facts that we now take for granted: "In the beginning, [we] did not believe that this plague could *chuanran* [spread by contagion or infection]." This assertion is intriguing because, in light of Angela Leung's recent work, we know that the Chinese term *chuanran* was regularly used to describe, as she puts it, "the transmission of the same disorder by contact with the sick" from the twelfth century on. By the seventeenth century, it was further used by Wu Youxing (ca. 1580–1669) in the specific context of febrile epidemics, which contaminated people "through the mouth and the nostrils."[56] Moreover, as Marta Hanson has cogently demonstrated, Wu dedicated an entire book *On Febrile Epidemics* (*Wenyi lun*) to arguing that the widespread outbreak of epidemic diseases was based on specific environmental rather than climatic pathogens.[57] Given this history, it is particularly difficult to comprehend Xi's assertion regarding the Chinese reluctance to accept the fact that "this plague could *chuanran*." Fortunately, a plausible answer to this puzzle can be found in Xi's historic speech at the International Plague Conference after the outbreak of the Manchurian plague.

This remarkable speech is much cited as evidence that China finally "acknowledged the superiority of modern medicine over Chinese medicine" because of the Manchurian plague.[58] Right before the paragraph quoted most often, Xi actually suggested that the Manchurian plague was not an ordinary "febrile epidemic": the disease was something that had been completely novel to the Chinese until three or four months before the epidemic began.[59] By no means did Xi refer to bubonic plague in his statement quoted above. Because the Hong Kong plague had also spread through other areas of China and other countries—including India, Japan, the United States, and Taiwan—there were numerous recent Chinese medical books focusing on the disease of bubonic plague (*shuyi*).[60] In light of this understanding, what Xi's ambivalent assertion really referred to was the pneumonic plague, which was little known at that time even in the West.[61]

In addition to being a disease previously nonexistent in China, there was another sense in which the Manchurian plague transcended the cognitive horizon of the late Qing Chinese—namely, the fact that the specific way in which this new disease spread was almost unthinkable in traditional terms. In her above-mentioned article, Angela Leung meticulously teases out layers of meanings for the term *chuanran*. According to my reading of her article, there were two distinct meanings of *chuanran*: it could mean an acute and widespread outbreak of epidemic or infection spread through direct and intimate person-to-person contact. In the first kind of *chuanran*, epidemic diseases were transmitted by way of environmental *qi*—particularly local *qi*, as Wu suggested in his book *On Febrile Epidemics*—which could enter people's bodies through the nostrils or mouth. By contrast, the second kind of *chuanran* was most often associated with nonepidemic, even chronic, diseases such as tuberculosis, leprosy, and venereal disease; it was rarely used to describe the spread of epidemic diseases. Hence we see two distinct kinds of meanings of the term *chuanran*, which were associated with two kinds of diseases—differentiable in both degree of communicability and perceived mode of communication. The two meanings of this term might help to explain the reason why people knew that many diseases tended to *chuanran* but had never grouped these diseases together into a category of infectious diseases.

To summarize, the Manchurian plague was beyond the horizon of the late Qing Chinese in two separate senses. First, pneumonic plague was a disease totally unheard of; second, the specific way in which the new plague spread was almost unthinkable in traditional terms. Combining two distinct meanings of *chuanran* in an unprecedented fashion, the pneumonic plague embodied a new conception of disease: an acute and widespread epidemic that was transmitted by direct and intimate human interaction.[62]

Nothing illustrates this novel connection more clearly than a handful of documentary illustrations called "diagrams on the system of *chuanran*" (*chuanran xitong tu*). Inserted in the official report on the epidemic, these diagrams were meant to display the systematic ways that the plague had spread in various locations in Manchuria.[63] Take, for example, the tragic incident in Mr. Sung's shop in Liaoyang. It began when Sung found two plague corpses left in his courtyard. To avoid interference with his business, Sung and his daughter-in-law hid the corpses under piles of snow. Through contact with the bodies, these two individuals became infected. In less than two weeks, the plague had spread to the neighboring Yang family and taken more than fifty lives within the two households. Starting with two anonymous corpses, the diagram on the system of *chuanran* for this case effectively linked fifty plague victims into a network of family members, relatives, neighbors, coworkers, and other residents of the area. As every victim was identified and linked to another, it appeared obvious that every one of them had become infected through contact with another person on the diagram. The existence of this type of disease, a virulent epidemic that spread by way of an extended network of infected individuals, was beyond the comprehension of Chinese people. No wonder that "in the beginning, [we—i.e., the late Qing Chinese] did not believe that this plague could *chuanran*."

It is worth pointing out that in this chapter I have intentionally left the term *chuanran* in Xi Liang's statement untranslated, instead of rendering it as either "infection" or "contagion." While the relationship between the concepts of contagion and infection is complicated and overlapping, "it would generally be assumed that contagion is direct, by contact, and infection indirect, through the medium of water, air, or contaminated articles," as the historian Margaret Pelling has clarified.[64] Drawing on this distinction, the most precise translation of what Xi intended to express is most likely that the Chinese people did not believe that "this plague could spread by contagion." While his intended meaning can be decoded with interpretive efforts, historians should bear in mind that his statement must have been very puzzling to his contemporaries. It was puzzling because in the Chinese language, one term—*chuanran*—was used to translate both infection and contagion. As a result, although Xi was trying to describe a novel phenomenon whose precise description demanded a distinction between infection and contagion, he was forced to use the term *chuanran*, whose meanings included both contagion and infection. Thus his language neither allowed the existence of the novel phenomenon nor was prepared to describe it. In order to capture the puzzling nature of Xi's assertion and to foreground the related issues of language and translation, I have chosen to leave *chuanran* untranslated.

No one took to this new knowledge more quickly than those who were physically involved with handling the sick. Having witnessed the horrific ways in which the plague had afflicted their colleagues who had dealt with patients and corpses, these people became deeply convinced of the contagious nature of the Manchurian plague. With this change of belief, according to Xi, resistance to cooperation in the medical management of the plague became a much thornier problem than previously, when people had not believed that the plague was transmitted contagiously through human contact.[65] Although the government tried to employ both carrots and sticks, many medical personnel, guards, and workers (for burying corpses) abandoned their posts until the military or police forced them to return. These parallel changes in both belief and behavior reflect the fact that, at least for people directly involved in containing the plague, a systematic understanding of epidemic contagion started taking hold.

Avoiding Epidemics

Fundamental differences in the recommendations of Western and Chinese medicine made it impossible to reconcile the two. The former recommended quarantine and isolation, whereas the latter recommended mobility. The systematic view of plague infection no doubt justified Dr. Wu Liande's unconventional measures: if doctors could isolate an individual patient before he or she came into contact with other people, they could prevent a string of infections and save potential patients from becoming part of the system. In contrast to confining the movement of people, an important traditional Chinese practice for coping with epidemics depended on individual mobility. Since Chinese people during the late Qing period considered plague to be caused by the "local *qi*" (*diqi*) of a certain place, they encouraged patients to avoid or even move out of the disease-causing location in the name of "avoiding epidemics" (*biyi*).[66]

In his 1910 study of the bubonic plague, Yu Botao, a famous practitioner and the founder of an important national association of Chinese medicine, documented two styles of avoiding epidemics in a special section devoted to the practice. The first style focused on avoiding the hot and damp spots in one's residence, since Chinese medicine visualized the plague as originating from "hot air rising from the earth like smoke pouring out of a chimney."[67] Even when a patient was found sick with plague, Yu suggested "moving him or her out of the house to a windy place under a big tree."[68] More importantly, the patient who was moved outside had to sit in an elevated chair or bed to keep a distance from the local *qi*.[69] Otherwise, moving outdoors would only worsen the situation. Rather than being concerned with the contagiousness of the patient, the author was preoccupied with the dangerous infection coming from the local *qi*.[70] The second style of avoiding epidemics derived from the writings of the famous late-Qing intellectual Yu Yue (1821–1907). Witnessing the practice of avoiding epidemics during plague outbreaks in the 1860s, Yu Yue described the following situation: "Once a patient was found in one household, the nearby dozen families all moved out of the neighborhood. Many were on the road but none could avoid [being infected]."[71] Instead of isolating the patient, dozens of people fled the dangerous place where that dying patient represented the first manifestation of a local epidemic.

While these two ways of avoiding epidemics differ substantially in terms of the scale of movement, they share the concern with escaping from the hot, disease-causing *qi* of the earth. Avoiding epidemics was a strategy designed to avoid a disease based on locality; it did not seek to avoid plague victims. As individuals abandoned their residences and joined an exodus of death, the practice of avoiding epidemics ran the serious risk of putting more people in danger.

In light of the systematic view of plague infection, modern quarantine (*jianyi*) and Chinese avoiding epidemics (*biyi*) were two practices in direct confrontation. As a result, forceful implementation of quarantine and isolation meant much more than adopting a new antiplague method. In the case of the Hong Kong plague, the state could be perceived as depriving people of their very last resort. As noted above, after the colonial authorities gave in to Chinese demands to grant the freedom to leave the colony, eighty thousand people—one-third to one-half of the Chinese population—fled Hong Kong for Guangzhou during the plague. By contrast, during the height of the Manchurian plague, a group of more than ten thousand coolies could not force their way past the Great Wall at Shanhaiguan because armed troops were on guard.[72] When a head-on confrontation seemed unavoidable, the medical officer there, a doctor named Xiao, sought approval from the central government: "If these coolies disobey the rules [against forcing their way through Shanhaiguan], [we want] to be allowed to treat them as bandits and execute them without being charged."[73] Although in the end this horrific policy was not approved by the Ministry of Foreign Affairs, Dr. Xiao's request made clear his determination to control the movement of people coming out of the plague-affected areas.

Unlike the individualist strategy of avoiding epidemics, blocking the network of infection was in the collective interest of the public and demanded governmental intervention. As the case of Sung's shop illustrated, just one infected patient could contribute to the spread of disease by failing to cooperate fully with the antiplague authorities. To highlight the lessons of this tragedy, the author concluded, "To those who hide corpses from the government and refuse to send contacted relatives to the detention camp, this [tragic incident] is their most serious warning."[74] According to the systematic view of plague infection, the task of containing infectious disease went beyond any individual's immediate self-interest but demanded full cooperation from everyone. Once the government realized that "the plague could *chuanran*" by way of extending a network of infected individuals, it had to take on the task of surveying, classifying, and controlling people, a task never called for in the traditional ways of coping with febrile epidemics.

Joining the Global Surveillance System

To signal the novelty of the Manchurian plague, in many places the official report referred to the plague not with traditional Chinese terms such as *yi* (epidemic), *dayi* (great epidemic), *wenyi* (febrile epidemic), or even *shuyi* (bubonic/rat epidemic), but as *baisituo*, the Chinese rendition of the Japanese *pesuto*, which itself came from the French *peste*, or "plague." Both the Chinese and the Japanese terms are transliterations (translations based on sound); the Chinese ideographs involved make no sense at all. In fact, Japan also used the same Chinese character *yi* (pronounced eki in Japanese) to refer to the plague until Kitasato and Yersin isolated the plague bacillus in Hong Kong. No one would have translated these terms in this way unless their intention was to highlight the complete foreignness of the concept.[75] Most importantly, *baisituo* was more than just a new disease to the Chinese; it introduced to the Qing state a new category of disease—*chuanranbing*, or infectious disease.

The Chinese government enacted no law concerning the notification of infectious disease at the beginning of the Manchurian plague. Since the Hong Kong plague had been transmitted from Guangzhou, the British colonial government afterward felt it crucial to be informed of future outbreaks of plague in China. The British decided to seek this information from medical customs at Chinese seaports, since the Chinese government did not concern itself with this task.[76] It was not until the end of the Manchurian plague that Xi Liang retrospectively issued tentative sanitary regulations concerning notifiable infectious disease. Since the regulations were specifically designed for the plague, they did not mention any other infectious diseases.[77] Nevertheless, notification of infectious disease lay at the heart of the interim report that the International Plague Conference submitted to the Qing court.

Thanks to William Summers' recent study, we now know in much greater detail why and how the late Qing government devoted itself to organizing China's first international scientific conference just months before its demise. More than anything else, the Qing government hoped to translate the pressure of international intervention—mostly from Japan and Russia—into "scientific investigation rather than administrative maneuvering."[78] It is remarkable how Qing governmental officials, especially Shi Zhaoji, displayed great diplomatic acumen in achieving this objective and fully appreciated the crucial role that cutting-edge scientific research played in this international power struggle. To help the Chinese and American scientists balance the anticipated Japanese domination of the conference, the Qing government invited American scientists to Manchuria more than a month prior to the conference and provided them with ample opportunities for performing autopsies on plague-infected victims.[79]

When the International Plague Conference opened in Mukden on April 3, it welcomed delegates from eleven nations, including Kitasato Shibasaburo, who had accepted the invitation at the last minute but "made several disparaging remarks about the ability and even the right of the Chinese to contribute to the conference."[80] In spite of such heightened tension, the conference managed to pass forty-five resolutions concerning measures that the Qing government should adopt to prevent any future plague outbreaks in China.

The conference participants discussed at length whether to recommend the official formation of a government organization to handle public health issues. They were fully aware that many suggestions would not be put into effect if there were no official with the designated responsibility of carrying them out. Some participants found it too idealistic to suggest that China should undertake an institutional reform and establish a central public health service. As one delegate put it, this recommendation impractically "proposes something better than we have in England."[81] Many participants also felt that, as medical scientists, it was beyond their expertise to pass a resolution concerning the reorganization of the Chinese government. On the other hand, the participants were very much concerned with making China an integral part of the global surveillance system over infectious diseases, especially the plague.

Starting with the first International Sanitary Conference in 1851, the fundamental objectives of these international meetings were "to protect Europe from disease importations and to lessen the burden quarantine placed on international trade."[82] While the majority of these conferences focused on protecting Europeans from "Asiatic cholera," the emergence of plague in India "encouraged States to conclude the International Sanitary Convention (ISC) of 1897 dealing with plague"[83] and thereby gave birth to the modern international surveillance system over infectious diseases. Echoing these efforts in monitoring the outbreaks of plague, the British representative, Dr. Reginald Farrar, pointed out during the Mukden Conference that "means of notification can only be had with a government department. Unless you have means of notification, you can not prevent plague."[84] After vigorous debate, representatives reached the following resolution: "With the view of giving effect to these recommendations, every endeavor should be made to organize a central public health department, more specifically with regard to the management and notification of future outbreaks of infectious disease."[85]

Instead of recommending an all-embracing project for constructing public health in China, as was often claimed by Dr. Wu Liande and some other scholars, the plague conference recommended that the Qing government should institutionalize the notification and management of infectious disease. Partially accepting this suggestion, the newly established Republican government created the North Manchurian Plague Prevention Service in the following year, with Dr. Wu as its director. This service was placed under the direct supervision of the Ministry of Foreign Affairs, and the foreign-controlled Maritime Custom Service financially supported it until 1929.

Conclusion: The Social Characteristics of the Manchurian Plague

Robert Koch, the father of the germ theory of disease, pointed out that cholera was "our best ally" in the fight for better hygiene,[86] but it apparently took a much more virulent epidemic to serve as the catalyst for public health in China. While previous studies have rightly emphasized the role of geopolitics in transforming the Manchurian plague into a sovereignty crisis for the Qing court, little is known about the factors that turned the Manchurian plague into such a powerful ally for promoting public health in China. To paraphrase the medical historian Charles Rosenberg, I would like to conclude this chapter by summarizing the unique configuration of "social characteristics" of this particular plague and the salient sociopolitical consequences that were, at least partially, constituted as responses to its social characteristics.

First of all, it is highly significant that the Manchurian plague was a pneumonic plague. Starting with its 100 percent mortality rate, the pneumonic plague effectively denied Chinese medicine any therapeutic value and demonstrated its powerlessness in containing infectious epidemics. The following statement is bound to strike the reader as ironic: Chinese medicine was demonstrated to be inferior to Western medicine in an event in which the latter did not cure a single patient. Second, to the late Qing Chinese population, the pneumonic plague exhibited the most puzzling combination of characteristics: it was very much like chronic or nonepidemic diseases in the specific way in which it spread, but in terms of speed and scope, its transmission was way beyond that of the most virulent epidemics they had ever experienced. Third, the existence of a purely pneumonic plague was also a new phenomenon to the international scientific community. It was during this outbreak that related knowledge was contested and then partially established.

The pneumonic plague, while known to plague experts, was greatly underestimated by the scientific community before this event. As mentioned previously, even after recognizing the pneumonic nature of the Manchurian plague, Dr. Kitasato Shibasaburo repeatedly warned that a bubonic plague might join forces with it when rats emerged from hibernation. After Japanese scientists dissected more than thirty-five thousand but found none infected with plague,[87] Dr. Kitasato was forced to admit, during the International Plague Conference, that "the Manchurian plague was a *pure* pneumonic plague—something that had not occurred in recent centuries (emphasis added)."[88] The plague conference opened with twelve scientific questions raised by a Chinese Imperial Commissioner, the majority of which were related to, if not directly concerned with, the following question: "Why should a bacillus that, as far as we know, has the same microscopic appearance and answers to the same bacteriological tests cause a pneumonic and septicemic epidemic here and only give rise to bubonic plague in India and other places where pneumonic cases occur only incidentally?"[89] In response to this question, which was based on the observation that the Manchurian plague was "strictly" pneumonic in form, there was much interest in the possibility that a variant strain of the plague was at work in Manchuria.[90] If Dr. Wu Liande had not been right about the existence of pure pneumonic plague and thus played a pioneering role in advancing scientific knowledge about its nature, it is hard to imagine that the Ministry of Foreign Affairs would have been interested in holding the first international scientific conference in China, let alone having the young Dr. Wu (then only thirty-two years old) serve as its chairman in the presence of international authorities on the plague.

Therefore, fourth, the pneumonic plague served as an ally for promoting public health, because it brought a scientific victory to the Qing court. Later on, in his preface to Dr. Wu's report on the plague, Liang Qichao, arguably the most prominent intellectual in late Qing China, did not hesitate to state that "science has been imported to China for more than fifty years. The only [Chinese] person who can face the world as a scholar is Dr. Wu."[91] Dr. Wu's pioneering research on this rare epidemic allowed China for the first time to face the world as a country performing cutting-edge scientific research. In this sense, to go back to the title of this chapter, the new knowledge that was discovered by means of microscopes—not just those used by Dr. Wu and his Chinese colleagues, but also those used by the Japanese scientists searching for the plague bacillus in rat corpses—was crucial in resolving the sovereignty struggle over Manchuria.

Fifth, as "the deadliest of all diseases," the pneumonic plague brought about "the most extreme and brutal policy ever seen in four thousand years."[92] As a rule, one extreme brings about the other. While the containment of the Manchurian plague has always been celebrated as a historic event for many different reasons, almost no one has noticed that it was also historic in such a dark sense—even though Xi Liang made its historic brutality the central message of his report to the Qing court.

It is very moving to see that after the plague had been successfully contained, Xi devoted the preface of his report to a public expression of his heartfelt reservation, puzzlement, and even dismay vis-à-vis the antiplague measures that he had authorized.[93] More than anyone else, Xi realized that China could no longer afford not to take control over epidemic disease; membership in the global surveillance system over epidemics was indispensable for the Qing state in order to maintain its sovereignty. On the other hand, as he shared with his fellow citizens the moral sensibility cultivated within the traditional cultural world, Xi felt dismayed by what he had witnessed during the process of containing the plague. He devoted all three pages of this preface to describing the ways in which antiplague measures had forced Chinese people to violate their most cherished ethical norms—to abandon suffering friends in detention camps, to fail to say goodbye to dying relatives, to burn their houses and belongings, and to cremate the bodies of their beloved family members.[94] "These are the most painful tragedies in the whole world. All benevolent fathers and filial sons would hardly bear to hear and witness this," stated Xi.[95] As Xi questioned the practitioners of Western medicine about their rationale, he was told that these apparently brutal measures represented real acts of humanism.[96] Xi could not but conclude that "there is no eternal measure for human affairs,"[97] leaving it for the reader to judge.

Xi was right to point out in this preface that these measures were "not only unheard of since antiquity [in China] but also rarely implemented by Westerners."[98] As historian Mark Gamsa points out, the antiplague measures implemented during the Manchurian plague "rank among the most intrusive of [those used in] Western-managed epidemics; mass cremation and forced quarantine on a similar scale would have been inconceivable in late Tsarist Russia, not to mention Western Europe."[99] In addition to such intrusive measures, the Manchurian plague polarized the dichotomy between the state medical elites and the general public. For the sake of containing the plague, state bureaucrats and medical officers were forced to treat the Chinese people as ignorant, unreasonable, and even immoral. Knowing the invisible bacillus and concerned about the well-being of the population, the state elites felt obligated to put into practice policies that even they regarded as brutal and extreme. They found themselves caught in a position of being simultaneously omniscient, benevolent, and brutal. If the Manchurian plague was indeed the watershed event that introduced public health into China, very unfortunately, part of its legacy was this dichotomized, hierarchic opposition between the state elites and the general public.

Finally, while the Manchurian plague has acquired the status of a historic event, its most radical aspect is so taken for granted that it often becomes invisible to the modern reader. That is, the containment of the Manchurian plague irreversibly made the Chinese state the subject of the history of Western medicine in China. This historic effect was not caused by the intrinsic nature of the pneumonic plague; instead it resulted from the historical contingency that this plague broke out in Manchuria, the border area over which Japan and Russia competed fiercely to extend their influence and encroach on China's territorial sovereignty. It was in this geopolitical context that public health was demonstrated to be of crucial importance to the Qing state. In response to the use of medicine as a political tool by Russia and Japan, the Qing state finally came to recognize the political function of public health and the superiority of Western medicine over traditional Chinese medicine for that function toward the end of the Manchurian plague. As suggested by this chapter's title, "Sovereignty and the Microscope," this medical event gave birth to the strategy of forging an alliance between the state and biomedicine by translating public health measures into tools for state building. As we will see in next chapter, this strategy was to have a profound influence on the development of both Chinese and Western medicine in China.

3 Connecting Medicine with the State: From Missionary Medicine to Public Health, 1860–1928

For believers in the universality of science and related ideas of the Enlightenment, the 1935 article "How to Popularize Scientific Medicine in China" might be puzzling. In this article, Yu Yan (1879–1954, also known as Yu Yunxiu), a practitioner of Western medicine and the most famous critic of traditional Chinese medicine, said, "I think that without the power of politics, there is no way to popularize scientific medicine in China. If we keep focusing on advertising [scientific medicine] to the masses, no one knows if there will be any effect at all in one hundred, or even one thousand, years."[1] Citing Japan's successful experience as an example, Yu went on to say, "The thriving development of scientific medicine in Japan since the [Meiji] Restoration has been based completely on political power. Lacking this political power, scientific medicine is not able to become popular in China. Politics and medicine are closely connected."[2]

In light of Yu's blunt statement, this chapter argues that what Yu Yan described in 1935 was the main strategy that practitioners of Western medicine had gradually developed and deployed since the turn of the century.[3] As Yu's statement reveals, when he and other practitioners of Western medicine strove to implement this state-centered strategy, the most challenging aspect was the creation of the vital link between politics and medicine, a link that did not exist in the public perception during the late Qing and even early Republican period. In their efforts to create this link between medicine and the state, these reformers adopted public health as the tool of choice in their strategy. They endeavored to make it an essential aspect of state building and thereby turned it into the most salient representative of Western medicine in the first half of twentieth-century China. Their state-centered approach to public health had a profound influence on the development of the health-care system in modern China.

Missionary Medicine

Among the various perspectives from which scholars regard the Manchurian plague as a landmark event in the history of Western medicine in China, one aspect, which its hero Wu Liande (Lien-teh Wu) strove to emphasize, is often ignored. In the *History of Chinese Medicine* coauthored by Wu, the chapter dealing with the period from 1911 to 1920 was entitled "Overthrow of the Manchurian Dynasty and the First Manchurian Plague Epidemic: The Beginning of Modern Public Health Work under Chinese Leadership."[4] With this seemingly straightforward chapter title, Wu highlighted the fact that a shift in leadership paralleled a shift in medical practice. While modern medicine previously had been represented by foreign missionaries and their curative medicine, the Manchurian plague marked the arrival of a new era in which modern medicine would be represented by new medical practices and new agents—public health practices and Chinese practitioners of Western medicine—in the equally new Republic of China.

For almost an entire century after its arrival, Western medicine in China was represented exclusively by missionary medicine, especially surgery. From the time Peter Parker (1804–88) opened his ophthalmologic hospital in Canton in 1834, medical missionaries had built up a reputation for their expertise in surgery and ophthalmology. As the Chinese increasingly abhorred the idea of intrusive therapies, even acupuncture came to be considered as "inappropriate to serve the Emperor" and was banned by the Qing dynasty's Imperial Medical College in 1822.[5] In the eyes of Western doctors, poor Chinese patients paid a heavy price for the national ignorance of the art of surgery. When Dr. Gillan accompanied Lord Macartney's historic embassy to Beijing in 1793, he could not help ridiculing the Chinese: "We did not see one person in all China who wanted either leg, arm, or any part of a limb."[6] Apparently, no one who needed a limb amputated had any chance of surviving the surgery.

A few decades later, when early missionaries arrived on the coast of China, where the Qing government had formally forbidden all forms of contact with the Chinese population, they discovered that the relative strength of Western surgery provided them a precious tool for expressing benevolence toward the Chinese in a personal setting. By the time that the China Centenary Missionary Conference was held in Shanghai in 1907, medical missions took major credit for overcoming the crucial obstacle that Protestant missionaries had encountered in China, namely "an atmosphere of suspicion, distrust, and hatred."[7] Later, Wu Liande went so far as to suggest that "it may even be said that modern medicine profited from the very distrust with which the Chinese looked upon the foreigners and their methods."[8] This profit came with an important limitation, however. Because "preventive medicine did not lend itself to evangelistic efforts as well as the personal contact between patient and doctors,"[9] medical missionaries established a pattern of individualized medicine that focused on curative treatment. Reflecting upon his experience in a 1925 article, "A Century of Medical Mission in China," Dr. James Maxwell, executive secretary of the China Medical Association, honestly admitted that "little indeed has been done here with preventive medicine."[10]

It is understandable that medical missionaries did not actively promote public health in China; missionaries had neither the appropriate training in advanced preventive medicine nor the necessary governmental authority behind them. In fact, as they concluded in their centenary conference in 1907, the badly needed breakthrough was "to reach the leading men of empire, the men of influence in China, intellectual, social, and political."[11] In addition, many people at that time, both Western and Chinese, suspected that the promotion of public health would do more harm than good in China because it would aggravate China's overpopulation problem, even though no one had any exact numbers.[12] As late as 1914, in the opening paragraphs of *Medicine in China*, the China Medical Commission of the Rockefeller Foundation still found it necessary to refute the apparently popular idea that "a high death rate is not on the whole undesirable [in China]."[13]

Along with their large-scale educational activities, missionaries were the dominant providers of formal health-care services in China, judged in terms of history, personnel, or hospital facilities. According to a survey done in 1916 by the China Medical Commission of the Rockefeller Foundation, Protestant missionaries supported 265 hospitals, 386 dispensaries, 420 physicians, and 127 nurses in China.[14] Protestant medical missions had a larger stake in China than in any other foreign country. In 1923, China had 53 percent of the missionary hospital beds and 48 percent of the missionary doctors in the world.[15] Moreover, the Medical Missionary Association, established in China in 1886, was the first medical missionary society in the world and was to remain the only medical association in China for decades to come.[16] Since this association required its members to be graduates of medical schools and to serve in Christian institutions, the vast majority of its members were foreign nationals.[17] Foreign medical missionaries were therefore the dominant organizing force in modern medical education and health services in China in the beginning decades of the twentieth century.

Western Medicine in Late Qing China versus Meiji Japan

It is not surprising that Western medicine was represented primarily by foreign missionaries until the end of the Qing dynasty, because medicine was not a major component of the late Qing reform movements. As historian Li Jingwei points out, the late Qing reformers did not consider medicine to be part of the Western formula for wealth and power. While the advocates of the Self-Strengthening Movement (*ziqiang yundong*, 1861–95) strove to adopt Western science, industry, weaponry, railways, and communication, they did not attribute comparable importance to medicine. Due to this lack of appreciation, only a few among the thousands of students who went abroad for a Western education chose to study medicine. It follows that, while modernizers and conservatives engaged in serious debate on many fronts of China's transformation, no substantial public debate took place concerning the issue of medicine before the twentieth century.[18] Even after the Qing government put an end to the imperial examination system in 1905, which had been the foundation of imperial ideology and gentry orthodoxy, and replaced it with the study of Western learning, Chinese medicine stood out as a unique field that many late Qing intellectuals considered as equal to, if not superior to, its Western counterpart.[19]

China's relative lack of interest in Western medicine—even though a significant number of people had by this time personally benefited from surgery offered by the missionaries—stands in sharp contrast to developments in Japan. It was medicine, anatomical drawings to be precise, that "inspired the rise of Dutch Studies (*Rangaku*) in 18th-century Japan, thus giving birth to one of the most decisive influences shaping modern Japanese history, namely, the study of Western languages and science."[20] A century later, when Nagayo Sensai (1838–1902), an enthusiastic student of Dutch Studies, visited Europe and the United States as a medical observer on the famous Iwakura Mission (1871–73), he was surprised to discover that "Germany offered a model for the central involvement of medicine—and medical experts—in the construction of a powerful and wealthy nation."[21] As a result of this continuous interest, medicine played a central role not just in the creation of a modern nation in Meiji Japan but also in the later expansion of the Japanese empire into its colonies of Taiwan, Korea, and Manchuria.[22]

The Chinese students who studied medicine in Japan were painfully aware that the significance of modern medicine was much greater in Japan than in China. As one student summarized the situation succinctly in 1907, because modern medicine in Japan followed the model of Germany, its medicine was based on science and strove to be a state-centered, collective enterprise. In contrast, as modern medicine in China was dominated by missionary medicine, it was oriented toward religion and confined to individual treatments. He saw this as the reason why medicine in Japan kept progressing, while it remained stagnant in China.[23] The author also pointed out that, after Japan's success in using medicine as a tool to colonize Taiwan, Goto Shinpei (1857–1929),[24] the president of the South Manchurian Railway Company, planned to apply the same strategy in Manchuria and demanded a huge budget for building hospitals and a public health infrastructure there. "It makes the relationship between medicine and the state evident," concluded the author.[25]

Although there had been efforts in China during the New Policy (*xinzheng*) period (1901–11) to implement sanitary measures,[26] it was not until the Manchurian plague that Viceroy Xi Liang finally offered the following definitive statement during the International Plague Conference in 1911: "We feel that the progress of medical science must go hand in hand with the advancement of learning. If railways, telegraphs, electric lights and other modern inventions are indispensable to the material welfare of this country, we should also make use of the wonderful resource of Western medicine for the benefit of our people."[27]

This statement lagged not only half a century behind Nagayo Sensai's discovery in Germany, but also half a century behind the late Qing reformers' efforts to learn Western technology and weaponry after the building of the Jiangnan Arsenal in 1865.[28] It is noteworthy that, in comparison to Japan's early adoption of Western anatomy, medicine was such a latecomer to China's reception of Western science and technology. In fact, even in comparison to most of colonial Southeast Asia, where "medical training provided the first, sometimes [the] only exposure of the colonized elite to science,"[29] the case of Qing China stands out as a salient anomaly because the Qing elite appreciated the natural sciences as a whole and engineering in particular much earlier than they appreciated Western medicine. Most importantly, the Chinese were late in adopting Western medicine not because they had less exposure to its benefits and efficacy than the Japanese, but because they were late in seeing the essential connection between medicine and their urgent pursuit of "wealth and power." The lesson that the Manchurian plague taught the Qing court was the crucial role that modern medicine played in defending a state's sovereignty. At last, medicine began to be perceived as linked to national affairs.

The First Generation of Chinese Practitioners of Western Medicine

The Manchurian plague radically transformed the careers of Wu and a whole generation of biomedically trained Chinese personnel as they jointly developed a new vision for developing modern medicine in China. Before the outbreak of the Manchurian plague, Wu's life at the Imperial Army Medical College in Tianjin had, by his own admission, been "not all too invigorating."[30] When Wu had proposed the creation of a model army hospital, his proposal had received the discouraging comment from the Qing government that "most of our military staff and citizens do not support Western medicine."[31] The Manchurian plague offered a badly needed opportunity for Wu, as he recalled, "to realize the ideas that had been dear to me all my life."[32] After the plague was brought under control, the newly established Republican government created the North Manchurian Plague Prevention Service in the following year, with Dr. Wu as its director. This service was placed under the direct supervision of the Ministry of Foreign Affairs and was financially supported by the foreign-controlled Maritime Customs Service until 1929. The lesson was clear: if one wanted to enlist the state's support, one should promote those aspects of biomedicine that were most "translatable" into perceived state concerns, such as protecting its sovereignty.

Encouraged by these new developments, Dr. Wu and other colleagues started organizing a professional association for Chinese practitioners of Western medicine. During the previous period when Chinese reformers had perceived little connection between medicine and national affairs, only a small group of Chinese students had chosen to study medicine. It is estimated that there were about eight thousand Chinese students in Japan during the peak time in 1905–6, but less than one hundred of these majored in medicine.[33] The lack of interest in medical education was the result of two related forces.

For one thing, because of the low status of medicine in traditional China, families of the gentry would not encourage their youngsters to study medicine,[34] even though the social status of the "scholar-physician" (*ruyi*) had been rising during the Ming and Qing dynasties.[35] Missionaries had long ago discovered the low social status of physicians; they considered it to be one of the most important obstacles to the introduction of modern medicine into China.[36] Because of the problem of social status, medical missionaries had had great trouble gaining acceptance from the Chinese gentry,[37] and the earliest Chinese assistants in the missionary hospitals came mostly from poor families.[38] By 1912, a Chinese medical journal still felt the need to encourage its readers to pursue a career in medicine.[39]

Second, since the Chinese government did not see medicine as part of the secret of the West's "wealth and power," it provided neither fellowships nor official careers for medical students. Even after the humiliating defeat in the Sino-Japanese war in 1895, when the late Qing government again decided to send students to foreign countries, its objective was to promote "practical learning" (*shixue*), which was consistently defined as "agriculture, engineering, business, and mining."[40] Moreover, when the United States returned the Boxer Indemnity fund in 1908, which was to be used for fellowships for Chinese students, the Qing government set up the official policy of allocating 80 percent of all fellowships for the study, again, of "agriculture, engineering, business, and mining."[41] In the end, the relative lack of medical students was largely a reflection of the government's policy, which was rooted in the general opinion that medicine was not essential to China's struggle for national survival.[42]

Although small in number, the Chinese practitioners of Western medicine established two associations at the turn of the century. Chinese medical students in Japan, especially those in Chiba Medical College, pioneered in organizing the Chinese Association for Medicine and Pharmacology (*zhongguo yiyao xuehui*) in 1906. Independently, biomedical practitioners in China, mostly trained in Europe and America, established the more influential National Medical Association of China (*zhonghua yixuehui*) in 1915, which began to publish the bilingual *National Medical Journal of China* (*Zhonghua yixue zazhi*).[43] Although Dr. Wu Liande had this plan of funding a professional medical organization before the outbreak of Manchurian plague, he could not garner enough support at that time.[44] Even when it was founded five years later, no more than a few dozen biomedically trained Chinese physicians attended the first meeting. Partially thanks to Wu's landmark success in containing the Manchurian plague, this small group of Chinese physicians started to foster a very different vision for China's medical development and for their own role in realizing that vision.[45]

Western Medicine as a Public Enterprise

The first issue of the *National Medical Journal of China*, opens with a revealing drawing (figure 3.1). The attached text explains, "Medical work in China is a Little Child trying to wrestle with the giant Disease." Standing in the middle of the picture, a threatening demon waves a chain of weapons, consisting of fifteen infectious diseases at a little child—tagged as "Medical Organization"—who strives to fight the giant demon with a toylike stick labeled "Public Interest." His head is clutched by the monster's hand, which carries the label "Public Indifference."

This picture represents a new vision of China's medical problems. First of all, no longer dominated by the horrifying tumors and surgically treatable diseases depicted in Lam Gua's famous series of paintings,[46] China's medical problems were now characterized by various kinds of infectious diseases. Second, the objects of medical intervention were no longer individual patients but these infectious diseases, personified as an unwieldy demon. Third, despite its personification as a demon, the target of medical work was by no means a pure biological entity. On the contrary, the demon of infectious disease was armed with weapons like "ignorance," "quackery," "no sanitation," "superstition," "patent medicine," and "poverty." Moreover, the intended medical work, as represented by the child, was insufficiently supported by the weak foundation of "no finance" and "no system." Finally, the problem of and solution to China's medical situation were two sides of the same coin: public indifference and public interest in the fight against infectious disease. As this illustration reveals, the new vision of medical work was remarkably social and collective in nature. When China's medical problem was framed in this way, the individualized curative medicine offered by the missionaries became almost irrelevant. The key to success was now to turn medicine into a public concern, or, as one of the objectives of the new association phrased it, "to arouse interest in public health and preventive medicine among the people."[47] Public health was prioritized on the association's agenda from the very beginning, as these biomedically trained doctors organized themselves and established the National Medical Association of China.

Historians have noticed that after the containment of the Manchu-

FIGURE 3.1 "Medical work in China," *National Medical Journal of China* (1915), cover page.

rian plague and the Revolution of 1911, even missionaries became interested in promoting preventive medicine in China,[48] and they launched China's first public health campaigns in 1915.[49] Nevertheless, most of the efforts were limited to "education" and spreading knowledge about hygiene.[50] At the conference of the China Medical Missionary Association in 1923, its Council on Public Health passed a resolution to place more emphasis on "the science and practice of preventive medicine." Revealingly, it moreover declared that "to this end, greater use should be made of Chinese doctors and trained helpers."[51] In the following year (1924), an editorial entitled "Preventive Medicine" was devoted to articulating a handful of reasons that "it is consonant with the work of the medical missionaries to support Public Health work."[52] Nevertheless, several months later, James L. Maxwell (1836–1921), a medical missionary in Taiwan, contested this new viewpoint, emphasizing that the fundamental principle for medical missionaries is "to follow their master, and through a method peculiarly His own. It is not for nothing that Gospels are more filled with the healing works of Christ than they are with anything else."[53] As this explicit resistance reveals, while the most visible emphasis of Western medical practice in China had started shifting to public health during the 1910s, it was by no means easy for medical missionaries to change their focus from curative medicine to public health.

In comparison, the shift to public health was surprisingly smooth for the members of the National Medical Association of China. In his meticulous study of the development of health-care services during the Republican period, Ka-che Yip found it unusual that the conflict over the proper boundary between public health agencies and the medical establishment, which was intense in the United States, never took place in Republican China. Yip reasoned that it was because "medical leaders were the architects and practitioners of modern public health and they fervently embraced its model."[54] Yip's important observation invites a further inquiry: Why did the leaders of the National Medical Association of China, many of whom were not trained in public health, offer strong and consistent support for public health? Or, to anticipate later historical developments, why would they later push the government to adopt policies of the State Medicine (*gongyi*), which aimed to minimize the role played by privately funded curative medicine? Again, the transformation of Wu Liande's career provides an illuminating answer to this question.

While Wu himself claimed that he had simply "seized the opportunity" made available to him by the Manchurian plague, he was keenly aware that his own career advancement signaled the opening of a new horizon of possibilities for his colleagues. Right after establishing the association in 1915, he argued, "Above all, let us pay more attention to the teaching of public health, for the *medical officers* [my emphasis] of the health of the future will exercise in China, as elsewhere, more influence in shaping the destinies of the profession and indeed of the country they serve than any body of men."[55] While Wu appeared to be predicting the future, he was recommending to his colleagues his newfound career as a medical officer of public health,[56] a role that, coincidentally, he himself had not been trained for.[57]

In addition to the respectable position of public health officer, Chinese practitioners of Western medicine had other good reasons for emphasizing public health. In his presidential address to the first National Conference of the National Medical Association of China in 1916, Yan Fuqing, also known as F. C. Yen (1882–1970),[58] emphasized public health as a field that should belong exclusively to the Chinese medical profession because "our government, rightly or wrongly, hesitates to grant this privilege to the foreign doctors. Indeed they have already done a noble share to educate our people along lines of Public Health and Preventive Medicine, but the health officers in our cities who are responsible for the effective carrying out of Public Health measures, are to be Chinese."[59]

In fact, it was because of this concern that the Qing court had urgently summoned Dr. Wu Liande to assist during the Manchurian plague. In other words, because of Chinese nationalism and the concern over China's sovereignty, the new emphasis on public health could help to cause a shift in medical leadership from foreign missionaries to Chinese nationals.[60]

"Public Health: Time Not Ripe for Large Work," 1914–24

Although practitioners of Western medicine made public health their official objective when they founded the National Medical Association of China in 1915, that year did not mark the start of the history of public health in China. On the contrary, one year earlier, the Rockefeller Foundation had decided to put off its involvement with large-scale public health construction in China. This was a historic decision, because no other institution had offered comparable support for medical development in China. Between 1913 and 1923, the foundation had spent a total of US$37 million, making China the recipient of the largest foreign donations.[61] In light of its influential role in the history of modern medicine in China, the foundation's carefully considered decision provides a valuable window into the daunting obstacles that preoccupied the early advocates of public health in the 1910s.

As the Rockefeller Foundation was considering its involvement with China in the early 1910s, public health was already listed as one of the three major fields of interest.[62] In 1914, the foundation established the famous China Medical Commission to study the health situation in

China. After a four-month survey in various parts of China, the commission returned to the United States and recommended that the foundation should undertake medical work in China on a large scale "with the understanding that it will involve a long time and that during that time the Foundation should be the most important factor in China in the development of medical instruction."[63] Most importantly, the commission decided to exclude public health as its core engagement with China and concluded as follows in its final report: "public health—time not ripe for larger work by foundation."[64]

When John B. Grant (1890–1962),[65] later widely acclaimed as the father of public health in China, strove to establish the first Chinese Department of Hygiene in the Peking Union Medical College in 1923, his main problem was that the China Medical Board had already considered this issue and issued a "definite statement" to put off its support for public health. The board provided four reasons for adopting this unconventional policy, which Grant quoted at length.[66]

To begin with, the China Medical Board made it clear that this policy was not in alignment with the general principle of "greater return" held by the Rockefeller Foundation. "In other countries preventive work in medicine first engaged the main efforts of the Foundation, since the fostering of public health presents the prospect of far larger results in the welfare of nations and individuals . . . than in the treatment of the sick. In China, however, several factors have contributed to defer the initiation of direct activities in hygiene and preventive medicine by the Rockefeller Foundation."[67]

Having acknowledged its China policy as an anomaly, the board proceeded to justify it by giving four reasons. First of all, "the systematic protection of public health is properly a governmental function. . . . In the disturbed political conditions now prevailing throughout China, with frequent changes in the government, . . . the prospects for the early development of public health work on a large scale have not been encouraging." Second, "confidence in scientific medicine is not sufficiently widespread to insure on the part of the people the co-operation necessary for the most effective work." Third, "it is important to precede any large effort in public health by a period of careful study of local conditions." Fourth, the number of highly trained personnel was not sufficient for any considerable extension of public health work. "The problem of medical education was therefore indicated as that which first demanded attention," concluded the China Medical Board.[68]

The lack of a stable government with a genuine interest in public health was widely considered to be the most serious problem. In fact, it had been identified as the decisive challenge for the following decades. With the containment of the Manchurian plague and the establishment of the Republic of China in 1912, people like Wu Liande at once celebrated the beginning of a new chapter in the history of modern medicine in China. Nevertheless, the first Chinese Republic turned into a terrible caricature in the hands of Yuan Shikai (1859–1916) and then disintegrated into local warlordism after Yuan's death in 1916. The fragmentation of the Chinese polity, the succession of civil wars, the deterioration of the rural economy, and the infringement of imperialist power made any nationwide medical campaigns impossible.[69] During its first meeting in 1915, the newly established National Medical Association of China had already urged the Chinese government to create a Central Medical Board and a national public health service.[70] However, little had materialized in the following decade; it was reported that the death rate in China had not decreased to any appreciable extent between 1911 and 1926.[71]

During the Warlord period, Wu Liande had become quite disillusioned with the Chinese government. In "A Survey of Public Health Activities in China since the Republic," written in 1923, he put it bluntly: "A Friend of mine especially interested in Public Health tells me that he is in China to sell Public Health. My reply to him is that he must sell it to the Chinese *people*, not to the Chinese *government* [Wu's emphasis]. As you know, China is really the least governed country in the world. . . . What may be feasible in Western countries is often difficult to practice in China and may even have to be discarded."[72] Faced with the chaotic situation in China, Dr. Wu, just like the China Medical Board, concluded, at least for the time being, that the Chinese government could not be counted on to promote public health.

The issue concerning the lack of qualified personnel went much deeper than it appears. Much more serious than the shortage of human resources, it concerned the problem created when large numbers of untrained personnel had been put into office during the New Policy Reform (1901–10). Borrowing the German concept of medical police from Japan, the late Qing government had established a Central Sanitary Bureau within the Ministry of the Interior and the Provincial and Local Departments of Health under the authority of the police. Because biomedically trained medical personnel were extremely rare at that time, people who knew little about modern hygiene often held these hygiene-related posts within the police department. Due to this serious drawback, in his efforts to contain the Manchurian plague, Wu Liande made it a goal "to replace, as far as possible, the untrained police in the routine inspection and report work with trained medical staff. The police, thus released, could return to their proper lay duties."[73] After the plague, Wu and his colleagues repeatedly criticized such sanitary police officers who knew nothing about modern hygiene other than street cleaning.[74] As late as 1926, when Shanghai's Department of Health was still under the leadership of the commissioner of police, more than 60 percent of its budget was spent on street cleaning.[75] The painful lesson to be learned here was never to create any governmental institution unless one could first secure the necessary qualified personnel.

For the Rockefeller Foundation, the lack of confidence in scientific medicine among the Chinese population meant much more than a serious obstacle to the construction of a public health system; it revealed the most serious of China's many deficiencies and the true nature of the foundation's mission in China. Explaining its objective in building the Peking Union Medical College (PUMC), the foundation's China Medical Board made it clear that "the chief result to be attained was the creation of an example on Chinese soil, of how a technique has become available of acquiring knowledge of natural phenomena by scientific methods."[76] Because the foundation adopted this narrowly conceived objective of instilling "the scientific spirit" into the Chinese minds with one elite institution, it consistently preferred medical research over more populist approaches to medical care, at least before it radically reevaluated and changed its policy in 1934 (see chapter 10).[77]

Almost ten years after the China Medical Board decided to defer initiatives in public heath, John Grant's "Proposal for a Department of Hygiene" (1923) was challenged severely on similar but more carefully articulated grounds. Having read Grant's proposal, Roger S. Greene (1881–1947), the resident director of the China Medical Board and director of PUMC, issued a number of highly unfavorable comments in his memorandum of 1924. The first of his ten comments begins, "It is doubtful whether the International Health Board should be asked to gamble on the development of public health in China." This statement is followed by five more "doubtful" or even "extremely doubtful" points. Concerning the lack of "confidence in scientific medicine," Greene appeared to be even more pessimistic than the China Medical Board; he pointed out that "demonstrations in curative medicine have been repeatedly demonstrated in China for over fifty years, and practically no impression has been made. Public health is a subject which is largely in the abstract and lacks the advantage of individual benefits which does so much to win friends for curative medicine."[78] Just like the medical missionaries before him, Greene was painfully aware of the crucial disadvantage of public health—its lack of direct benefits to individual Chinese patients.[79]

More importantly, by now Greene had a much deeper understanding of China's horrific situation, in which medicine, even public health, could not offer much help. At that time, more than 80 percent of the Chinese population was illiterate, and rural China was on the verge of collapse. Poverty was claiming more lives than disease. In his comment on Grant's proposal, Greene was quite right to doubt "whether very much can be accomplished beyond the control of dangerous communicable diseases in a country . . . [in] which economic conditions are such that more people freeze to death and die from tuberculosis than in Europe and America."[80] If the only practical objective was to control epidemic diseases, there was no need for a huge hygiene department to conduct research, because the root problem was "the lack of interest and economic resources to take [advantage] of the existing knowledge."[81]

Confounding the issue of poverty, Greene found five deep-rooted obstacles against launching large-scale public health measures in China. After finally agreeing to support the limited goal of "control of epidemic diseases," Greene nevertheless issued the following strong conclusion: "The International Health Board can not admit its responsibility to create a sentiment in China by health demonstration in order to induce the Government to organize a health service," and predicted that "this will probably have to come very slowly . . . through efforts of smaller voluntary groups."[82]

Subsequent events turned out to be very different from Greene's prediction,[83] even though the five obstacles were widely accepted as posing real difficulties for the creation of a public health system. In sharp contrast to Greene, who had advised against "induc[ing] Government to organize a health service," within four years Grant and his Chinese colleagues convinced the Nationalist government to build an independent Ministry of Health and to consider promoting their ambitious project of State Medicine. In terms of coping with the five daunting obstacles individually, a whole new horizon opened up, because Grant and his Chinese colleagues endeavored to create a new relationship between the state and health care.

The Ministry of Health and the Medical Obligations of Modern Government, 1926–27

In light of the fact that by the mid-1920s many advocates of public health identified the government (or the lack thereof, to be more precise), as the number one obstacle to establishing public health in China,[84] it is very remarkable that by 1928, when the Nationalist government established its capital in Nanjing, the state suddenly rose to become the solution to all the problems raised by Greene and many like-minded leaders of medicine. The question is, When and how did this dramatic change of attitude take place?

While Roger Greene, on behalf of the Rockefeller Foundation, concluded in 1924 that the time was still not ripe for larger work in public health, some practitioners of Western medicine persisted in pushing for its realization. It is useful to compare two of their proposals for building public health in China: *Memorandum on the Need of a Public Health Organisation in China: Presented to the British Boxer Indemnity Commission* (1926) by the Association for the Advancement of Public Health in China, and John Grant's *Memorandum for Provisional National Health Council* (1927). Written just one year apart, these two proposals came from almost the same group of public health advocates.[85] Remarkable similarities therefore exist between the two documents in terms of the understanding of China's health problems and the proposed solutions. Against this remarkable similarity, however, salient differences surface concerning whether there was a need to establish an independent Ministry of Health.

On behalf of the Association for the Advancement of Public Health in China, Liu Ruiheng (1890–1961) prepared a systematic proposal entitled "On the Need of a Public Health Organization in China" in April 1926.[86] This remarkable proposal was prepared for the British Boxer Indemnity Commission, because four years previously, in 1922, the British government had decided to follow the example of the United States and other countries that had received indemnities for the Boxer Rebellion (1900) and return those funds to China. These returned funds would be used for specific purposes, not as desired by China but as the colonial powers decided.[87] In response to this opportunity, in February 1926, when the Sixth Biennial Conference of the National Medical Association took place in Shanghai, its members heatedly discussed the possibility of lobbying for the British indemnity fund.[88] By the end of this conference, they had decided to urge the British government to allocate a substantial portion of this fund "for the specific purpose of promoting public health activities in China."[89] In order to confer with the British Boxer Indemnity Commission, the association established a Committee for the Advancement of Public Health in China, which prepared the proposal during the following two months. In this sense, this proposal represented the collective consensus of the National Medical

Association of China in 1926. Because Liu was not only the president of this association but had also been the superintendent of the PUMC since 1925 and a long-term ally of John Grant, the two proposals came from people with close ties and similar views about medicine in China.

Within the former proposal for the British government, Liu provided answers to a series of questions that were very similar to the ones raised by Greene. Apparently these questions occupied the best minds among the Chinese advocates of public health at that time. Liu did not consider "the Problem of Stable Government," to be serious, because "the present instability of government is limited to national and provincial organization and does not affect local communities and cities."[90] More importantly, Liu argued that public health had already secured for itself a "nominal position" in Chinese government; as in Japan and Germany, local health administration was an integral part of the police. "This circumstance obviates the difficulty of creating a new branch of government to undertake public health functions as would be necessary were provision [made] for the latter along lines found in Anglo-Saxon countries."[91] Instead of creating a new ministry, the crucial task was to staff these public health positions under the police with trained personnel. The lack of a stable central government in the foreseeable future did not constitute a serious problem, because Liu saw "the necessity of restricting most of any proposable health work for the next decade to its establishment in local centers."[92] In accordance with this judgment, the whole proposal focused on local endeavors to promote public health.

Less than one year later, John Grant prepared his memorandum for a very different audience and with different objectives. After a decade of suffering at the hands of the warlords, the Nationalist Party launched the Northern Expedition in 1926, determined to reunify China by military force. As its Nationalist Revolutionary Army succeeded in gaining control over the central Yangzi Valley in late 1926, the Nationalist government moved its capital from its southern location in Canton north to Wuhan in central China. Encouraged by the military progress, at this point Nationalist leaders started planning the organization of a government that would realize their stated objective of state building.[93] In Grant's view, the political advancement of the Nationalists carried with it two distinct implications for medical development in China. In terms of the development of public health, "the establishment of Nationalist authority in 1927 proved a turning point";[94] on the other hand, "1927 saw the suspension in South China of a large proportion of medical institutions receiving foreign cooperation."[95] Grant's second point is a major understatement to describe the catastrophe that befell missionary medicine. Because of the manifested hatred and social disruption caused by the Nationalists and their xenophobic sentiments, by July 1927, only five hundred of the former eight thousand missionaries remained in China.[96] Even Yan Fuqing, the Chinese dean of the Xiangya Medical College, was seen as part of this foreign infringement and was forced to leave Changsha in December of 1926.[97]

Instead of feeling threatened, Grant stayed and interpreted the Nationalist advancement as an unprecedented opportunity for promoting public health. To join forces with the Nationalists' attempt to realize a self-aggrandizing, modernizing agenda,[98] Grant suggested in his memorandum that a Provisional National Health Council be established that would prepare for the creation of an independent Ministry of Health within one year. At the end of the memorandum, Grant added a note explaining that this document had been translated by Dr. Yan Fuqing into Chinese and was to be presented by Sun Ke (1891–1973), the minister of communication and the son of Sun Yat-sen, to the National Council of the Nationalist Party in March 1927. "The proposal has been discussed with several members of the Council and received their approval. The Minister of Finance has agreed to provide the funds,"[99] Grant pointed out explicitly. On March 13, the Nationalist Party's Central Executive Committee passed a resolution to add five new ministries to the government, including the Ministry of Health.[100] Several days later, Mme. Sun Yat-sen nominated Liu Ruiheng to be the director of this new ministry.[101] Grant's and Yan's connection and their memorandum likely played a direct role in creating the first Ministry of Health in China.[102]

In addition to its practical influence in persuading the Nationalists to establish a new ministry, Grant's memorandum constituted an ideological breakthrough in terms of the discourse about the relationship between medicine and government. "The country is now passing through a Revolution which extends also to the conception and the carrying out of governmental obligations to the people."[103] Opening his memorandum this way, Grant proceeded to argue that the Nationalist Revolution should be a revolution of "governmental obligations" and that "not the least of the obligations of modern government to the people it represents is that of medical protection."[104] This deceptively straightforward statement marked a crucial reversal of logic. For Roger Greene, Liu Ruiheng, and many of their Chinese colleagues, the primary concern was the popularization of scientific medicine and public health in China. They saw the unfortunate situation of the Chinese state as an obstacle that they had to endure, requiring them to circumscribe (Liu in 1926), or to put off their otherwise ambitious project of establishing public health (Greene in 1924).

Grant did not start out with the objective of popularizing medicine and then trying to cope with the problem of the state. On the contrary, his starting point was the state. Situating medicine in the context of the recent development of Chinese government, Grant argued that "medical protection" was just a new extension of a series of "governmental obligations," following police and national education, neither of which had been viewed as "governmental obligations" in China just twenty-five years ago. In view of this reversal of logic, the question was no longer how the medical profession should best deal with the problem of an unstable state, but whether the Nationalist government would commit itself to taking on another new "governmental obligation" so it could function more successfully as a modern state. In this sense, Grant's state-centered discourse signaled the beginning of a crucial change in conceptualizing the relationship between government and medicine.

As the foundation of Grant's theoretical framework, the idea of "governmental obligation" also provided the key rationale for his practical suggestion that China needed an independent Ministry of Health. Once the "recent extension of governmental obligation" was accepted, it followed that "preventive and curative medicine cannot be administered separately without detracting from the efficiency of both. Effort in the execution of this principle is seen in the elevation of health administration to a position of a separate ministerial rank in some twenty national governments, almost all within the past decade."[105] As this statement shows, Grant did not base his argument exclusively on the importance of public health. On the contrary, curative medicine was equally important for the fulfillment of governmental obligations, especially in China, where modern curative medicine was so sparse.[106] In addition to providing services of preventive medicine, public health, and medical administration, the Ministry of Health was created to make available to the Chinese people a "minimum degree of the benefits of curative medical science at a relative small cost."[107]

For the sake of fulfilling this novel objective, Grant explicitly suggested that the Nationalist government should break away from the German and Japanese model of medical administration, in which the health department was placed under the Ministry of Interior or Home Affairs.[108] Not having high regard for the American system either, Grant urged the Nationalist government to adopt the most advanced health policy that had recently been developed in England and Russia after the First World War,[109] especially the British establishment of an independent Ministry of Health in 1919.[110] The objective was to create "a Ministry of Health, to include eventually *full state control* [author's emphasis] of all medical matters both preventive and curative."[111] It is remarkable that, once a state-centered logic had been adopted and health care had been accepted as a governmental obligation, the radical idea of "full state control" became an inevitable, even logical, conclusion for medical development in China. This was the very beginning of so-called State Medicine in China; to its architect, Grant, it was this ambitious objective of State Medicine that justified the need for a separate Ministry of Health. Moreover, to highlight the logical nature of this reasoning, when Grant explicitly promoted State Medicine one year later, after the Ministry of Health was sure to be established, the title he chose for his article was "State Medicine: A Logical Policy for China."[112]

Nevertheless, the Chinese translation of Grant's proposal that was later published in the *National Medical Journal of China* differed from the English original in almost all the historic points that I have just highlighted.[113] First of all, instead of being presented as a translation, the published article was presented as an original paper authored by Yan Fuqing; Grant's name and his memorandum were not mentioned at all. Second, the Chinese article consistently downplayed the role of "governmental obligation," replacing Grant's high-minded opening paragraph with a list of practical and nonmedical—economic, military, and political—benefits that medicine could bring to the nation. While most references to "governmental obligation" were retained in the Chinese article, the term was translated with the Chinese term *zhize,* whose meaning is closer to "responsibility" than to "obligation" (*yiwu*). Moreover, right after the section entitled "Reasons for a Ministry of Health," Grant's version had contained the capitalized subtitle "Medical Obligations of Modern Government." In Yan's Chinese article, this subtitle was replaced by "Medical Administrations of Various Modern Countries," even though most of the other contents were faithfully translated and preserved. Finally, while Yan's Chinese article also strongly recommended the establishment of a Ministry of Health, it carefully deleted the attached elaboration "to include eventually full state control of all medical matters both preventive and curative."[114] The crucial link that Grant had constructed between the Ministry of Health and State Medicine just disappeared from the Chinese translation.

There is no way of telling whether these significant revisions were made by Yan to win the support of the Nationalists or whether they were the result of negotiation with the Nationalist leaders, because the Chinese article was published after the Nationalist government's decision to establish the Ministry of Health in March 1927. Therefore, we cannot know for sure who—medical leaders, Nationalist leaders, or both—demanded the change of these crucial points in Grant's memorandum or for what reasons. Either way, the Chinese proposal deliberately jettisoned both the key rationale (governmental obligation) and the crucial objective (full state control of medical matters) of Grant's memorandum, but it realized Grant's practical suggestion—the Nationalist government agreed to establish a Ministry of Health.

On the surface, the birth of China's first Ministry of Health signaled great hope for the advancement of medicine. As documented by Ka-Che Yip, however, this unprecedented government endeavor to promote medicine was more a compromise of political convenience than a well-thought-out commitment to health. The idea of setting up a Ministry of Health did not come from the core members of the Nationalist Party but from the newly joined warlord Feng Yuxiang (1882–1948), the famous "Christian General." For the Nationalists, the main concern was to keep General Feng in the alliance; for that purpose, they considered adding a new ministry to the cabinet, for the sake of his supporters. Very revealingly, it was a follower of Feng's, Xue Dubi (1892–1973) who, prompted by John Grant and Yan Fuqing, proposed to General Feng the idea that the new ministry to be created should be a Ministry of Health and should be headed by him.[115] As the result, Liu Ruiheng became the viceminister of health. Apparently, neither the idea for nor the interest in a Ministry of Health came from the Nationalists. The lack of long-term commitment became evident when the Nationalist government downscaled the ministry into a Health Administration just one year after its establishment, because the new ministry no longer served its political purpose. Nevertheless, with this medical administration, which was staffed mostly with faculty and alumni from PUMC, the first generation of Chinese practitioners of Western medicine finally succeeded in creating a space for their profession in the emerging state machinery.

Conclusion

This chapter has aimed to provide a historical account of China's response to the problem that Yu Yan retrospectively formulated in 1935, that is, how to popularize Western scientific medicine in China. By the time that the Ministry of Health was established by the Nationalist government in 1928, the answer to this question had taken on a recognizable form. In order to popularize scientific medicine in China, its promoters had no choice but to rely on the political power of the state. While the state appeared to emerge as the subject of medical history, the real actors of this history were mostly the Chinese practitioners of Western medicine discussed in this chapter, such as Wu Liande, Liu Ruiheng, Yan Fuqing, and their Western allies, John B. Grant, and the leading players at the Rockefeller Foundation.[116] It was this group of medical practitioners who actively strove to create an unprecedented new relationship between their profession and the state.

In the process of recruiting the state, these men developed a new vision for Western medicine in China that strongly emphasized public health. Ironically, in the early twentieth century, the Chinese state was often considered to be the major barrier, rather than a solution, to promoting public health. To show how risky and uncertain this approach was as a political strategy, this chapter has documented the Rockefeller Foundation's changing policies toward the promotion of public health in China. This radical shift of policy was possible because in 1928, when the Republican government put an end to the warlord period and started committing itself to building the first modern Chinese state, John Grant and his Chinese colleagues turned this enduring set of difficulties into reasons why a state-centered medical system was both an absolute necessity and a valuable opportunity for China's medical advancement. Assuming the state to be the answer to all their daunting problems, public health advocates strove to make their innovative State Medicine an integral part of the Nationalists' state-building efforts, thereby making themselves dual agents tasked with "popularizing Western medicine while constructing the state."

This strategy had important and long-lasting effects for the development of modern health care in China, because, as a result, almost all the leaders of the Chinese medical profession during this period had assumed the role of the architects and promoters of public health. This feature is in sharp contrast with the situation in the United States, where the medical profession displayed little interest in salaried public health positions. As a result, "public health evolved into a somewhat separate professional specialty, staffed by biologists, statisticians, engineers, and others with specialized training."[117] Nothing better illuminated this important feature of the Chinese medical profession than its ultimate endorsement of State Medicine in 1937, a crucial development that is discussed in chapter 10.

On the other hand, however, the strategy of promoting modern medicine by way of the state had an unintended, confining effect on medical development as well. While the original intention of the Chinese doctors had been to use the state as a tool for popularizing scientific medicine, in their efforts to recruit the state they could easily be directed to use medicine as a tool for realizing the state's political objectives. For example, Wu later devoted himself to helping the government to regain control of the seaport Quarantine Service,[118] which had been administered by foreigners since the latter half of the nineteenth century, because the establishment of a Chinese-run Quarantine Service could pave the way for the return of the treaty ports and the recovery of tariff autonomy for the Chinese government. As criticized by another practitioner of Western medicine, the new Ministry devoted a disproportional share of its resource to issues related to foreign affairs.[119] As Chinese practitioners of Western medicine developed their strategy to recruit, even colonize, the emerging Nationalist state, they gradually accepted the state's established concerns as their own prioritized medical problems.

Since the foremost political objective of the Nationalists was state building, these practitioners of Western medicine could not help serving as dual agents, being simultaneously pioneers of public health and agents of the state. To adopt sociologist Pierre Bourdieu's analytic framework, as they endeavored to recruit the state for medical development, the practitioners of Western medicine committed themselves to the dual process by which "[the agents of the state], under the guise of saying what the state is, caused the state to come into being by stating what it should be."[120] In the processes of instituting the state, they "constituted themselves into state nobility," becoming the powerful figures that Wu promised—state medical officers. Emerging in conjunction with the Nationalist state and Chinese nationalism, the first generation of biomedically trained Chinese doctors swiftly took over the leadership positions from the medical missionaries who had dominated modern medicine in China since the mid-nineteenth century.[121]

The Nationalist state did not greet the various medical visions with equal enthusiasm. Within one year, the state downsized the Ministry of Health into a Health Administration (*weishengshu*). While the ministry actively improved seaport quarantine to help recover China's sovereignty, it paid very little attention to the idea of State Medicine, only mentioning it once in passing in its first year's twelve volumes of gazettes. The governing elites had no problem deciding which medical project best served the state's pressing concerns while demanding less input of its precious resources. It was not until the end of the second phase (1929–47) that they succeeded in making the state, at least on paper, embrace its medical obligation to the people and commit to implementing State Medicine. Meanwhile, as they succeeded in establishing the Ministry of Health and securing for some of their colleagues the relatively powerful positions of state medical officers, these practitioners of Western medicine became increasingly trapped in their strategy of making the state the subject of medical history. The next chapter shows how wrong this strategy could turn out to be. Instead of successfully resolving "the problem of Chinese medicine" through the power of the state, they ended up teaching practitioners of Chinese medicine their strategy of success, that is, to build the link between medicine and the state.

4 Imagining the Relationship between Chinese Medicine and Western Medicine, 1890–1928

The present book opens with the containment of the Manchurian plague and the historic effect this event had on establishing "the superiority of Western medicine over Chinese medicine." Nevertheless, it is crucial to point out that the landmark status of this event was completely due to—but at the same time also reinforced—a state-centered perspective on medical history. If we do not focus so exclusively on the relationship of the state to Western medicine but on the response of the general public to the medical and political measures taken in response to the plague, the landmark status of this event becomes less obvious. Decades before the outbreak of the plague in 1911, the population in late Qing China had already recognized certain weaknesses of Chinese medicine and had developed visions for incorporating some of the strengths of Western medicine into the theoretical framework of Chinese medicine. Innovative medical thinkers had imagined various ways of relating Western medicine to Chinese medicine long before the Manchurian plague forced the Qing state to draw the historic conclusion of the superiority of Western medicine.

A revealing snapshot of such appreciation in Chinese medical circles for the contributions of Western medicine appears in the lecture notes compiled by the newly established China Medical Society (*zhongguo yixuehui*)—one of the earliest societies organized by practitioners of Chinese medicine—for its schools of Chinese medicine in 1909. In its attempt to create a new educational institution for Chinese medicine, the society emphasized the urgent need for two subjects:

> What the physicians of Western medicine are most dissatisfied with in Chinese medicine is its lack of anatomy and experimentation. We should consult Western and Chinese books on anatomy and physiology so as to produce lecture notes on the subject of anatomy. On the other hand, what practitioners of Chinese medicine are most dissatisfied with in Western medicine is its ignorance about yin-yang and *qi*-transformation (*qihua*). Therefore, we should also prepare lecture notes on *qi*-transformation so as to preserve the essence [of Chinese medicine].[1]

As this statement reveals, at the end of the Qing dynasty, practitioners of Chinese medicine in this society had agreed that they needed to learn Western medical knowledge, especially anatomy. At the same time, they still held high regard for their own medical tradition, characterizing Chinese medicine as based on a unique theory of *qi*-transformation that reached beyond the horizons of Western learning.

This paired characterization of Chinese medicine as based on *qi*-transformation and of Western medicine as based on anatomy was first conceived of in the 1890s by Tang Zonghai (1851–1908), the widely acclaimed founder of the School of Converging Chinese and Western Medicine (*zhongxiyi huitong xuepai*). The statement issued by the China Medical Society testified to the long-lasting popularity of Tang's paired characterization of these medicines as well as his pioneering efforts to combine the strengths of both traditions.

However, this vision of medical syncretism had encountered serious challenges since the late 1910s. Instead of praising Tang and other figures for their pioneering appreciation of Western medicine, beginning in the latter half of 1910s, Chinese practitioners of Western medicine—represented by Yu Yan (1879–1954, also known as Yu Yunxiu)[2]—felt the strong need to completely discredit such efforts to merge Chinese medicine with Western medicine. To break away from this imagined syncretism, through which Chinese medicine and Western medicine could learn from and benefit each other, Yu Yan and others strove to impose a modernist framework in which it was no longer possible for either traditional practitioners or the general public to imagine the coexistence of these two medical traditions, let alone a mutually complementary relationship between them. This proposed modernist framework involved a partitioning of Chinese medicine into three categories: theory, Chinese drugs (*zhongyao*), and experience (*jingyan*). Furthermore, theory was to be abandoned once and for all, while Chinese drugs and experience were to be subjected to scientific research. Based on a globally circulated discourse of modernity, this tripartite characterization of Chinese medicine has played a dominant role in shaping the modern history of Chinese medicine ever since. After documenting Tang Zonghai's vision for medical syncretism, this chapter traces the rise of this tripartite characterization of Chinese medicine, the local and regional forces that gave it credibility, and the resistance to it by traditional practitioners until the eve of the 1929 confrontation.

Converging Chinese and Western Medicine in the Late 1890s

Because Tang Zonghai is currently remembered as the founding master of the School of Converging Chinese and Western Medicine, historians view him mainly as a medical practitioner. If we do not impose this categorization upon him, it becomes clear that Tang was more notably a Confucian scholar, and one who was as anxious about the fate of Chinese civilization as many other reform-minded scholars of the late Qing. In other words, his project for medical eclecticism represented a Confucian scholar's appreciation and response to late nineteenth-century Western civilization.

Lacking any familial background in medicine, Tang devoted the first half of his life to preparing for the civil service examination. Unlike many other scholar physicians who turned to medicine only later in their careers (often after repeatedly failing the civil service examination), Tang achieved at the relatively young age of thirty-eight the prestigious rank of "presented scholar" (*jinshi*).[3] He was a rare exception in that he chose to pursue a medical career rather than a lucrative and highly regarded career as government bureaucrat. As a result of his clinical skills as a traditional practitioner, and especially after the publication of his 1884 book, *On Blood Disorders* (*Xuezhenglun*), he made a name for himself in Sichuan Province. In the 1880s, after traveling to the Jiangnan area (most notably Shanghai), he became strongly interested in Western science and medicine. Finally, in 1892, he published a groundbreaking book with the title *The Essential Meanings of the Medical Canons: (Approached) through the Convergence and Assimilation of Chinese and Western Medicine* (*Zhongxi huitong: Yijing jingyi*), abbreviated below as *The Essential Meanings of the Medical Canons*.

It was in *The Essential Meanings of the Medical Canons* that Tang Zonghai first presented the following famous formula: Western medicine is good at visible configuration (*xingji*); Chinese medicine is good at *qihua* (*qi*-transformation).[4] As many scholars have noticed, Tang's formula had resulted in a long-lasting dichotomy between Chinese *qi*-transformation and Western anatomy, as shown by the statement made by the China Medical Society in 1909. Ironically, as I have argued in detail elsewhere,[5] scholars are often unaware that Tang's conception of *qi*-transformation was built upon and therefore heavily influenced by newly imported technology from the West, namely, the steam engine. For Tang, the *qi* of *qi*-transformation was closely modeled upon steam, whose Chinese counterpart, *zhengqi*, was a newly coined term to translate this modern concept into Chinese. Counter-intuitively, it was the steam engine that inspired Tang to create a new understanding of *qi*-transformation in the human body and a new characterization of Chinese medicine as being in opposition to Western anatomy.

Moreover, this new understanding of *qi* enabled Tang to reform Chinese medicine by incorporating into it both the new knowledge and illustrations of Western anatomy. Up to this point, most scholars have overlooked the intriguing fact that the visual representation of the body, especially of the visceral organs, was the medium that Tang Zonghai adopted to present his vision of converging Western and Chinese medicine. In the brief preface to *The Essential Meanings of the Medical Canons*, he listed and commented on five kinds of medical illustrations: (1) traditional Chinese diagrams of the viscera; (2) illustrations from *Correcting the Errors of Medicine* (*Yilingaicuo*), authored by the famous critic of Chinese medicine Wang Qingren (1768–1831) and published in 1830; (3) Western anatomical illustrations, such as those included in *Treatise on Physiology* (*Quanti xinlun*), authored by the Scottish missionary physician Benjamin Hobson in 1851; (4) traditional diagrams of the meridian channels; and (5) two illustrations that he had created himself. For Tang, the central task of integrating Chinese and Western medicine thus consisted of discerning the appropriate ways of reconciling these illustrations by preserving or abandoning them, adding to or subtracting from them, combining them, and, ultimately, by creating new ones.

In terms of Chinese history, the Song dynasty "marks a turning point in the graphic representation of the body,"[6] but Tang Zonghai nevertheless advocated abandoning all traditional Chinese diagrams of the viscera because "they mostly did not match the real configurations of the human visceral organs."[7] Tang came to this radical conclusion after serious engagement with the many criticisms laid out by Wang Qingren and Benjamin Hobson, whose works, as Benjamin Elman emphasizes, "represented the first sustained introduction of the modern European sciences and medicines in the first half of the nineteenth century."[8] A good example of Tang's engagement with both authors appears in his response to their criticism that Chinese medicine was ignorant about the urinary function of the kidney.[9] After admitting that it amounted to an embarrassing crisis for Chinese medicine, Tang included in his book an illustration of the kidney that foregrounds a pair of urinary tubes (*niguan*).[10] This image of urinary tubes, an anatomical structure unknown to the Chinese until this time, was perhaps the first of its kind in a book on Chinese medicine.

In addition to their importance in helping to rectify the errors of Chinese medicine, Western anatomical illustrations provided crucial tools with which Tang could elaborate the "essential meanings of medical canons." Nothing better explains the implication of Tang's book title than his claim to have resolved a long-term controversy over the question whether the Triple Burner (*sanjiao*) is a theoretical construct or a visible, material entity. In the process of arguing for the latter position, Tang went so far as to present an illustration of the Triple Burner (figure 4.1, left) which, as I discovered and document elsewhere,[11] was in fact a close reproduction of an illustration of the peritoneum from the authoritative *Gray's Anatomy* (figure 4.1, right). To be more precise, Tang's illustration of the Triple Burner appears to have been copied from a Chinese book called *Anatomy, Descriptive and Surgical* (*Quanti chanwei*), the first Chinese translation of *Gray's Anatomy,* which had been completed in 1881 by Dauphin W. Osgood. From Tang's point of view, the detailed lines of the peritoneum provided the crucial evidence that the Triple Burner in the *Inner Canon of the Yellow Emperor* (*Huangdi neijing*) was in fact a visible material entity and thereby resolved the controversy once and for all.

The detailed fashion in which Tang discussed these five kinds of illustrations confirms that Tang was an avid student of Western science. Not only did he make the crucial decision to replace traditional Chinese-style diagrams of the internal organs with modern anatomical illustrations, but he also appropriated the concept of steam to invent a new conception of *qi*-transformation and thereby created a new interpretation of the *Inner Canon* that was presumed to be consistent with Western anatomical illustrations. This three-part relationship reveals the true meaning of his book title: *The Essential Meaning of the Medical Canons: (Approached) through the Convergence and Assimilation of Chinese and Western Medicine.* It was in this process of communicating

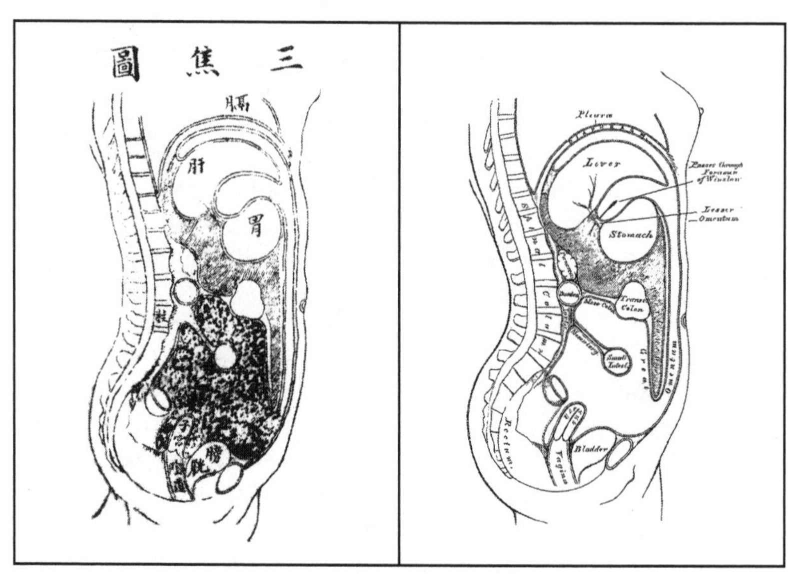

FIGURE 4.1 Left: "Illustration of the Triple Burner" (*sanjiao tu*) from Tang Zonghai, *Zhongxi huitong: Yijing jingyi* [The essential meanings of the medical canons: (Approached) through the convergence and assimilation of Chinese and Western medicine], 2 vols. (Shanghai: Qianqingtan shuju, 1908), 1:24; orig. pub. 1892. Right: "The reflections of the peritoneum, as seen in a vertical section of abdomen" from *Gray's Anatomy* (1858), 599.

between the two styles of medicine, by means of the conception of *qi*-transformation, that Tang believed to have recovered, or newly created, a correct understanding of the *Inner Canon*. *Qi*-transformation could serve as a tool to bridge the gap between the two styles of medicine because it was reconceptualized as the transformation of steam.

Non-Identity between the Meridian Channels and the Blood Vessels

Having elaborated how Tang used anatomical images to serve as a medium for integrating two styles of medicines, I should emphasize that Tang also used anatomical images to draw boundaries between them. Ironically, in the late 1920s their boundary-drawing function became so prominent that it foreshadowed Tang's efforts to integrate the two styles of medicine by way of his new conception of *qi*. In order to trace the radically changed ways in which people understood Tang's project in the late Qing and early Republican periods, we have to examine closely Tang's original thesis of the fundamental "non-identity" (*butong*) between the meridian channels (*jingmai*) and the blood vessels (*xueguan*).

In the eyes of modern biomedicine, the Chinese concept of meridian channels is a chaotic and confusing amalgamation of the vascular, nervous, and endocrine systems, along with other concepts that have no counterpart at all in the modern scientific understanding of the human body. In the words of Manfred Porkert (written in the 1980s), the concept of *jingmai* (meridian channels) should be translated as "arterial pathways" because it is formed by compounding two Chinese words: *jing*, meaning "course" or "path," and *mai*, meaning "pulse" and "artery." According to Porkert, the *jingmai* "are conceived as the conduits through which various forms of physical energy are transmitted, [but] these pathways in themselves are purely theoretical constructs."[12] While Porkert thus did not hesitate to treat the meridian channels as theoretical constructs, many historical Chinese sources suggest that the system of meridian channels includes physical elements rather close to the blood vessels. Because the courses of the meridian channels do in fact mirror the network of blood vessels in many respects, Wang Qingren famously suggested that meridian channels were meant to represent the physical system of the blood vessels, only in a rather crude and incorrect fashion.[13] Although the anatomical status of meridian channels remains an unsolved mystery,[14] the concept is crucial for many of the practical therapies of Chinese medicine, such as acupuncture, moxibustion, and drug therapy in the Cold Damage (*Shanghan*) tradition.[15]

When Tang articulated the thesis of non-identity in response to the criticism of the meridian channels made by Hobson and Wang, he presented his argument alongside a Western-style illustration called "Diagram of blood vessels":

> On the basis of this diagram, Westerners have decided that there is no such thing as the twelve meridian channels and eight extraordinary channels. The book *Correcting Errors of Medicine* also judges the meridian channels to be baseless. They do not realize that because they have always dissected dead bodies, they surely have been unable to identify these channels and acupuncture points. Besides, the meridian channels are not identical with the blood vessels. As suggested in the *Inner Canon*, certain meridian channels contain an abundance of blood but a scarce amount of *qi* while others contain an abundance of *qi* but a scarce amount of blood. On the basis of this fact, we know that the meridian channels contain both blood and *qi*; they should not be equated with either blood vessels or trachea. The Western view of the human body includes a separate entity

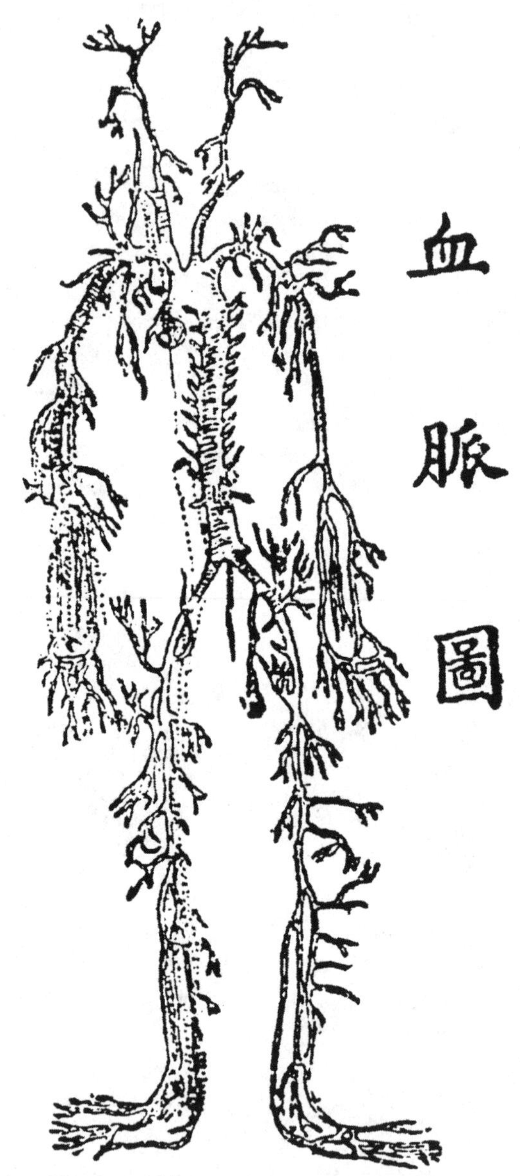

FIGURE 4.2 "Diagram of blood vessels" (*xuemai tu*), from Tang Zonghai, *Zhongxi huitong: Yijing jingyi*, [The essential meanings of the medical canons: (Approached) through the convergence and assimilation of Chinese and Western medicine] (Taipei: Lixing shuju, 1987), 110; orig. pub. 1892.

> called the autonomic nervous system *(zihe naojin)*. . . . This idea sounds similar to what the *Inner Canon* referred to as meridian channels. It is a pity that Westerners do not know the Chinese language so they never seriously study the *Inner Canon*.[16]

In comparison to the five kinds of illustrations that Tang discussed in his preface, the way that this Western "Diagram of blood vessels" was used is a salient anomaly. Instead of using anatomical images to elaborate on the teachings of the *Inner Canon,* he used this diagram to demonstrate that certain aspects of the *Inner Canon* were beyond the knowledge of Western anatomy. Rather than achieving the stated objective of converging Chinese and Western medicine, this diagram served as unique visual evidence for a breakdown in integration. Secondly, Tang provided an important methodological foundation for this breakdown of integration—"because they have always dissected dead bodies, they surely have been unable to identify the meridian channels and acupuncture points."[17]

Even at this moment of failed integration, Tang did not think that the differences between the two medicines were insurmountable. On the contrary, he mentioned that the Western concept of the autonomic nervous system sounded quite similar to that of the meridian channels as discussed in the *Inner Canon*. Tang's reference here to the autonomic nervous system is noteworthy. He appeared to be very concerned with the Western idea,[18] made famous by Hobson in his *Treatise on Physiology*, that the whole body is governed by the brain instead of by the heart-mind (*xin*), as Chinese medical theory stipulates.[19] Instead of embracing it, Tang considered this new knowledge that sensation and movement were functions that belonged exclusively to the brain as only "partially true."[20] In light of this reservation, it becomes understandable why Tang found the concept of the autonomic nervous system so appealing. It is quite possible that Tang had learned from Osgood's *Anatomy, Descriptive and Surgical* that the autonomic nervous system is supposed to "spread all over the visceral organs and blood vessels and to function by itself without the involvement of human volition."[21] Thus, just like the Chinese conception of meridian channels, the autonomic nervous system was linked to the internal organs; moreover, it had a life of its own, independent from the voluntary control of the brain. Perhaps because this concept of the autonomic nervous system undercut the dominance of the brain and gave some "autonomy" to the internal organs, Tang concluded, with some reservation, that "this idea [of the autonomic nervous system] sounds similar to what the *Inner*

Canon referred to as meridian channels."[22] As in his conception of the Triple Burner, which included the material structure of the peritoneum, the meridian channels could be based on the equally material structure of the nervous system.

Finally, Tang did not think that the meridian channels were in principle beyond the cognitive horizon of Western medical doctors; they would be able to learn about these structures if they only mastered the Chinese language and studied the *Inner Canon*. There was no effort on Tang's part to construct an incommensurable relationship between the two systems of medicine. This accords with Tang's statement in the preface that his objective was to "combine Chinese and Western doctrines and not to be biased by any delimiting difference, but to strive to find the *one* true knowledge."[23] Nevertheless, as we will soon see, Tang's famous comparison between Chinese *qi*-transformation and Western anatomy, especially the "non-identity" of the meridian channels and the blood vessels, was to be turned into a dichotomy of two incommensurable worlds when the intellectual climate radically changed in the 1920s.

Yu Yan and the Tripartition of Chinese Medicine

If taken against the backdrop of this popular idea of "converging Chinese and Western medicine"—an idea that survived the abolishment of the civil service examination in 1905 and the overthrow of the Qing dynasty in 1911—one can appreciate how radically Yu Yan intended to transform the relationship between Chinese and Western medicine when he came to center stage in the late 1910s.

In order to understand the struggle between practitioners of Chinese medicine and Western medicine during the Republican period, no one is more central than Yu Yan. For both traditional and biomedical practitioners in the 1920s and 1930s, the notion of a "medical revolution" (*yixue geming*) was inseparable from him. According to Yu's own recollection, he started developing a critical stance toward Chinese medicine while studying Western medicine in Japan's Osaka Prefectural Medical College (which, in 1931, became the Osaka Imperial University) during 1908–11 and 1913–16. Although he was a lifelong student of the philologist and revolutionary Zhang Taiyan (1868–1936),[24] who supported the reform of Chinese medicine, Yu became an increasingly determined critic of it—so much so that he proposed to abolish it altogether during the first National Public Health Conference held in China in 1929. After his proposal fell through and unexpectedly provoked the National Medicine Movement, Yu actively participated in every major debate between the two factions of medicine. As one of the very few physicians trained in both Western medicine and traditional Chinese medicine during this period, Yu played a crucial role in setting the agenda of these debates. In the 1950s, after the Communist government decided to promote traditional Chinese medicine, the Nationalists' medical policy was often criticized as being Yu Yan's policy of "abolishing the [Chinese] medicine but preserving [Chinese] Drugs" (*feiyi cunyao*).[25] For practitioners and supporters of Chinese medicine, Yu personified the oppressive forces of Western medicine.

When Yu Yan started composing *Critique of the "Divine Pivot" and "Basic Questions"* (*Lingsu shangdui*)[26] in 1914, while he was studying in Japan, he was apparently ahead of most of his compatriots. By the time that it was published, in 1917, he had already returned to China. In the opening sentence of this book, Yu declared that his purpose in writing it was to "disclose all the fallacies of the *Divine Pivot* (*Lingshu*) and *Basic Questions* (*Su-wen*)," the two extant parts of the most ancient Chinese medical classic, the *Inner Canon*.[27] On the basis of modern anatomy and physiology, Yu systematically repudiated almost every fundamental concept of Chinese medicine: Yin and yang, the five phases, the five viscera and six bowels,[28] and the twelve meridian channels. Having done so, Yu expected that "as soon as this article was published, many practitioners of the old medicine would rise up expressing their objections."[29] But that was not the case. "Who would have known that their spirits were so low and there wouldn't be any response at all?" recalled Yu.

While the publication of *Critique of the "Divine Pivot" and "Basic Questions"* in 1917 was meant to serve a destructive function, in 1920 Yu published a long article to articulate his view about a constructive project for Chinese medicine called Scientific Research on Nationally Produced Drugs (*Guochan yaowu de kexue yanjiu*). Although Yu's two works appeared to come from opposite directions, the latter project in fact was completely built upon the *Critique* published three years earlier. As a follow-up to his *Critique*, Yu's article took a previously unproblematic question as its departure point: Since Chinese medical theories were unscientific and nothing but far-fetched speculation, and since traditional Chinese anatomy, physiology, and pathology had all been proven to be wrong, how could Chinese medicine be effective in treating certain diseases?[30]

While at this moment Yu Yan seemed to be alone in his concern with this question, it would soon become hotly debated by both styles of medical practitioners in the late 1920s.[31] Regardless of whether one was searching for the "scientific value" or the "national essence" of Chinese medicine, one first had to provide a convincing answer to this question. As a result, the question about the real "location" of Chinese medical knowledge—which elements of Chinese medicine should be credited for their curative efficacy—emerged as a great enigma.

Yu offered four options as possible answers to this most frequently asked question about Chinese medicine.[32] Not surprisingly, the first was the use of Chinese drugs (*zhongyao*). The second was traditional practitioners' experience (*jingyan*) in prescribing drugs. Third, he noted that patients occasionally recover by themselves without any effective medical intervention. Yu's fourth explanation was what we now call the placebo effect—the psychological effect of doctors' authoritative diagnosis and prediction. Since the last two possible answers were no more than suggestions that Chinese medicine was ineffective, Yu attributed the actual efficacy of Chinese medicine to what he called its "facts," namely, Chinese drugs and experience. Yu further asserted that, with regard to the curative efficacy of Chinese medicine, "its theories and facts are two distinctly different things."[33] Thus, at the same time that Yu singled out Chinese drugs and experience as the valuable essence of Chinese medicine, he decoupled them from its medical theories.

In two important senses, Yu argued, the experience of Chinese medicine was unrelated—and should be considered as unrelated—to its theories. First, rather than being guided by theory, experience was in fact acquired through the instinctive testing of Chinese drugs. In order to emphasize the instinctive, pretheoretical nature of experience, Yu went so far in other writing as to compare the experience of the local population to the instinctive behavior of animals.[34] For Yu, because Chinese drugs were based on this instinctive behavior, the so-called "Chinese drugs" in fact had nothing particularly "Chinese" about them except that they happened to be produced in Chinese territory. Being "grass roots and tree bark" (*caogen shupi*), the so-called Chinese drugs were raw materials from nature, having nothing to do with either Chinese culture or Chinese medical theories. According to Yu, therefore, these so-called Chinese drugs should be more correctly referred to as "nationally produced drugs" (*guochang yaowu*). In this sense, Yu's conception of experience was instrumental in creating a conceptual basis for the project of Scientific Research on Nationally Produced Drugs.

Second, Yu suggested that the history of Chinese medicine included a special sort of experience that preceded any theoretical formulation. Only this specific kind of experience deserved serious study because it had not been polluted by the speculative theories of Chinese medicine.

According to Yu, the Song dynasty was the historical turning point for Chinese medicine, because thereafter speculative theories had begun to be added onto practical experience.[35] By recasting the history of Chinese medicine in this way, Yu Yan further decoupled Chinese drugs from medical theory along a temporal axis. Instead of progressing by accumulating more facts, Chinese medicine had actually regressed since the Song dynasty.[36] With regard to conducting scientific research, Yu Yan therefore suggested that researchers should focus on studying those formulas collected before the Song dynasty.

Ironically, the Song dynasty probably contributed more official support to Chinese medicine than any other imperial government in Chinese history, particularly during the Northern Song dynasty (960–1127 CE). According to Asaf Goldschmidt's recent studies, the Northern Song government implemented a series of revolutionary policies to reform and support medicine, including the establishment of an Imperial Medical Service and Imperial Pharmacy, the systematization of medical curricula and examinations, the standardization of acupuncture and moxibustion therapy, the expansion of medicinal therapy, and, most importantly, the reprint of the ancient medical canons (one of which was *The Treatise on Cold Damage* [*Shanghan lun*]),[37] which had previously been inaccessible to the majority of physicians. Moreover, due in part to the Song dynasty emperors' strong interest in medicine, scholar-officials also became more interested in medicine as a profession—a discipline they had previously considered less than respectable. In fact, in order to attract talented scholars to the field of medicine, emperor Huizong (1082–1135 CE) even coined the new term "scholar-physician" (*ruyi*).[38] Because of these revolutionary developments, the social status of medical practitioners rose significantly.[39]

While practitioners of Chinese medicine during the Republican period considered the Song dynasty the golden age in terms of governmental support for Chinese medicine, from the viewpoint of Yu Yan and other critics, this social elevation of medicine had resulted only in its catastrophic deterioration. Yu elaborated as follows:

> I think that the medicinal formulas that postdate the Song and Yuan periods are adulterated by scholarly doctrines. Therefore, they are very suspicious. . . . On the contrary, the formulas circulating among the vulgar healers and villagers are much less deceptive than those prescribed by the scholar-physicians. We should try our best to collect their [i.e., the vulgar healers' and villagers'] discoveries, which could be used as raw material for

> our research. If we pay due respect to the facts discovered by human instinct, we can uncover the truths of experience and confirm their validity by means of experiments.[40]

Thus, on both epistemological and historical grounds, Yu Yan had demonstrated that Chinese drugs and experience were separable—and *should* be separated—from the allegedly speculative theories of Chinese medicine. According to this picture of Chinese medicine, atomized bits of theory-free experience were supposed to have accumulated in a mechanical fashion, with no systemic relation to each other. Just like the Baconian facts of natural history that had served to overthrow the system of Aristotelian natural philosophy,[41] Yu's notion of experience collapsed the system of Chinese medicine into atomized objects ready for scientific research. Instead of being merely a repressive ideology, Yu's conception of experience stimulated traditional practitioners to take action—to discredit their own medical theories, to treat their own medical knowledge as if it were a collection of atomized experiences, to view the time-tested Chinese drugs as nothing but raw materials of nature, to trust those who hitherto had been considered untrustworthy, and to value and collect common medicinal formulas from among the general population. Clearly, the more practitioners of traditional medicine pursued this line of action, the more Chinese medicine disintegrated.

To Avoid the Place of Confrontation

Although in the beginning Yu Yan was very disappointed to find that no traditional practitioners rushed to the defense of Chinese medicine, he became even more dismayed when the traditional practitioners Yun Tieqiao (1879–1935) and Yu Jianquan responded to his *Critique* seven years after its publication. Yu Yan found it hard to believe that they both had merely adopted Tang Zonghai's tired strategy, which had been developed more than three decades earlier:

> The smart ones then thought about adjusting their old theories to avoid my criticism. Last year, Yun Tieqiao wrote *Record of Perceptions of the Classics* (*Qunjing jianzhi lu*), suggesting that the five viscera in the *Inner Canon* are not identical with the five physical viscera but with the five organs of seasonal *qi*-transformation. This is actually a strategy to avoid the place of confrontation [*bi di zhi ju*]. Yu's [Yu Jianquan's] argument is another example of "a strategy to avoid the place of confron-

> tation." Recognizing the weakness [of Chinese medicine], they have escaped to a no-man's-land.[42]

It is very revealing that Yu used a spatial metaphor—"to avoid the place of confrontation"—to describe the strategies of both Yun Tieqiao and Yu Jianquan as they appropriated *qi*-transformation to defend Chinese medicine. This spatial metaphor vividly reveals that the struggle between the two styles of medicine had shifted to a new arena—ontology. Compared to the general idea of science, the greatest challenge to Chinese medicine was posed by the widely accepted idea that science—and, by extension, Western medicine—held a monopoly on the truth of ontology. Since there was seemingly only one such ontological world, there seemed to be no way that Chinese medicine and modern medicine could coexist. More than anything else, the rise of ontology radically restructured the field of struggle between the two styles of medicine and turned it into a zero-sum game. As a result, in the 1924 debate between Yu Yan and Yu Jianquan over the relationship between blood vessels and meridian channels, Tang's conception of *qi*-transformation was radically transformed from a tool of communication, as he had conceived it in the 1880s, into a tool "to avoid the place of confrontation" in the new space of ontology.

When Tang argued in the 1890s that meridian channels and the biomedical blood vessels were far from identical, he regretted that Westerners were not familiar with the Chinese language and had never studied the *Inner Canon* in depth, as if they could have otherwise appreciated the Chinese doctrine of meridian channels. In sharp contrast to Tang's optimistic expectation, however, when Chinese experts well-versed in both medicines did appear for the first time several decades later, they brought forth a much more daunting challenge to Chinese medicine than that from its Western critics. As one of those rare individuals who had mastered both systems of medicine, Yu Yan remarked that, among all the points he had raised in his *Critique*, he was especially proud of his criticism of the theory of meridian channels. Unlike the Western critics of Tang's time, who had simply claimed that "meridian channels were nonexistent," a claim that Tang and like-minded traditional practitioners could easily attribute to their ignorance of Chinese medicine, Yu Yan's objective was not to demonstrate the nonexistence but rather the wrongfulness of the doctrine of meridian channels.

Because Yu Yan's objective was to demonstrate how incorrect the Chinese theory of meridian channels was as a representation of the system of blood circulation, his criticism presupposed that meridian channels and blood vessels could be compared, if not considered identical. To demonstrate the comparability between the two, Yu devoted twelve long paragraphs to discussing the routes of each of the twelve meridian channels in detail. During this process, he pointed out the numerous mistakes that became apparent when one compared them with the corresponding blood vessels. While we might be overwhelmed by the number of discrepancies between the two systems, the common points that established the condition of comparability between the two are equally as important. Instead of dismissing the meridian channels as nonexistent, Yu Yan displayed his erudition in both styles of medicine to argue for the specific conclusion that the term "meridian channels in the *Inner Canon* was meant to refer to arteries. When the deeply-situated vessels were taken to be meridian channels, [the *Canon*] wrongly confused veins with arteries."[43] Once the fact of the comparability of meridian channels and blood vessels was established, it followed that the system of meridian channels was a terrible misrepresentation, marked by numerous mistakes. The logical conclusion was that medicine based on such erroneous knowledge was bound to fail.

In response to Yu Yan's specific critique of the theory of meridian channels, Yu Jianquan based his argument on Tang Zonghai's conception of *qi*-transformation. In an essay pointedly entitled "On the Non-Identity of Meridian Channels and Blood Vessels," published in 1924, Yu Jianquan strove to refute the identification of the two concepts on which Yu Yan's criticism had been based. For this purpose, Yu Jianquan appealed to Tang Zonghai's famous formula, referring specifically to his comments made along with the "Diagram of the blood vessels" (figure 4.2).[44] To be sure, Tang did provide statements that could be interpreted as suggesting a dichotomy between the two styles of medicine. He had said, for example, "Western anatomy also recognizes layers and folds but not meridian channels; only physical configurations but not *qi*-transformation."[45] Nevertheless, unlike Tang, who had endeavored to depict *qi*-transformation and physical configuration as mutually complementary concepts, Yu Jianquan strove to separate them into two ontologically independent worlds. On the basis of a macro-ontological dichotomy between *qi*-transformation and physical configuration, he argued for a micro-dichotomy between meridian channels and blood vessels.

In order to better explain Yu Jianquan's new interpretation of meridian channels, I refer to it by its Chinese term, *jingmai*, in the following two paragraphs. Pressured by the compelling evidence that Yu Yan had quoted from the *Inner Canon*, such as the phrase "what is called *mai* ("vessels") is the house of blood," Yu Jianquan admitted, "It is beyond doubt that the *mai* contained blood."[46] In order to challenge the direct identification of *jingmai* with the blood vessels, Yu Jianquan now strove to decouple *jingmai* from *mai*, which seemed hopelessly similar to the biomedical system of the blood vessels. In other words, instead of interpreting *jingmai* as a compound term comprised of both *jing* and *mai* (in the sense of either pulse or arteries), as suggested by Porkert, Yu Jianquan invented an interpretation of *jingmai* that was unrelated to either *mai* ("vessel") or *xuemai* ("blood vessels").

To achieve this objective, Yu Jianquan developed a series of mutually supporting points. First of all, unlike the *xuemai*, which contained streams of blood, *jingmai* was said to refer to the routes of *jingqi* ("channel *qi*") and, therefore, belonged exclusively to the realm of *qi*-transformation.[47] Secondly, according to Yu Jianquan, the term *mai* in *jingmai* "referred to the pathway of *jing*; it is the route of *jing* and by no means refers to the stream of *xuemai*."[48] In this sense, *jingmai* should be, and could be, differentiated from *xuemai*: the latter contained blood and could be equated with the blood vessels, while the former contained only *jing qi*. Finally, and most importantly, Yu Jianquan added a new feature to the doctrine of *qi*-transformation, and to the concept of *jingmai* in particular: he emphasized that "the way that the twelve *jingmai* spread out all over the body is like a seamless cloth that is invisible."[49] With the addition of this new feature of invisibility, *qi*-transformation was transformed in Yu's hands into an invisible and immaterial entity, completely beyond the gaze of modern anatomy and thus independent of the material realm of "physical configuration." Since his objective was to turn *qi*-transformation and physical configuration into two independent, unrelated, and even unrelatable, entities,[50] it is no wonder that Yu Jianquan never mentioned Tang's innovative efforts in relating the two by means of the conception of *qi* as steam.

Although Yu Jianquan admired Tang Zonghai greatly and praised him as the father of the School of Converging Chinese and Western Medicine, his argument for the "non-identity of meridian channels and blood vessels" directly collided with Tang's. For Tang, the two concepts had not been identical because the concept of meridian channels covered more than just the blood vessels. Tang never denied that meridian channels included the blood vessels, or at least a portion of them. In sharp contrast, Yu Jianquan believed that the meridian channels did not include the blood vessels but represented a completely independent system. Pressured by Yu Yan's materialist arguments, Yu Jianquan's response strove to create a space detached as far as possible from the material world of modern anatomy. In this sense, it was very insightful that Yu Yan characterized his argument as "a strategy to avoid the place of confrontation."

This strategy became a necessity when Chinese medicine was forced to accept a modernist ontological space in which meridian channels and blood vessels could not coexist. In response to this modernist epistemic hegemony, an equally modernist defensive strategy arose, represented by Yu Jianquan, which involved a historic retreat. To "avoid the place of confrontation," Chinese medicine had to carefully retreat from all potentially dangerous zones of contact, thereby turning itself into a mirror-image of its oppressor. As Yu Jianquan decided to abandon the potentially arguable point of the blood vessels, he had to make the meridian channels completely invisible and immaterial. Despite the apparent continuity between Tang Zonghai and Yu Jianquan in terms of upholding the doctrine of *qi*-transformation, its very conception and function went through a fundamental change by being shifted into an immaterial and invisible realm that was completely outside the space occupied by modern science. This shift was closely related to the rise of scientism, which had become a popular ideological force in public discourse since the May Fourth Movement in 1919. Threatened by the idea that modern science held a monopoly on ontological truth, advocates of Chinese medicine developed this strategy in order to circumvent the authority of science. It was in this specific context of defense that *qi*-transformation became the commonsensical characterization of Chinese medicine that has persisted to this day. As indicated by this case, the ultimate price that Chinese medicine paid for this defensive strategy was a distorted self-image determined largely by its enemies.

Although people emphasized different dimensions and understandings of *qi*-transformation, it nevertheless ceased to be a tool of communication, as Tang had conceived it. Instead, from this time on, the concept of *qi*-transformation functioned as "a strategic stronghold," as Yu Yan bitterly complained, to resist the attacks of science and Western medicine. Through Yu Jianquan, Yun Tieqiao, and other like-minded defenders of Chinese medicine, this incommensurable opposition between the two styles of medicine began taking shape in the early 1920s and persisted until the historic confrontation in the late 1920s.

Ephedrine and Scientific Research on Nationally Produced Drugs

Although Yu Yan's *Critique* had been met with resistance by traditional practitioners in the early 1920s, his call for scientific research on Chinese drugs had—within a mere decade of the publication of his original article—developed into a national consensus. However, as a result of the Communist government's criticism in later years, which referred to Yu's policy as "abandoning Chinese medicine but preserving [merely] Chinese drugs" (*feiyi cunyao*), scholars often fail to remember that Yu Yan deserves credit for championing scientific research on Chinese drugs in the early 1920s. If not for the unexpected assistance and scientific breakthroughs of the Peking Union Medical College (PUMC), it is uncertain whether scientific research on the Chinese materia medica could ever have become the national consensus, a consensus that has had a long-lasting influence on the development of the modern medical sciences in the twentieth and twenty-first centuries in China.

By the late nineteenth century it took unusual courage for any scientist to disregard the global trend of medical research and suggest that traditional materia medica was still worthy of serious research. In spite of its roots in traditional materia medica, the development of modern pharmacology since the latter half of the nineteenth century had revolutionized the field, making the study of materia medica "the least progressive of all medical disciplines."[51] Nothing illustrates this drastic change of attitudes better than Oliver Wendell Holmes's oft-quoted 1860 statement that, with a few exceptions (such as opium) "if the whole materia medica, as now used, could be sunk to the bottom of the sea, it would be all the better for mankind—and all the worse for the fishes."[52]

In light of this prevailing shift from traditional materia medica to modern pharmacology, it is highly plausible that Yu Yan's unusual interest in traditional drugs had its origins in his Japanese educational background. The Japanese tradition of modern pharmacology can be traced back to Nagai Nagayoshi (1844–1929), who was one of eleven people making up the first of several publicly funded groups to be sent to Europe by the Meiji government.[53] Under the supervision of August Wilhelm von Hofmann at Berlin University, Nagai had received his PhD in pharmacology—the first to hold such a degree in Japan. From the time he returned to Japan to serve as a professor of pharmacology at Tokyo Imperial University in 1884, Nagai devoted most of his career to the chemical analysis of traditional drugs. His most famous achievement was the isolation of an alkaloid—which he named *ephedrine* in 1885—from the Chinese herb *mahuang* (ephedra). A widely admired founder of Japanese pharmacology and the first president of the Pharmacological Society of Japan, Nagai played a crucial and long-term role in making the study of traditional drugs a legitimate branch of modern pharmacology in Japan.

It should be mentioned that this strong interest in scientific research on traditional drugs was perceived at the time as a Japanese obsession—one not shared by the European and American scientific communities. As John Grant later recalled about his visit to Japan in 1924, he had been appalled to find that "not only in each medical school but in each large city you would find a municipal laboratory spending thousands and hundreds of thousands of yen each year on the investigation of indigenous Chinese drugs."[54] Not surprisingly, Grant was doubtful that this research had produced anything worthy of such a hefty investment.

It was this global intellectual context that had made the credibility of Chinese drugs the center of the controversy with which the present book begins. As a result of such suspicious attitudes toward the efficacy of Chinese drugs, when the dying Sun Yat-sen finally decided to be treated with Chinese drugs on February 19, 1925, the administration of the Peking Union Medical College (PUMC) demanded that he be moved out of its hospital. While this dramatic action was later widely remembered as an expression of PUMC's uncompromising attitude toward the use of Chinese drugs, it is all too easy to forget that the school had concurrently been at the forefront of conducting scientific research on Chinese drugs. In fact, several months before this incident, researchers at PUMC had enjoyed a landmark success by "discovering" an alkaloid from the Chinese herb ephedra—that is, ephedrine, the compound that Nagai had named four decades earlier but that had long since been forgotten—to be effective for asthma relief. Thanks to the success of this research at the US-sponsored PUMC, the Japanese-inspired vision of researching Chinese drugs became nationally approved in China on the eve of the 1929 confrontation.

What made this landmark success possible was the unusual decision by PUMC to devote its Department of Pharmacology to research on Chinese materia medica.[55] In order to justify the decision to support what was commonly seen as a dying science, Henry S. Houghton—soon to become the director of PUMC—spelled out his rationale in May 1920 in a letter to Edwin R. Embree, secretary of the Rockefeller Foundation and China Medical Board: "It would appear fundamental to the planting of western medicine in China that we find out and recognize the worth of her own material, hoping that our own may in turn receive just recognition from her. . . . It is earnestly hoped that this department may be so developed that full justice may be given to it."[56] Apparently, PUMC made this unprecedented decision in order to demonstrate its good intentions in recognizing and developing appreciation for the value of traditional Chinese culture, just as it had done in designing its famous buildings, which deliberately incorporated some characteristics of Chinese architecture.[57]

To promote the study of Chinese drugs, PUMC appointed the faculty members for its Department of Pharmacology with utmost care. The first full-time faculty member—and later director—of this department was Bernard E. Read (1887–1949), a chemist by training, who was, however, better-known for his strong interest in Chinese materia medica and natural history. To assist Read in catching up on modern pharmacological research, PUMC financed his advanced education at Yale, where he received his PhD in pharmacology in 1924. At around the same time, PUMC also invited Carl F. Schmidt, professor of pharmacology at the University of Pennsylvania, to serve as a visiting professor from 1922 to 1924 and initiate research on the Chinese materia medica. Upon Schmidt's arrival in Beijing, he received "a list of supposedly important Chinese drugs" from Read as well as a recommendation to investigate the herb astragalus (*huangqi*) from Victor Heiser, then president of the International Board of the Rockefeller Foundation.[58] It is worth pointing out that astragalus was the Chinese drug that had allegedly helped Hu Shih (1891–1962), the famous leader of the New Culture Movement, to recover from a serious diabetic condition in 1920–21. Because of this positive experience, Hu had—albeit with some hesitation—recommended his physician, Lu Zhong'an, to Sun Yat-sen. Incidentally, as soon as Sun was removed from the hospital at PUMC, one of the first two Chinese drugs that Lu prescribed to Sun was astragalus.[59]

To Schmidt's disappointment, he found neither a trace of a definite active constituent nor significant pharmacological actions in any of the recommended Chinese drugs, including astragalus. Around the time when Schmidt was considering terminating the investigation altogether, K. K. Chen (Chen Kehui, 1898–1988) accepted a position as senior assistant at PUMC, returning to China from the United States in August 1923. Following is Schmidt's recollection of what happened next:

> If it had not been for a series of coincidences that led us to ephedrine (my writing to Chen before he came to Peking, telling him of my unpromising experiences with Chinese drugs, his telling something of this to an uncle at a family reunion in Shanghai and getting recommendation to look into *mahuang*, and our trying a hasty aqueous extract of this drug in a preparation for a student experiment), my two years in Peking would have been highly interesting but scientifically unrewarding.[60]

When K. K. Chen triggered this "series of coincidences," he was only twenty-five years old. After receiving his bachelor's degree in chemistry from the University of Wisconsin in 1920, Chen had spent two years (1921–23) studying in its medical school before returning to China to work in the Department of Pharmacology at PUMC. Chen and Schmidt coauthored their original paper on ephedrine in 1924,[61] Schmidt being so generous as to insist that Chen be listed as its primary author. At this point, Chen and Schmidt were completely unaware of the studies that Nagai had performed on *mahuang* four decades previously. As Chen recalled, they were preparing "to give a new name to this alkaloid," only to learn from the literature that it was already "a well-known chemical compound called ephedrine."[62]

When Schmidt left PUMC for Philadelphia in October 1924, he told Houghton, "The more I think of it, the more certain I feel that pharmacology stands second to none, and is equaled possibly only by parasitology, as a field for productive work in Peking."[63] In the United States, the first publication on asthma soon stimulated what could be called an ephedrine "gold rush."[64] Within five years, research on *mahuang* had grown into a small industry that resulted in the publication of five hundred research papers on ephedrine worldwide. To summarize this enormous amount of research, Chen and Schmidt again coauthored a 117-page monograph, "Ephedrine and Related Substances," in the American journal *Medicine* in 1930.[65]

As Schmidt later recalled, the research on ephedrine provided much more than just a paradigmatic example for the potential benefits of research in indigenous Chinese drugs: it helped to move modern pharmacology in China from a "negative phase" to a "positive phase,"[66] by transforming the goal of research from aiming to understand the fundamental mechanisms of individual medicinal herbs to promoting the development of new and effective drugs. Thanks to his successful research on ephedra during the early stages of his career, K. K. Chen was invited by Eli Lilly and Company to serve as its director of pharmacological research in 1929, with the understanding that he would have "complete freedom of research, particularly in Chinese materia medica."[67] Chen went on to become a world-renowned scientist, working at Eli Lilly and Company for thirty-four years (1929–63) and serving as president of the Society for American Pharmacology and Experimental Therapeutics in 1952.[68] Most important for the history under investigation in the present book, ever since the publication of Chen's groundbreaking research in 1924, it has become standard practice for any discussion about Chinese medicine to cite the case of ephedrine. Even those who had originally held conflicting attitudes toward Chinese medicine and had been its biggest critics consequently began to share the opinion that Chinese drugs, unlike other aspects of Chinese medicine, were worthy of serious scientific research.

Inventing an Empirical Tradition of Chinese Medicine

While Yu Yan's concept of experience was a tool with which he attempted to disintegrate Chinese medicine into an object for scientific research, it nevertheless helped foster the notion that there existed an "empirical"—and therefore more valuable—subtradition of Chinese medicine. In other words, although the notion of a completely theory-free pharmaceutical tradition in Chinese medicine—or perhaps in any medical tradition—is questionable on a general level, the concept of experience, understood as an actors' category and a relational concept,[69] had played an important role in reassembling the tradition of Chinese medicine. The most salient efforts to reassemble such a tradition were those of Yu Yan and his mentor Zhang Taiyan (1869–1936), who sought to elevate *The Treatise on Cold Damage* into the canon of this empirical tradition.

Among all the advocates of Chinese medicine, Zhang Taiyan perhaps enjoyed the greatest respect among his contemporaries both as a great scholar and as a political figure in the revolution of 1911.[70] Starting out as an activist in the late Qing reform movement, Zhang later switched positions to join the Chinese United League (*Tongmenghui*) and developed the first formal expression of Chinese nationalism in opposition to the Manchu-controlled Qing dynasty. He coined the phrase "*Zhonghua Minguo*," which eventually became the name of the Republic of China. Concurrent with his political efforts to overthrow the Manchu regime, Zhang championed the National Essence Movement (*guocui yundong*) to foster a national cultural revival and to preserve the "national essence" of China. After 1918, Zhang withdrew from politics and lived the life of a scholar and teacher, educating many leading scholars of the Republican period. With such a remarkable reputation, Zhang was widely respected among practitioners of Chinese medicine as their leader. In 1927, Zhang accepted the invitation to serve as honorary president of the Shanghai Institute of Chinese Medicine (*Shanghai zhongguo yixueyuan*), formally becoming an icon for reforming Chinese medicine.

Just like his student Yu Yan, Zhang Taiyan singled out *The Treatise on Cold Damage* as the representative canon for empiricism in Chinese medicine. The *Treatise* was clearly more empirical when compared with both the *Inner Canon of the Yellow Emperor* in ancient times and the four great masters of the Jin and Yuan dynasties whose medical doctrines were deeply embedded in neo-Confucian philosophy.[71] Zhang and Yu Yan came to this important conclusion because their scholarly orientation was deeply rooted in Han learning (*Hanxue*), the evidential scholarship that emerged in the late Ming and Qing and was critical of neo-Confucianism, or Song learning (*Songxue*).[72] As scholars of Han learning considered neo-Confucianism to have distorted the true meaning of the foundational Confucian texts, they strove to trace upstream to the earliest layer of interpretation of these texts. For Zhang and Yu, the uncorrupted canon in the field of medicine was *The Treatise on Cold Damage* compiled by Zhang Zhongjing at the end of the Han dynasty.

Moreover, Zhang and Yu's appreciation of *The Treatise on Cold Damage* was very much influenced by the Japanese "ancient formula current" (*gufang pai*), a style of Chinese medicine popular in Edo Japan. Beyond just revealing their personal inclinations, their approval of it reflects a then-emerging general trend in Republican China. As Jia Chunhua points out, the Japanese "ancient formula current" played a crucial role in inspiring and shaping the modern study of *The Treatise on Cold Damage*.[73] This Japanese approach was highly critical of the medical developments in China during the Jin-Yuan period and thereby became a distinctive Japanese tradition of Chinese medicine on the basis of the *Treatise*. Among other objectives, Japanese scholars of this tradition fundamentally challenged the conventional wisdom that the *Treatise* was based upon the theoretical framework of the *Inner Canon*, even though Chinese scholars since the Jin-Yuan period had devoted much interpretative effort to substantiating this connection. In a nutshell, the "ancient formula" scholars strove to establish *The Treatise on Cold Damage* as a practical book of clinical practice that was independent of the theoretical framework of *Inner Canon*. Moreover, as Elman points out, the Japanese approach was a part of larger efforts by Tokugawa scholars "to detach classical learning and ancient medicine from China and make it Japanese."[74]

Zhang Taiyan admired this Japanese tradition of Chinese medicine, especially the Japanese scholarship on *The Treatise on Cold Damage*, so greatly that he once openly asserted: "If [Zhang] Zhongjing were to return to life, he would invariably say, 'my way prospers in the East [i.e., Japan].'"[75] Zhang's scholarship on the *Treatise* helped to elevate this text to a prominent status in the Republican period, and his enthusiasm for Japanese scholarship helped create a lucrative market for the translated Japanese works on traditional medicine.

Among these translated Japanese texts, the one that most directly influenced the concept of experience and the notion of an empirical tradition was *Japanese-Style Chinese Medicine* (*Kôkanigaku*), written by Yumoto Kyushin (1876–1941). Formally trained in Western medicine, Yumoto nevertheless published this book in 1927 to defend the value of traditional East Asian medicine. As Yumoto made clear in the preface, "I deeply believe in the ancient formula current, and the core content of this book is based on Zhang Zhongjing's *Treatise on Cold Damage*."[76] As a result of the strong interest in Japanese scholarship, Yumoto's book was translated into Chinese independently by two authors within only three years of its publication.

In this much-appreciated book, Yumoto proposed the notion of "experience with the human body" (*renti jingyan*) as an alternative basis for traditional East Asian medicine. Inspired by Yumoto's writing, advocates of Chinese medicine advocated an opposition between Chinese and Western medicine rooted in the view that Chinese medicine was supposedly based on "experience with the human body," while Western medicine was held to be based on "animal experiments." As a result, they argued that Chinese medicine was more appropriate for treating humans than Western medicine. More than just a tool for scholarly debate, the concept of "experience with the human body" also had practical implications for the protocol for scientific research on Chinese drugs, a crucial issue that I analyze in chapter 9.

Despite the fact that Zhang and Yu shared a background in Han learning and enthusiasm for *The Treatise on Cold Damage*, the master and his disciple held very different opinions regarding the value of Yumoto's contributions. After reading Yumoto's book, Zhang reportedly encouraged his student Zhang Cigong (1903–59) to pursue advanced studies in Japan.[77] In contrast, Yu Yan later devoted a whole booklet to refuting Yumoto's notion of "experience of the human body" point by point.[78] As I document in detail elsewhere, it was in the engagement with Yumoto's book that the notion of experience gradually took shape, becoming the crucial, although problematic, epistemological concept for modern Chinese medicine.[79] In light of the remarkable common ground shared by Zhang and Yu, their radically different attitudes toward Japanese-style Chinese medicine and the related concept of experience suggested strongly opposed visions with regard to reforming, reassembling, or dismantling the tradition of Chinese medicine.

Conclusion

At the time that the idea of converging Chinese and Western medicine was still popular, Yu Yan had championed a partitioning of Chinese medicine into three categories: theory, Chinese drugs, and experience. With a few exceptions, such as Ding Fubao, Yu Yan was a solitary pioneer when he proposed this tripartite characterization of Chinese medicine in the late 1910s, but by the time he debated with opponents in the early 1920s, he was no longer alone in his thinking. Once the confrontation broke out in the spring of 1929, this characterization of the medicine at once became the basic framework for the struggle between advocates and opponents of Chinese medicine. As a discourse of modernity, this characterization of Chinese medicine effectively put an end to the late Qing vision of converging Chinese and Western medicine as conceived by Tang Zonghai in the 1890s.

I emphasize that this characterization of Chinese medicine is a modernist discourse because its tripartite division was based on a modernist divide between nature and culture, with Chinese drugs being characterized as raw material from nature, an allegedly erroneous theory viewed as a purely cultural construct. In the opposition between theory and Chinese drugs, which neatly filled the positions on each side of this great divide, experience played the role of an ambiguous mediator between the poles of the modernist framework. Understood in the sense of the quasi-Darwinian concept of instinct, experience was taken as the empirical basis of Chinese medicine and hence as one of two elements, along with Chinese drugs, worthy of serious scientific research.

Although this tripartite characterization of Chinese medicine was clearly structured by globally circulated modernist discourses, by no means was it merely a local duplication of that "universal" discourse in China. To state the obvious, the concept of experience was not created exclusively for the sake of mediating the relationship between modern and indigenous medicines, even though using it in this way was a strategy shared by advocates of non-Western medical traditions around the world.[80] Rather, it was also created to shed light on the internal differences among the subtraditions of Chinese medicine and to valorize the relatively more "empirical" tradition of medicine that developed before the Song dynasty vis-à-vis the style of medicine during the latter Jin-Yuan period. By elevating the *The Treatise on Cold Damage* over the *Inner Canon of the Yellow Emperor* as a more experience-oriented and thus more valuable canon of Chinese medicine, advocates of Chinese medicine could claim to have revived the "empirical tradition" of Chinese medicine and thereby associate Chinese medicine with Japanese-style Chinese medicine and with the image of Japan as a model for China's modernizing efforts. Thus, because the concept of experience had to serve such specific functions in these local and regional contexts, it could not possibly have remained a decontextualized, general discourse of modernity. The once abstract notion of experience became increasingly materialized—and reconfigured—as *jingyan* when it was connected to and thus influenced by the concrete local tasks that shaped the identity of modern Chinese medicine.

Even more localized than the notion of experience was the valorization of Chinese drugs as promising objects for scientific research. This valorization of traditional materia medica was a local or, more correctly, a regional feature of modern East Asia. Although the conception of traditional drugs as the "raw material of nature" was clearly rooted in the modernist divide between culture and nature, pharmacologists in Europe and North America had ceased to consider these raw materials as promising research objects long before this time. Because of the inspiration of Japanese pharmacology, Yu Yan actively promoted the study of Chinese drugs even as he strove to repudiate the credibility of Chinese medicine; and it was for the sake of appealing to Chinese nationalism that PUMC was willing to devote some tentative efforts to the research of Chinese drugs. Thanks to "a series of coincidences," these otherwise tentative efforts achieved important success in the study of ephedrine, consequently giving scientific validation to this local and regional valorization of Chinese drugs. It is worth mentioning that even with the noteworthy success of ephedrine, this local valoration was viewed as being mainly an expression of nationalist sentiment by many Western observers until the 1970s—and perhaps is still so regarded to this day.[81]

What practitioners of Chinese medicine in the 1920s found most unacceptable among the three categories of this tripartite characterization of Chinese medicine was the idea that Chinese medical theories were erroneous representations of the world that should be abandoned once and for all. Most practitioners of Chinese medicine simply ignored Yu Yan's *Critique*; those of them who cared to debate with Yu adopted a defensive strategy of "avoiding the place of confrontation," thereby transforming the concept of *qi*-transformation from a tool of communication into one of incommensurability. There was little sign that the community of Chinese medicine would recognize the critical flaw of their doctrines and launch a fundamental reform from within. Partially because of this frustrating experience, Yu Yan later decided to radically change his strategy when history appeared to have turned a new page, namely, when the Nationalist Government established China's first Ministry of Health in 1928.

5 The Chinese Medical Revolution and the National Medicine Movement

The Chinese Medical Revolution

Very few scholars of modern China have heard of the so-called Chinese medical revolution (*zhongguo yixue geming*) that occurred in the late 1920s, and with good reason: it utterly failed to achieve its objective of abolishing the practice of Chinese medicine. As a great historical irony, however, its efforts had a revolutionary consequence in the opposite direction, for it gave birth to the National Medicine Movement (*guoyi yundong*), and, in the eyes of C. C. Chen, it had the most unfortunate long-term effect of creating a bifurcated medical field in China.[1] By way of documenting these two crucial events—the Chinese medical revolution and the National Medicine Movement—this chapter traces the rise of a decade-long collective struggle between the practitioners of the two medicines, a struggle that fundamentally transformed the modern history of Chinese medicine.

Yu Yan is widely credited with creating and promoting the idea of a Chinese medical revolution.[2] When he claimed in 1928 that he had been advocating for such a revolution for more than a decade, he was referring to the publication of his *Critique of the "Divine Pivot" and "Basic Questions"* in 1916. The problem is that among the forty articles collected in *Mr. Yu's Essays on Medicine* (*Yushi yishu*), published in 1928, none included the term *medical revolution* in its title. As soon as Yu published his 1928 article "Construction and Destruction in Our National Medical Revolution,"[3] however, "medical revolution" became one of his favorite catch-phrases, to be used frequently in future writings.[4] Moreover, in 1932, four years after the publication of *Mr. Yu's Essays on Medicine*, Yu changed its title to *Collected Essays on Medical Revolution* (*Yixue geming lunwenxuan*), and in the following year he published volume two under the same new title. In light of such a clear-cut adoption of this concept from 1928 on, we can ascertain with confidence that it was the Nationalist Revolution (*guomin geming*) that directly evoked, and was invoked by, the "medical revolution."[5]

As Yu Yan explicitly pointed out, his conception of medical revolution drew upon the ideas promoted by German liberal pathologist Rudolf Virchow (1821–1902), the founder of social medicine. Quoting verbatim Virchow's famous slogan, "Medicine is a social science, and politics is nothing but medicine on a larger scale,"[6] Yu promoted the dual projects of "socialization of medicine and hygienization of politics."[7] In this regard, Yu joined a larger trend in colonial East Asia and Southeast Asia. As a recent study by Anderson and Pols demonstrates, Virchow's vision of social medicine inspired many native medical practitioners in colonial Asia to go beyond clinical practice and to fashion themselves as leaders of the nationalist movement.[8] It was in the context of this larger endeavor to realize a revolutionary vision for biomedicine—and for politics—in China that Yu Yan pushed for his long-held agenda to abolish the practice of Chinese medicine.

It was no accident that Yu suddenly raised the banner of a medical revolution around 1928. At the end of 1928, the Nationalist Party's Revolutionary Army finally brought the Warlord period to an end and nominally unified China.[9] Partially because of their self-proclaimed commitment to building a modern state and partially because of political convenience, the Nationalists established a Ministry of Health in China's new capital of Nanjing. As a result of these developments, several practitioners of Western medicine—including Yu—had secured the important post of state medical officer. This strategic position provided Yu with more effective ways to marginalize Chinese medicine than simply through attacking its theoretical foundations. Now Yu and his colleagues recast their attack on Chinese medicine as part and parcel of their effort to realize a national medical revolution. Yu said,

> Is there any other reason that I have shouted out to promote a medical revolution and tearfully appealed to my people? What has deeply agonized me are the following: the old medicine does not obey science, the medical administration is not unified, the construction of public health has stagnated in many respects, and the shameful name of "Sick Man of East Asia" (*dongya bingfu*) has not been deleted.[10]

Thus, the medical revolution, as Yu conceived it, integrated the task of abolishing Chinese medicine into the projects of building a national medical infrastructure and overcoming the "Chinese deficiency" in terms of hygienic modernity.[11] Once the "problem of Chinese medicine" was recognized as a problem of the state, Yu found it quite natural to propose using extraordinary "political means" to solve it.[12]

Controversy over Legalizing Schools of Chinese Medicine

During the period when biomedical practitioners developed the strategy to "popularize Western medicine with the state," practitioners of Chinese medicine gradually learned the necessity of associating one's medicine with the state. The first time that traditional practitioners learned this lesson occurred when the government promulgated a series of ordinances for modern schools in 1912 but did not include any ordinances for Chinese medicine. Given that no schools of Chinese medicine (*zhongyi xuexiao*) had existed in the imperial period, the "omission of Chinese medicine from the national educational system" did not necessarily suggest a governmental policy against Chinese medicine. In response to this ordinance, however, traditional practitioners launched their first campaign with a meeting in Shanghai that included representatives from nineteen provinces. The government finally capitulated to their demands, and this meeting paved the way for the establishment of approximately a dozen schools of Chinese medicine, including the most influential Shanghai Technical College of Chinese Medicine (*Shanghai zhongyi zhuanmen xuexiao*) in 1916.[13]

A much larger campaign took place a decade later when the respected Society for the Advancement of Chinese Education (*zhonghua jiaoyu gaijinshe*) held a national meeting in Taiyuan in 1925. During the meeting, the Jiangsu Chinese Medical Society submitted a proposal that schools of Chinese medicine be incorporated into the national education system. Although the Society for the Advancement of Chinese Education was famous for its leading role in promoting modern education, this conference nevertheless passed a resolution to adopt this proposal and referred it to the Ministry of Education.[14]

Instead of resisting intervention by the state, traditional practitioners asked the Chinese government to take on the unprecedented task of certifying their schools. Their unconventional request was far from an isolated event; rather, it was a reflection of a much larger trend during the Republican period. As Marianne Bastid insightfully points out, "In education, more than any other area, the bourgeoisie, gentry, and former constitutionalists were best prepared to occupy the territory they had been unable to conquer under the monarch and which the revolution had just opened up by overthrowing the Qing. The territory was the machinery of the state."[15]

Just like the bourgeoisie and the gentry, proponents of Chinese medicine considered education to be a newly opened channel through which they could gain access to the state. Clearly, this earliest campaign by traditional practitioners did not concern itself with medicine per se, but with access to the interests newly created by the state.

It follows that traditional practitioners, just like advocates of modern medicine, strongly criticized the Chinese government for not realizing the importance of medicine and for its consequent failure to intervene sufficiently in medical matters. To emphasize the precedent of enlightened government in Chinese history, they praised the Song dynasty (960–1200) for its unusual commitment to promoting medicine.[16] Some went further to argue that the decline of such governmental commitments was the root cause of Chinese medicine's lagging behind Western medicine in recent times.[17] One such practitioner stated, "With regard to medicine, the imperial governments belittled it as being just a craft, society treated [practitioners] as hired labor, and the government never promoted it. These are the reasons why [Chinese] medicine has not progressed. It is not because [Chinese doctors'] skill was crude. If the state had placed a high value on [Chinese] medicine—as it has done in establishing schools for Western medicine and awarding [its practitioners] many privileges, [Chinese medicine] would never have declined to the present level."[18]

In comparison, to recognize the "superiority" of modern Western medicine, it was much easier for traditional practitioners to dwell on the fact that Western medicine had enjoyed privileges, professional interests, and governmental support that had never been accessible to them. Instead of resisting the expansion of the state into the medical sphere, some practitioners of Chinese medicine saw quite early on the potential benefits of having an "unconventional state" that concerned itself with medical matters.

This campaign for legalizing schools of Chinese medicine provoked practitioners of Western medicine to establish the Shanghai Union of Physicians (*Shanghai yishi gonghui*) in 1925, which brought together the most vocal and determined critics of Chinese medicine: Yu Yan, Wang Qizhang (1885–1955, also a graduate of Osaka Prefectural Medical College), Pang Jingzhou (1897–1966, a graduate of the German-founded Tongji University),[19] and Fan Shouyuan. The opposition between advocates and opponents of Chinese medicine further intensified a few months later when the National Alliance for Education (*quanguo jiaoyu lianhe hui*) held its conference in October 1925 in Hunan. During that conference, two provincial educational associations again proposed to assimilate the schools of Chinese medicine into the school system. Again, the conference voted to support the proposal.

While this was occurring, Yu Yan published "Denouncement of the Proposal for Old-Style Medical Schools"[20] in 1926, in which he refuted nearly all the points listed in the proposal for assimilating schools of Chinese medicine into the national school system. Unlike his prior theoretical attacks on the *Inner Canon*, this time Yu's article was immediately showered with counterattacks from traditional practitioners. In spite of drawing such a different response, the arguments against Chinese medicine found in Yu's "Denouncement" did not go much beyond his *Critique of the "Divine Pivot" and "Basic Questions"* published in 1916. Neither work mentions Chinese medicine as an "obstacle" to medical administration and public health, although that argument became his key rationale when Yu Yan started promoting the idea of a Chinese medical revolution in 1928.

Abolishing Chinese Medicine: The Proposal of 1929

In the spring of 1929, the state finally took action against Chinese medicine. The Central Board of Health held its first Public Health Conference on February 25, five months after the establishment of the Ministry of Health. At that conference, the board unanimously passed the proposal to regulate traditional medical practice. The Central Board of Health consisted exclusively of biomedical physicians, including Liu Ruiheng (J. Heng Liu, vice minister of health and president of Peking Union Medical College), Yan Fuqing (dean of Central University Medical College, Shanghai), Wu Liande (director, Plague Prevention Service),

Hu Dingan (commissioner, Nanjing), and Yu Yan (president, Medical and Pharmaceutical Association of China, Shanghai branch).[21] The proposal passed at this conference drew on, but also modified, the proposal drafted by Yu Yan, which required traditional practitioners to register with the government and to attend government-sponsored supplementary education courses in order to continue practicing medicine. According to this proposal, registration would end on the last day of 1930, and the supplementary classes would be offered for only five years. Traditional practitioners were not allowed to organize schools, nor could they publish promotional materials in newspapers. Since Yu Yan's proposal left little doubt that registration of traditional practitioners would be a one-time event,[22] its ultimate objective was clearly the abolition of Chinese medicine.

In this proposal, entitled "Abolishing the Old Medicine to Sweep Away the Obstacles to Medicine and Public Health,"[23] Yu explained his reasons for abolishing Chinese medicine in great detail. Yu was by no means the first person to come up with these criticisms of Chinese medicine. Nevertheless, due to his position as a member of the Central Board of Health, Yu's arguments crystallized enduring themes for future struggles between the two groups of practitioners. After quoting the proposal, I will consider his four arguments point by point.

> Reasons: The medicine of today has advanced from the curative to the preventive stage, from individualized to socialized medicine, from aiming at curing individuals to curing the masses. The modern public health service is based entirely on scientific medical knowledge with corresponding political backing. I beg herewith to submit four reasons for the advisability of abolishing the old medical practices.
>
> First: The old medicine of China adopts the doctrines of Yin and Yang, the Five Phases, the Six Qi, and the systems of the viscera and channels. These are pure speculation and completely untrue.
>
> Second: In diagnosis [the practitioners of old medicine] depend completely on pulse palpation. This method arbitrarily divides one portion of the artery into three parts—inch, cubit, and bar—that supposedly correlate with all the internal organs. With such absurd theories, they deceive themselves more than anybody else. It is ridiculous, just like astrology.
>
> Third: Since they do not know even the fundamentals of diagnosis, it is impossible for them to certify the cause of death,

> classify diseases, or combat epidemics, not to mention engage in eugenics and racial improvement, which are completely beyond their reach. Concerning the important issues of people's well-being and national survival, the old medicine is totally useless for civil administration in realizing those goals.
>
> Fourth: The evolution of civilization is from the supernatural to the human, from the philosophical to the practical. Meanwhile, the government is trying to combat superstition and abolish idols so as to bring the people's thoughts to proper scientific channels, while the old-style physicians, on the other hand, are daily deceiving the masses with their faith healing. While the government is educating the public as to the benefits of cleanliness and disinfection and the fact that germs are the root of most diseases, the old-style physicians are broadcasting such theories as falling ill with typhoid in spring after one catches cold in winter, and falling ill with malaria in autumn after one suffers from the heat in summer. These reactionary thoughts are the greatest obstacle to the scientization [*kexuehua*] of people's medical beliefs. In short, as long as the old medicine is not eliminated, people's thoughts will never change, new medical reforms cannot be successfully introduced, and public health measures will be impossible to carry out.[24]

While his first two arguments rehashed his earlier attacks on Chinese medical theories, this time Yu's argument brought a teleological vision of medical development into the foreground. As he states in his opening, "The medicine of today has advanced from the curative to the preventive stage, from individualized to socialized medicine, from aiming at curing individuals to curing the masses."[25] Moreover, as he assumed that "curing the masses (*qunzhong*)" was the direction of medical evolution, such medical development demanded an alliance between scientific medical knowledge and modern political theory.[26] Yu's argument appeared persuasive because he was, in fact, articulating the strategy that his modern-trained colleagues had been implementing in order to create the structural linkage between the state and Western medicine.

We may begin with Yu's third argument, the first novel argument, which concluded that Chinese medicine was totally useless in addressing the important concerns of "people's well-being and national survival." It is noteworthy here that even when Yu tried to demonstrate the "uselessness" of Chinese medicine, he avoided explicitly denying that it was effective in healing individual patients. On the contrary, as discussed in chapter 4, Yu elaborated in great detail on the question why "unscientific Chinese medicine can be effective in treating certain illnesses," giving four explanations there.[27] It is revealing that Yu and other biomedical doctors argued relentlessly against evaluating the merits of the two medicines in terms of their relative efficacy in curing individual patients,[28] which is the defining characteristic of what Yu called individualized medicine.[29]

For Yu, Chinese medicine was useless in the specific sense that it was unable to "certify the cause of death, classify diseases, combat epidemics, not to mention engage in eugenics and racial improvement."[30] What is "useful," however, largely depends on who is evaluating the "usefulness." For Yu and his biomedical colleagues, the crucial importance of "certifying the cause of death" and of "classifying diseases" was rooted in their responsibility for constructing "national vital statistics," that is, a specific kind of information useful and available primarily to the state medical bureaucrats, that is, practitioners of modern biomedicine like themselves.

Yu's fourth argument claimed that, worse than being useless for medical administration, the "reactionary thoughts [of traditional medicine were] the greatest obstacle to the scientization of people's medical beliefs."[31] Predictably, Yu took aim at these "reactionary thoughts" in the context of infectious diseases and opposition to germ theory. In light of the performance of traditional practitioners in the Manchurian plague campaign, Yu's argument was more than justifiable. In order to justify the unprecedented exercise of governmental power, the state had a strong interest in assisting the proponents of biomedicine to monopolize all legitimate medical knowledge in the nation. In Yu's language, the state's monopolizing project was called "the scientization of people's medical beliefs."

Arguments three and four exemplify what sociologist Pierre Bourdieu characterizes as the "symbolic violence" of the state.[32] Before practitioners of Western medicine had occupied the position of state medical officers, their disease classifications and causes of death had simply been one among many competing visions of the medical world. Once the state promulgated a series of ordinances on the basis of this knowledge,[33] Western medicine could defeat Chinese medicine on the simple ground that now it shared operational codes—that is, standardized classifications of diseases and causes of death—with the state. From then on, what the proponents of Chinese medicine were struggling against was no longer another knowledge tradition of medicine but the *official* knowledge sanctioned by the state. As a result, Chinese medicine not only appeared useless to the state but, moreover, became an "obstacle to medicine and public health."[34]

Without engaging in any serious debate with traditional practitioners, practitioners of Western medicine had successfully transformed the agenda of medical enterprise, redefined the priority of medical problems, and transformed their medical doctrine into the official knowledge of the state. All of these results were accomplished by virtue of what Yu called the alliance between scientific medical knowledge and modern political theory. Because they were threatened by this alliance, it did not take long for the leaders of Chinese medicine to realize that what Chinese medicine needed most urgently was "to develop its own political perspective."[35]

The March Seventeenth Demonstration

No one, not even the traditional practitioners themselves, seemed to have expected that they could block the tidal wave of the medical revolution.[36] While Yu and other biomedical practitioners had enthusiastically welcomed the coming of the Nationalist Revolutionary Army, some traditional practitioners had sensed that the unification of the nation could mean an end to their schools of Chinese medicine. After Yu's proposal was revealed to the public, Wang Qizhang, another vocal critic of Chinese medicine, wrote a strongly worded essay to warn the students of these medical schools to think about changing their careers.[37] Wang had good reasons to feel confident that Yu's proposal would win out. First of all, the Nanjing government had apparently committed itself to an agenda of modernization,[38] and his colleagues overwhelmingly dominated the newly established Ministry of Health.[39] Furthermore, even in the previous Warlord period, traditional practitioners had failed in most of their attempts to manipulate national policy to their benefit.

Two crucial factors can account for the failure of previous political campaigns undertaken by traditional practitioners: they had neither an organized infrastructure nor a shared vision of common interests. Moreover, the fact that the Japanese Meiji government had successfully suppressed traditional medicine in Japan by implementing a medical licensing system in the 1870s must also have been in the back of everyone's mind at the time.[40] Many traditional practitioners therefore doubted that it was a good idea to deal with the government at all, not to mention to actively pursue the interests of the state. When some of them mobilized a political campaign to assimilate the schools of Chinese medicine into the educational system, many of their colleagues purposely ignored this endeavor.[41] To prevent the Chinese government from emulating the strategy of the Japanese government, it was often suggested that traditional practitioners should resist governmental regulation on the ground that "for four thousand years, the profession of Chinese medicine had never adopted a licensing system."[42] In short, traditional practitioners had so far not been very successful in political campaigns because many of them did not recognize their collective interests in associating Chinese medicine with the state.

Ironically, Yu's proposal provided an instant solution to these two problems. Being preoccupied with the Japanese experience, traditional practitioners saw Yu's proposal as demanding the wholesale abolition of their practice. Consequently, organizing a national federation to resist this proposal immediately became a shared goal.

While Yu's proposal was designed to eliminate traditional practitioners as a group, at that time they had not yet established a communication network that would have constituted them as a group per se. Even though they had rallied for political campaigns in the past, they had always had trouble forming a permanent national association. "Once there was a move to outlaw Chinese medicine, [traditional practitioners] joined the association one after another. As soon as there was no immediate danger of forced registration, they disappeared like smoke," one traditional doctor sadly complained.[43] Therefore, when some traditional practitioners in Shanghai tried to mobilize a mass protest against this proposal, they had to start almost from scratch.

The protest was first organized by the All-China Medical and Pharmaceutical Associations (*Shenzhou yiyao zonghui*), which had organized the earlier campaign aimed at assimilating schools of traditional medicine into the educational system in 1913. Many of the organizers of this demonstration were colleagues, teachers, and students from the social network of its vice president, Ding Ganren. For example, Chen Cunren (1908–90), one of the initiators of the March 17 demonstration, was the most trusted student of Ding Zhongying (1886–1978), Ding Ganren's second son. As Chen recalled, he initially had no idea how to send out appeals for a mass meeting to traditional practitioners throughout the country. Fortunately, Chen Cunren was in charge of a medical weekly, and another initiator named Zhang Zanchen (1904–93) was the editor of a popular journal of traditional medicine, *Annals of the Medical Profession*. From the list of their subscribers, Chen and Zhang randomly picked two people from every county in China, mailed them the petition, and asked them to carry the petition to their local associations of Chinese medicine, if any existed.[44]

To their surprise, 262 delegates representing 131 organizations attended a three-day convention hosted by the powerful Shanghai General Chamber of Commerce (*Shanghai zongshanghui*). Since the Shanghai Chamber of Commerce had acted as the political voice of the Chinese business elite since the fall of the Qing dynasty, its support for Chinese medicine particularly upset the proponents of Western medicine.[45] More than two thousand practitioners of Chinese medicine closed their clinics for half a day to support this demonstration, and the nationally renowned drugstore *Huqingyu* protested especially fervently.[46] Full-page advertisements appeared in the leading dailies, spreading a rumor that Yu's proposal had been bolstered by a $6 million dollar bribe from foreign pharmaceutical companies. The ensuing demonstration by the proponents of Chinese medicine was described as the largest mass movement since the Nationalist state had put an end to the warlords.[47] Inside the assembly hall, a pair of giant posters was hung on the wall that read "Promote Chinese medicine to prevent cultural invasion!" and "Promote Chinese drugs to prevent economic invasion!"[48] To recruit other allies, traditional practitioners had thus not only adopted the rhetoric of cultural nationalism, as Ralph Croizier has convincingly argued,[49] but also the rhetoric of the National Goods Movement (*guohuo yundong*).[50]

Founded just before the 1911 revolution, the National Goods Movement encouraged Chinese people to buy domestically manufactured goods in order to aid China's economic independence.[51] By responding to national humiliation with concerted economic action, it signaled the rise of a new kind of popular movement in Chinese history. As a result, governmental officers and newly emerging Chinese capitalists collaborated to associate patriotism with the purchase of domestic commodities. By translating Chinese drugs into "national goods" (*guohuo*), the advocates of Chinese medicine intended to recruit not only people involved in the Chinese pharmaceutical industry, but also people who were already committed to the National Goods Movement—in other words, people who would otherwise have had little interest in the medical struggle. This turned out to be a very successful strategy. In addition to the traditional practitioners themselves, the National Business Association, the National Goods Maintenance Association, and the National Labor Union of the Pharmaceutical Industry (mostly pharmaceutical workers) were among those who immediately committed to supporting their demonstration.[52] In fact, the Chinese pharmaceutical associations not only joined this protest from its very beginning but also hosted delegates who had come from areas outside of Shanghai.

Before the end of the first day, a representative of the Hangzhou Association of Chinese Medicine voiced his feeling for all participants: "It is just unprecedented that we have this national assembly and the consolidation of all the [local] associations of Chinese medicine. It is happening because we have been downtrodden to such as heart-broken situation. Today is the first day of our gathering. We should take the day as a day of commemoration, that is, to mark March 17 as the day that we will commemorate forever."[53] His proposal passed with passionate applause, and March 17 was adopted as the day "in commemoration of the grand consolidation of Chinese medicine and pharmacy."[54] What exhilarated the participants of this convention more than anything else was the simple fact that they were all gathering under one roof as representatives of a national assembly, thereby turning traditional Chinese medicine into a national entity. This objective certainly could not be realized with just one gathering, and many subsequent efforts were necessary in order to turn this inspiration into reality. One hundred and five proposals were discussed and passed in the ensuing three-day convention.[55] In addition to defying Yu's proposal, advocates of Chinese medicine resolved to establish a permanent national organization—representing not only practitioners of Chinese medicine but also the pharmaceutical industry—in order to "consolidate power and defend against invasion."[56]

Many delegates proposed to articulate the conditions for a permanent alliance between the professions of Chinese medicine and the pharmaceutical industry.[57] As a result, a new organization, the National Federation of Medical and Pharmaceutical Associations (NFMPA; *quanguo yiyao tuanti zonglianhehui*), was created that consisted of three subfederations: the associations of [Chinese] medicine, the [Chinese] pharmaceutical industry, and a third that consisted of pharmaceutical workers.[58]

It is important to clarify here that the conference participants deliberately avoided speaking of "Chinese medicine" but instead emphasized "national" (*quanguo*) in the names of their federation and subfederations. Presumably this was done to create the impression that they represented the majority of medical and pharmaceutical practitioners in China. Under the federation, branches were established at the provincial, county, and district levels. Facing the immediate threat of complete abolition, practitioners of Chinese medicine were eager to become members. Within three years, the number of member associations increased from 242 to 518, including affiliates in Hong Kong, the Philippines, and Singapore.[59] Thanks to Yu's proposal, an international network of Chinese medical practitioners was taking shape for the first time in history.

During this first conference, a dramatic moment came when the representatives had to select an official name for their profession. Not only did they reject pejorative names such as "non-scientific medicine" or "old medicine," but, surprisingly, they also rejected the name "Chinese medicine" (*zhongyi*), even though some of them had been using that name for quite a while. Ralph Croizier's groundbreaking study concludes that the National Medicine Movement was largely motivated by the psychological need to preserve a uniquely Chinese identity in the midst of a sweepingly changed cultural setting.[60] If that had been the major motivation, though, it would be hard to imagine why its advocates preferred the name "national medicine" (*guoyi*) to "Chinese medicine" (*zhongyi*).[61]

The Ambivalent Meaning of Guoyi

The answer to this crucial question lies in the ambivalent meaning of *guoyi*. First of all, the English translation of *guoyi* as "national medicine," while accurate in most contexts, eliminates one important connotation of the Chinese term.[62] The two English concepts of "nation" and "state" are both commonly translated with one Chinese term, *guojia*. Consequently, when Chinese people put *guo* as an adjective in front of a noun such as *yi* (medicine), we simply do not know if the word *guo* in the compound *guoyi* is meant to characterize *yi* (medicine) as being "national" (i.e., as a part of Chinese culture) or as "belonging to the state." This undifferentiated use of one Chinese word for both "nation" and "state" might have a historical origin: What took place as two separate historical processes in Europe, namely the construction of states and the construction of nations, became intertwined as a single process in early twentieth-century China. To be precise, as Prasenjit Duara put it, in twentieth-century China "state making was proclaimed within the framework of nationalism and related ideas of modernization."[63]

This lack of differentiation in Chinese between the concepts of "nation" and "state" produced the ambivalent meaning of *guoyi*. On the one hand, as Croizier in the 1960s and medical modernizers in the 1930s argued, *guoyi* could be understood as something particularly Chinese, that is, the "national essence" of Chinese culture. On the other hand, just as in the case of the "national calendar" (*guoli*), a reference to the Gregorian calendar promulgated by the Republican government, *guoyi* could be understood as the medicine that was officially sanctioned by the Chinese state. The term "national calendar" had been in use to refer to the Gregorian calendar since the first day of 1912, when Sun Yat-sen was inaugurated as the first president of the new Republic.[64] In the present context, the meaning of *guoyi* embraces that of "Chinese medicine," but not vice versa.

Strong evidence supporting this interpretation of *guoyi* (i.e., as "state-sanctioned medicine" and not just "national medicine") comes from an unexpected source—the practitioners of Western medicine. After traditional practitioners strategically labeled their practice as *guoyi*, biomedical practitioners in a subsequent annual convention of their own passed a resolution to prohibit traditional practitioners from adopting this name.[65] Most surprisingly, after the Nationalist government founded the Institute of National Medicine in 1931, the Shanghai Union of Physicians, an organization of biomedical doctors, officially requested the government to designate Western medicine as *guoyi*, because they considered it the only style of medicine worthy of the designation "national medicine."[66] From their point of view, as in the case of the "national calendar," the *guo* in *guoyi* did not refer to the Chineseness of the medicine at all but to the fact that it was officially sanctioned by the state.

By taking advantage of the ambiguous meaning of *guoyi*, traditional practitioners managed to associate their style of medicine simultaneously with cultural nationalism and with statism. In other words, they succeeded in identifying it with "national essence" (*guocui*), strengthening its link to Chinese culture, and with "national drugs" (*guoyao*), which linked it to the advancement of the state's economic autonomy. The intention to directly recruit the state—instead of involving it indirectly through the ideology of cultural nationalism—was, in fact, precisely the reason why the proponents of traditional medicine chose the name *guoyi* (national medicine). Taking inspiration from the National Goods Movement, a delegate pointed out,

> The National Goods Association never called itself the Chinese Goods Association. The Chinese medicine of our country is a national essence. Chinese drugs are produced in China. Therefore, we should take the opportunity in this Federation to replace the word "Chinese" with "national." The names "national medicine" and "national drugs" fit the modern trend very well.[67]

Like those in the National Goods Movement, what the advocates of Chinese medicine wanted most to recruit was not the cultural China but the emerging Chinese state. Symbolically, March 17, the first day of their demonstration, was designated as "national medicine day" (*guoyi-*

jie), which was observed by supporters of Chinese medicine every year until the Communists took over China. In response to being oppressed as a group by the state, traditional practitioners strove to organize themselves into a national entity. The terms *national drugs*, *national medicine*, and especially *National Medicine Day* testify that the modern history of Chinese medicine began on March 17, 1929, when a unified group of traditional practitioners encountered the first modern Chinese state.

The Delegation to Nanjing

On March 21, 1929, five traditional practitioners boarded the night train for Nanjing. As a delegation representing the National Federation of Medical and Pharmaceutical Associations, these five men carried a petition to be presented at the Nationalist Party's Third National Conference (*guomindang disanci quanguo daibiao dahui*), which was in session at the time. The petition contained four main points: The government shall (1) officially declare its commitment to promoting Chinese medicine and drugs, (2) rescind the proposal by the Board of Health to abolish Chinese medicine, (3) incorporate the schools of Chinese medicine into the national school system, and (4) reserve membership on the Board of Health to practitioners of Chinese medicine.[68] This last request was an entirely novel one for traditional practitioners. It shows that they had gradually learned about both their "obligations" and "rights" in relation to the state through their collective struggle against biomedical doctors.

On the eve of their departure, the delegates still felt uncertain, if not gloomy, about their mission.[69] To their surprise, however, many top-ranking politicians responded positively to their petition, including the secretary-general of the Nationalist Party, Ye Chucang (1887–1946), the prime minister of the Executive Yuan, Tan Yankai (1880–1930), and the director of the Examination Yuan, Dai Jitao (1891–1949). In addition, many of these figures emphasized the importance of Chinese medicine to the national economy.[70] This would never have been the case if the traditional practitioners had not succeeded in recruiting the Chinese pharmaceutical industry.[71] In a meeting with the delegates in his private residence, Tan Yankai, prime minister of the Executive Yuan, said,

> Governmental policy should never betray the people's need. Therefore, the resolution of the Central Board of Health can by no means be put into practice. Taking Hunan Province as an

> example, even in big cities there are only very few practitioners of Western medicine. In the counties, even traditional practitioners, let alone biomedical practitioners, are terribly lacking. If the resolution were really to be put into practice, patients would have no other option but to wait for their death, and the peasants, workers, and businessmen in [the Chinese] pharmaceutical industry would all lose their jobs.[72]

With news reporters standing by his side, Tan further sought help from Xie Guan, the leader of the delegation and a widely admired senior practitioner of Chinese medicine, by asking him for a pulse diagnosis and a prescription. Tan's willingness to receive the consultation showed his unreserved trust in Chinese medicine, which was made public when Xie's prescription was published in the major newspapers the next day.[73]

Tan Yankai's diary allows us to view this episode from a different angle. Tan indeed wrote about his meeting with the delegates on the morning of March 22. Their meeting was attended by Lu Zhong'an, the famous traditional practitioner from Nanjing, who had allegedly helped Hu Shih to recover from diabetes and had been involved in the controversy over treating Sun Yat-sen. While Tan did not record his thoughts about this visit, he described Xie Guan's style of prescription as "so gentle that it belongs to the local style of the lower Yangtze River."[74] Tan was clearly interested in Chinese medicine and trusted it in his personal life. In his diaries from 1895 to 1930, he mentions Chinese medicine twenty-seven times, mostly in a positive sense, and he includes reference to a miraculous cure that he had heard about from friends.[75] Most revealingly, he agreed with traditional practitioners that Sun Yat-sen had passed away because he "had been hurt by [Western] surgery."[76] Tan himself passed away in September of the year following his meeting with the delegates, but from the day of that meeting onward, he provided crucial—though brief—political support to the National Medicine Movement and thereby influenced its course of development during the Republican period.[77]

With enthusiastic support from political leaders, the delegates deliberately delayed their visit to the Minister of Health, Xue Dubi (1890–1973) in order to expose him to more criticism from the Nationalist Party and the public. It was not until March 24 that Xue finally succeeded in hosting a dinner banquet with the delegates. According to an interview that Xue conducted with a journalist from the popular newspaper *Shenbao*, Xue pointed out that "among two thousand counties in

China, less than one-tenth or two-tenths have health services provided by Western medicine."[78] He also stated that "a so-called proposal to abolish Chinese medicine indeed existed but was merely a proposal."[79] As an immediate response to the demand for the "participation of Chinese medicine in the medical administration," Xue invited Xie Guan and Chen Cunren on the spot to serve as consultants for the Ministry of Health.[80] Before leaving the meeting, Xue assured the delegates that "as long as I am minister, I will never approve this proposal."[81] The next morning, the delegation returned to Shanghai with the first major political triumph for traditional practitioners: Yu's proposal had been blocked.

Envisioning National Medicine

Because it directly triggered the rise of the National Medicine Movement, Yu Yan's proposal is seen as a crucial historic document and is included in almost every book concerning the twentieth-century history of Chinese medicine. Nevertheless, one fact is often neglected: the Nationalist government never adopted it—not even during the meetings of the Central Board of Health in which it was proposed. The whole process was not kept secret; rather, it was widely popularized in Shanghai's newspapers and in various journals of Chinese medicine. During a meeting of the Central Board of Health on February 24, 1929, the board members deliberated on four proposals concerning Chinese medicine, one of which was Yu Yan's proposal. After a marathon session, the board decided to combine all four of them into one and to name this new proposal "Principles Concerning the Registration of Practitioners of Old Medicine." It consisted of only three items: (1) the deadline for registration was December 31, 1930; (2) so-called schools of Chinese medicine were prohibited from being established; (3) with regard to the issue of regulating news magazines and advertisements for the "old" medicine, the Ministry of Health would pursue these objectives at the appropriate time.[82]

The board members were keenly aware of the danger of backlash from supporters of Chinese medicine. Unlike Yu's original proposal, the new proposal did not require traditional practitioners to attend supplementary training courses to secure their right to practice medicine: the government would instead grant registration to anyone who had ever practiced medicine. In addition, it pushed governmental control of medical advertisement into the remote future. Most symbolically, it changed the name of the proposal from Yu Yan's "Abolishing the Old

Medicine" to the positive-sounding "Registration of the Old Medicine." This change of title was emphasized when the Ministry of Health responded to the shower of protests from various levels of society. In the several dozens of brief telegraphs that it sent out, the ministry stated: "Concerning Chinese drugs, we very much want to promote them. Concerning Chinese medicine, we should try to improve it, even to scientize it. Within the resolution of the Board of Health, there is no such thing as 'abolishing Chinese medicine.'"[83]

In spite of the public denial from the ministry, advocates of Chinese medicine strove to frame their campaign as one against the threat posted by Yu's "Proposal to Abolish Chinese Medicine." For example, while all four proposals and the final proposal were included in the official documents of the board meeting and resolution, advocates of Chinese medicine did not bother to reproduce the contents (except for the titles) of the other three proposals when they published this document in their journals and magazines.[84] Instead, they published Yu Yan's original proposal verbatim, even though the Central Board of Health had not endorsed it.[85] In the end, the traditional practitioners intentionally misrepresented the official document by naming it the "Original Text of the Proposal to Abolish the Practice of Chinese Medicine, Discussed and Decided by the Central Board of Health."[86] It is a testimony to the success of the traditional practitioners in framing this campaign that Yu's otherwise short-lived proposal thus received wide circulation and became a symbol for this historic event.

This widely shared perception obscured one crucial dimension of the National Medicine Movement. While the traditional practitioners were no doubt mobilized by the threat of abolishment, their concern over the issue of legalizing the schools of Chinese medicine was equally important. As I have mentioned above, what mobilized traditional practitioners to organize their first nationwide campaign in 1912 was the protest against the "omission of Chinese medicine from the school system."[87] By the time that the historic confrontation took place in March of 1929, people from both sides took this issue to be the core of the struggle. While relaxing both the policies and the wording of Yu Yan's proposal, the Central Board of Health saw these schools as the most worrisome development. Therefore, the final proposal kept intact the recommendation from Yu Yan's proposal to strictly prohibit the establishment of schools of Chinese medicine. As a pointed response, the proponents of Chinese medicine made the "participation of Chinese medicine in the [national] education system" their number one demand when the three-day assembly concluded in Shanghai.[88] Nevertheless, among the four requests that the delegates carried to Nanjing, this was the only one that was not accepted by the apologetic Minister of Health Xue Dubi. This unfulfilled request reveals that the core of further struggle would be about the assimilation of Chinese medicine into the national education system.

Once the campaign had successfully blocked Yu Yan's proposal—or, more precisely, the revised "Proposal for the Registration of Old Medicine," the traditional practitioners were thrilled with their victory and the social support backing it. This sentiment of thrill and surprise is apparent in many articles published immediately after the March campaign. Many people were keenly aware that this event represented a turning point in the history of Chinese medicine. Within a month after the March 17 campaign, for example, the journal *Annals of the Medical Profession* published a special issue entitled "The Struggle by the Chinese Medical and Pharmaceutical Societies." In the preface, Zhang Zancheng, the journal's editor-in-chief and the organizer of this campaign, pointed out that "we have collected these documents for the reference of future historians of medicine."[89] Clearly, participants in the March 17 campaign knew that they were participating in an event that would make history for Chinese medicine.

Unfortunately, this political victory proved to be short-lived. Less than two months after the March 17 campaign, the Ministry of Education demanded that schools of Chinese medicine be renamed as "Facilities for Practical Training" (*chuanxisuo*) arguing that these so-called schools "are not based on science and have no qualification standard for the students. It is not appropriate to characterize them in terms borrowed from the national education system."[90] In response, the National Federation of Medical and Pharmaceutical Associations (hereafter, "the federation") not only organized lobby groups but also established two special committees for preparing curricula and for compiling and editing textbooks. It also held a national conference in July in 1929 to standardize curriculum and teaching materials. In spite of these efforts, the Ministry of Education proceeded to ban schools of Chinese medicine in August. In addition, it demanded that so-called hospitals of Chinese medicine be renamed as "medical rooms" (*yishi*) and promulgated regulations prohibiting traditional practitioners from using stethoscopes, syringes, and even the name "Chinese-Western medicine" (*zhongxiyi*).[91]

In order to resist the multiple waves of suppression, the federation again organized a three-day national conference on December 1 in Shanghai, with 223 associations and 457 representatives in attendance. Their first objective was defined as "the participation of Chinese medicine in medical administration." Moreover, the representatives formally approved the proposal presented at the last conference in March to replace the term "Chinese medicine" with "national medicine" as their official designation.[92] Another important resolution was to contact and seek alliances with practitioners of the Japanese style of Chinese medicine (*huanghan yixue*) on behalf of the federation.[93] Conference participants from all over the country were so eager to consolidate their alliance that they extended the conference for two more days and then again sent a delegation to Nanjing on December 7. On December 13, Chiang Kai-shek (1887–1975), then chairman of the Nationalist government, openly issued a personal instruction to withdraw all "suppressive regulations" issued against Chinese medicine by the Ministry of Health and Education.[94]

While traditional physicians were thrilled to have Chiang's personal support, they also came to realize that it might not be in their best interest to incorporate the schools of Chinese medicine into the national education system. For one thing, the governmental regulations for schools of modern medicine were beyond their means. For example, regular schools were required to have an endowment of one hundred thousand dollars. In addition, medical schools could not operate independently but had to be affiliated with a university. In practical terms, it was simply impossible for the schools of Chinese medicine to meet these requirements.[95] It became clear that practitioners of traditional medicine faced a serious dilemma. As one author put it, "If the schools of Chinese medicine did get incorporated into the national education system, they would all be forced to close within six months, without a single exception."[96] In order to solve this dilemma, traditional practitioners came up with the idea to establish an independent administrative organization, the Institute of National Medicine (*guoyiguan*).

According to the recollections of its originator, the traditional physician Jiang Wenfang (1898–1961) and other core members of the federation came to the conclusion that they needed a governmental agency specifically dedicated to the administration of Chinese medicine. In order to create such an agency to meet the needs of Chinese medicine, the proposal deliberately left the nature of the Institute of National Medicine ambiguous. According to the original proposal submitted to the government, the Institute of National Medicine had three major tasks: improving national medicine, conducting research on national drugs, and managing affairs related to national medicine and pharmacy.[97] Since the proposal indicated that the budget for the institute came from the federation and had no governmental funding, the institute had to be a civil society devoted to scholarly purposes. Yet, as its objectives included "managing affairs related to national medicine and pharmacy," the institute somewhat resembled a branch of governmental administration. As Jiang Wenfang recalled, in order to create this ambiguity, they deliberately included language in the proposal suggesting that "the Institute of National Medicine was modeled after the Institute of National Martial Arts (*guoshuguan*)."[98] According to its organizing guidelines, the Institute of National Martial Arts had been endowed with the power to oversee the affairs of martial arts all over the country, and Jiang therefore felt that this model could pave the way for a future extension of the Institute of National Medicine from a scholarly organization to an administrative agency. As one might expect, the Ministry of Health was strongly opposed to entrusting administrative power to a "scholarly society" and therefore refused for several months to approve the institute.[99]

In April 1930—independent of the proposal submitted to the Executive Yuan by the federation—Tan Yankai and some leading political figures raised awareness about, and received approval for, the Institute of National Medicine within the Central Political Conference of the Nationalist Party. While Tan's new proposal was based upon the one drafted by the federation, it had been transformed in the hands of the Nationalists in a few crucial aspects. On the one hand, the government would provide the budget for the institute and entrust it to organize hospitals and medical schools. On the other hand, the three tasks mentioned in the original proposal were deleted and replaced with one new objective, namely, "to use scientific methods to put Chinese medicine into order and to improve therapeutics and the manufacture of pharmaceuticals."[100] The new proposal thereby highlighted the crucial importance of science and deleted the third task of "managing affairs of Chinese medicine." Generally referred to as the "scientization of Chinese medicine" (*zhongyi kexuehua*), the institute's official objective forcefully imposed upon Chinese medicine the dominant ideology of scientism and thereby conditioned the development of Chinese medicine in the twentieth century.

Conclusion

The National Medicine Movement was neither a passive movement of resistance nor a conservative movement aimed at preserving a fragile tradition, even though it is conventionally remembered as a response to the governmental threat of "abolishing Chinese medicine." Instead of resisting the state, the proponents of this movement developed the vision of National Medicine and actively struggled to create a closer alliance between Chinese medicine and the Nationalist state. As demonstrated by their efforts to incorporate the schools of Chinese medicine into the national education system, they strove for the new professional interests that had been created and sanctioned by the state, including professional licensing, the educational system, and the power associated with medical administration.

No matter how limited the real practical benefits were that the Nationalist state was able to offer, practitioners of Chinese medicine were thereby brought into a movement for collective social mobility. This social mobility became possible because practitioners of Western medicine had succeeded, at least partially, in transforming medicine into a modern profession that played a crucial role in national affairs. In this sense, it is ironic that practitioners of Western medicine not only posed the most serious threat to Chinese medicine, but also introduced the unprecedented possibility for collective upward mobility to practitioners of Chinese medicine. This newly created state capital and the associated possibility for collective social mobility are essential for understanding the modern history of Chinese medicine. At least during the Republican period, traditional practitioners were not passively responding to the state but actively and collectively pursuing this upward mobility by way of the state.

When the Institute of National Medicine was inaugurated on March 17, 1931, it at once symbolized and embodied the vision of "national medicine" that had been developed since the 1929 confrontation. First of all, as the only institute to include this key term in its name,[101] the institute reinforced the vision that the objective of this movement was to make Chinese medicine thrive alongside the modern state. Second, because the Nationalist government had identified the central objective of the institute to be the reordering of Chinese medicine by means of scientific methods, traditional practitioners were forced to embrace and cope with the concept of science—one of the two most powerful ideologies of the May Fourth Movement. This is a key development that I discuss in chapter 7. Third, because the government eventually outlawed the National Federation of Medical and Pharmaceutical Associations and ordered it to be dissolved in 1930, the Institute of National Medicine became the official successor of the first nationwide federation of Chinese medicine. As a consequence, it inherited from the federation the urgent tasks of further consolidating the alliance among all the participating associations of the National Medicine Movement and of creating a nationally unified profession of Chinese medicine. In order to appreciate the efforts of those involved in creating the national entity that we now know as Chinese medicine, it is valuable to first examine the complicated and diversified medical landscape in which they launched this historic effort in the early 1930s. Finally and most importantly, the president of the Institute of National Medicine was keenly aware that a crucial element was wanting in their efforts to build National Medicine. Given that Chinese medicine had just barely survived the threat posed by the emerging power alliance between biomedicine and the Nationalist state, the president warned his fellow practitioners that what they needed most urgently was "to develop a political perspective" so as to create a similar alliance for Chinese medicine—a crucial topic that will be addressed in chapter 10.

6

Visualizing Health Care in 1930s Shanghai

Reading a Chart of the Medical Environment in Shanghai

Four years after the 1929 confrontation, the famous critic of Chinese medicine Pang Jingzhou (1897–1965) produced a detailed chart (figure 6.1) to provide a visual representation of the chaotic medical environment in contemporary Shanghai. Consisting of more than forty items, from formal medical associations and governmental organizations to sidewalk plaster vendors and spirit mediums, this chart represents Pang's effort to provide a "bird's-eye view" of Shanghai's landscape of health-care services. While Pang's chart is extremely detailed and includes many seemingly insignificant items, surprisingly, it does not include a single item that carries *Chinese medicine* in its name. Moreover, although his chart does include one label that includes the term *Western medicine*, namely, Association of Western Medicine [12], Pang refused to accept this association as a qualified member of Western medicine. It is thus puzzling that neither of the two styles of medicine can be easily mapped onto Pang's detailed portrayal of the medical landscape in 1930s Shanghai. The reason is actually quite simple: no such things as "Chinese medicine" and "Western medicine"

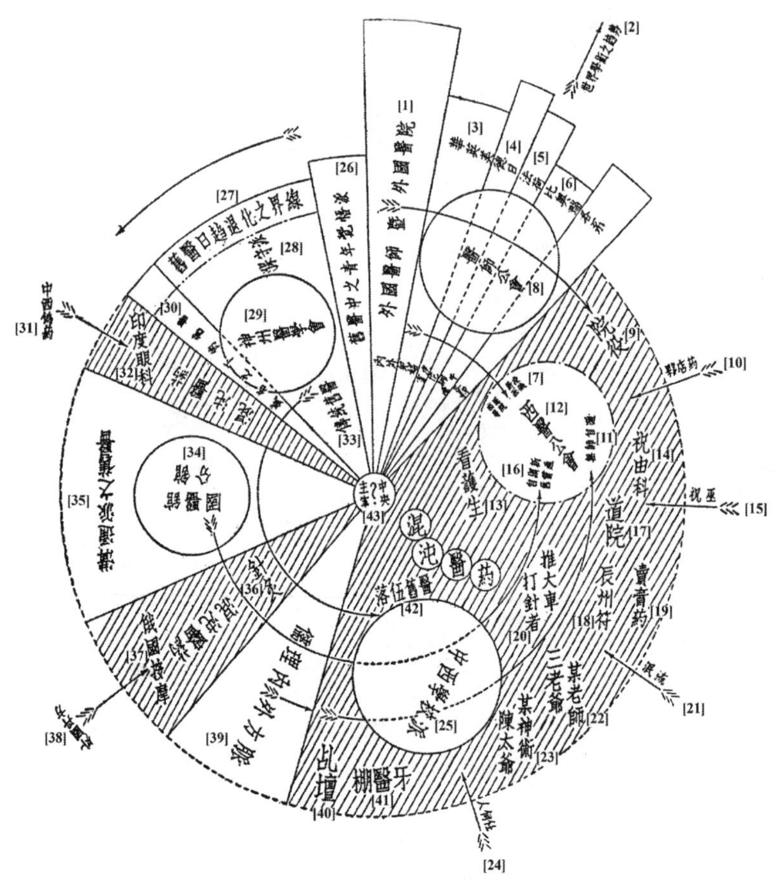

FIGURE 6.1 "Medical environment in Shanghai," from Pang Jingzhou, *Shanghaishi jinshinian lai yiyao niaokan* [*A Birds-Eye View Report of Medicine and Pharmacy in Shanghai in the Last Decade*] (1933), p. 11. English translations of the terms in this figure follow: [1] foreign doctors and hospitals; [2] global academic trends; [3] China, England, United States; [4] Germany and Japan; [5] France and Holland; [6] Belgium, Austria, Switzerland; [7] opportunistic practitioners of the New Medicine; [8] Association of Physicians; [9] hospital staff; [10] drugstore attendants; [11] self-taught practitioners; [12] Association of Western Medicine; [13] nurses; [14] specialists of *zhuyou* exorcism; [15] shamans and exorcists; [16] self-proclaimed experts in both the New and the Old Medicine; [17] Daoist temples; [18] talismans from *Chenzhou*; [19] vendors of medical plasters; [20] injection givers; [21] gangsters; [22] various masters; [23] divination; [24] anybody; [25] faction of Chinese-Western medical schools; [26] faction of awakened young practitioners of the Old Medicine; [27] the receding frontier of the Old Medicine; [28] conservative faction; [29] Shenzhou Medical Association; [30] reputable practitioners of the Old Medicine; [31] fake Chinese and Western pharmaceuticals; [32] Indian ophthalmology; [33] traditional practitioners of the Old Medicine; [34] Institute of National Medicine (Shanghai Branch); [35] communication faction within the Old Medicine; [36] acupuncture; [37] Russian massage; [38] foreign prescriptions; [39] Neo-Confucian medicine; [40] altars for spirit mediums; [41] dentists' huts; [42] outmoded practitioners of the Old Medicine; [43] center in charge. A larger version of this diagram can be downloaded at http://www.press.uchicago.edu/sites/Lei.

existed in 1930s China, even though I have been documenting a struggle between them.

While I have been forced in the interest of textual flow to employ these two categories to refer to the two styles of medicine, there was really no such thing as a unified medical system, whether biomedical or Chinese, before the historical confrontation that took place in 1929. On the contrary, the gradual consolidation of early modern Chinese medicine into more organized groups and standardized systems of practices was partially the hard-won result of the historical struggle under consideration in this book. To exaggerate a bit, I have been telling a story in which the two main actors—Chinese medicine and Western medicine—are not fully present until the later parts. To treat them uncritically as if they had been distinct groups all along would obscure a crucial dimension of this history—that is, how the leading figures selectively reconstituted these two styles of medicine by way of drawing boundaries against other practices and excluding some health-care providers from their proper domains. In order to trace this process of transformation, we have to look beyond these two labels and consider the internal diversity, heterogeneity, and fluidity of both Chinese and Western medicine. For this purpose, the present chapter takes a step back in chronology to make visible the internal heterogeneity of what I have in previous chapters been calling "Chinese medicine" and "Western medicine" for the sake of convenience.

Instead of tracing the chronological development of a historical entity or a central theme, however, this chapter focuses on making sense of Pang's chart, which appears in his book *A Birds-Eye View Report of Medicine and Pharmacy in Shanghai in the Last Decade.*[1] Pang received his medical training at the German-influenced Tongde Medical College in Shanghai and became president of his alma mater in 1925 at the young age of twenty-nine.[2] He was deeply concerned with public affairs and subsequently devoted himself to the Red Cross Society of the Republic of China, becoming its secretary general when the Sino-Japanese war intensified in 1936. The book was originally published in 1932 as a series of forty-six essays in the *Medical Weekly* section of Shanghai's most popular newspaper, *Shenbao*. From this origin, it is clear that Pang wanted to report to the general public both the "chaotic medical environment" in Shanghai and the dynamic restructuring process that had been taking place since the 1929 confrontation.

While historical writing is mostly concerned with textual narratives, this chapter has the unusual objective of re-creating in the reader's mind a "bird's-eye-view" of the health-care landscape in 1930s Shanghai by way of a close reading and interpretation of a visual representation. I emphasize the visual representation here because some crucial information embedded in Pang's chart was not articulated in the accompanying text. For the purpose of retrieving nontextual, visual information, we should understand Pang's chart in the way that Francesca Bray urges her readers to decipher a special kind of Chinese chart that is, like Pang's chart, made up of almost nothing but words. Bray says that we should understand such a chart as one that "offer[s] spatial encodings of factual information, structures, processes and relationships, translating temporal or intellectual sequences into purely spatial terms, and encrypting dynamic process as static layout."[3] With the help of textual information provided by Pang and his contemporaries, this chapter strives to make sense of the spatial encodings in Pang's chart. To guide readers in deciphering this chart and forming a meaningful image in their minds, I have organized this chapter not along the axis of temporal progression but along the two-dimensional spatial expansion of this chart.

This visual representation is valuable for the present book because it captures the formation of these two styles of medicine as they struggled against each other in the Republican period. Including many items that are never explicitly discussed elsewhere in this book, Pang's chart effectively reminds us how much was excluded during the formative process of these two styles of medicine. To point out the most obvious, we might wonder what happened to the numerous religious practices depicted by Pang. By focusing on the seemingly value-neutral category of "medicine," the present book unavoidably bypasses many important healthcare services of that time. As we examine and interpret this detailed chart, I hope that this chapter can make readers aware of the limitations of the present book and at the same time encourage them to engage with, if not challenge, Pang's worldview, which was surely biased because of his critical stance toward Chinese medicine. To provide an example of such critical engagement with Pang's chart, I visualize a different mental image in the concluding section of this chapter—one that might have originated in the mind of a practitioner of Chinese medicine.

Western Medicine: Consolidation and Boundary-Drawing

Several features of Pang's chart immediately demand our attention.[4] First of all, the circumference of the largest circle and the boundaries between the fields within the circle are formed by two sorts of lines. The nature of the lines that demarcate a specific slice—solid or dotted—indicates whether the boundaries between that group and the neighboring groups were either rigorous and clear-cut or permeable and vague, respectively.[5] Apparently only foreign doctors and hospitals, licensed practitioners of Western medicine, and a certain conservative faction among the practitioners of Chinese medicine managed to police the boundaries of their groups. Second, the largest circle includes shaded and nonshaded areas. The shaded sections denote what Pang described as "chaotic medicine and pharmacy," which in his view constituted almost half of all medical services provided in Shanghai.

Third, there are five small circles, which represent five formally organized institutions: clockwise from the top, we find the Shanghai Association of Physicians (*yishi gonghui*, founded in 1925) [8], the Association of Western Medicine (*xiyi gonghui*, founded in 1929) [12], the "faction of the Chinese-Western medical schools" [25], the Institute of National Medicine [Shanghai Branch] (founded in 1931) [34], and the Shenzhou Medical Association (founded in 1912) [29]. Among these five institutions, the only one circumscribed by a dotted line is the Association of Western Medicine [12]. Pang considered this group to be more loosely constituted than the other organizations; it allowed unrestricted entry and had a particularly diversified membership.

What interested Pang more than sorting out the group boundaries was monitoring the flow of members between the various factions. With a dozen or so arrows representing these flows, Pang's chart captures these dynamic relationships. Three arrows directly extend into the dotted circle of the Association of Western Medicine [12]. These are identified with three different types of practitioners: The "opportunistic practitioners of the New Medicine" [7] (from the slices representing practitioners of Western medicine [3] [4] [5] [6]), "self-taught practitioners" [11] (from both the "faction of Chinese-Western medical schools [25] and "neo-Confucian medicine" [39]), and the "self-proclaimed experts in both the New and the Old Medicine" [16] (from the Shanghai Branch of the Institute of National Medicine [34]). Moreover, since the boundary around the Association of Western Medicine was not rigorously guarded, as represented by the dotted line enclosing this circle, anyone from the surrounding environment—for example, "hospital staff" [9] (from the slice representing "foreign doctors and hospitals" [1]), nurses [13], drugstore attendants [10], specialists of *zhuyou* exorcism [14], vendors of medical plasters [19], and so on—could freely move in and out of this association.

Given the fact that this organization identified itself as a group associated with Western medicine, the nature of its membership comes as a surprise: According to Pang, many of its members were prescribing traditional formulas of Chinese medicine. Moreover, none of its members practiced medicine in any certified hospital of Western medicine or taught biomedicine in any medical school. Most revealingly, the majority of its members claimed expertise in the area of venereal disease, of all things![6] It was widely reported that many former assistants in Western hospitals marketed themselves as self-proclaimed experts in venereal diseases, specializing in injecting the newly invented drug 914, which was in high demand in China. According to Dr. Wu Liande's estimation, "as much as 50–60 percent of the [Chinese] adult population was afflicted with venereal diseases," and therefore "Harbin alone with its 300,000 inhabitants has over 200 such 'hospitals.'"[7] No wonder that Pang felt the need to include in his chart this awkward-sounding group of medical providers—the "injection givers" [20].

Moreover, the concern over this group of minimally trained practitioners explains why licensed practitioners of biomedicine demanded that the government rigorously prohibit any other practitioners from employing hypodermic needles. Pang described in graphic detail how a well-respected hereditary practitioner of Chinese medicine treated one hundred patients a day, giving many of them injections of vaccines for diphtheria and scarlet fever.[8] According to the official document that the Shanghai Association of Physicians [8] submitted to the Shanghai Bureau of Hygiene, some traditional practitioners went so far as to insert hypodermic needles through their female patients' underwear to avoid physical contact.[9] Because of the development of effective biomedical treatments, the hypodermic needle became a valuable instrument, and thus a symbol, for the competing parties of medical practitioners. When the critics of Chinese medicine thought about their opponents' lofty idea of integrating the two styles of medicine, their first thought must have been about this dangerous appropriation of the hypodermic needle. That is perhaps the reason why Pang puts the "self-proclaimed experts in both the New and the Old Medicine" [16] right above the "injection-givers" [20].

Apparently, the members of this so-called Association of Western Medicine did not belong among what I have been calling "practitioners of Western medicine" in the past chapters of this book. The majority of them had no formal medical training at all; some of them had learned the rudiments of medicine through hands-on assistant work in the missionary hospitals. As a rule, they came from poor families and were deficient even in general education.[10] In his chart, Pang therefore purposely excluded the Association of Western Medicine from the section of the pie representing the practice of real Western medicine [1–6] and placed it in the shaded section of "chaotic medicine." Without going further into this chart, we see clearly that our standard notion of Western medicine cannot comprehend the much more complicated situation of biomedical services that we encounter in Shanghai in the 1930s.

It is not surprising that Pang clearly distanced his profession from the Association of Western Medicine, whose members employed rudimentary techniques borrowed from biomedicine without having received proper training. Biomedical physicians with a medical diploma had always complained that their Chinese patients distrusted Western medicine partially because there were too many unqualified, self-proclaimed practitioners of "Western medicine."[11] Consequently, these less-than-qualified, unlicensed practitioners of Western medicine also lived under the constant threat of governmental regulation.[12] In terms of the struggle under investigation, these less-than-qualified practitioners did not side with what I have been calling "practitioners of Western medicine" in the standard use of that phrase. On the contrary, inspired by the traditional practitioners' political victory in March 1929, they decided to establish their own separate Chinese Association of Western Medicine (*Zhonghua xiyi gonghui*) within a matter of months after this event.

The inclusion of "Western Medicine" in the name they chose for their association is highly revealing. It conceded both the foreign origin of biomedicine and the relatively legitimate status of Chinese medicine. Because of its negative connotations, biomedical practitioners had consistently resisted using the term *Western medicine* to describe biomedicine in China.[13] It is ironic that the group I have referred to as "practitioners of Western medicine" in previous chapters did not want to associate their practice with the name *Western medicine* and that what I have been calling "practitioners of Chinese medicine" preferred to use the name *National Medicine*. Eight years after this struggle over names, neither of them got to be called by the name they had chosen for themselves.

Both the "foreign doctors and hospitals" [1] and the Chinese-born biomedical doctors trained in various foreign countries [3–6] maintained relatively rigorous qualification requirements for their members. Consisting of about 265 members, the "foreign doctors and hospitals" group was most advanced in following "global academic trends" [2], a fact that Pang emphasized visually by the unusually long radius of its slice. Nevertheless, Pang differentiated between foreign doctors and Chinese-born doctors of Western medicine (consisting of 660 members), the group that he himself identified with. As Pang put it, "The foreign doctors and hospitals have no relation at all to the [Shanghai] Association of Physicians. Therefore, they are separated into another group in the chart. While they have advanced facilities, only a minority of people have access to their medical services—capitalists, the wealthy and the powerful, and foreign residents. They seem to have only a minimal relation to the citizens."[14]

Let us finally proceed to the key group, which I have characterized as practitioners of Western medicine in the past chapters. The members of this group of Chinese biomedical practitioners are further separated into subgroups according to the country where they received their medical education: China, England, and the United States [3]; Germany and Japan [4]; France and Holland [5]; and Belgium, Austria, and Switzerland [6]. Because the Chinese language during this time still lacked a standardized scientific and biomedical terminology, Chinese practitioners of biomedicine continued to use the foreign languages in which they had been trained, even though efforts toward standardization had started as early as 1915. There were serious tensions among these subgroups, especially between the members of two major medical associations: the National Medical Association of China (*zhonghua yixuehui*), whose members had been trained mostly in the United States, England, and in English-speaking biomedical institutions in China, and the Medical Association of the Republic of China (*zhonghua minguo yiyao xuehui*), which consisted of physicians who had been trained in Japan and Germany. The divide between these associations was so sharp that out of the 794 members of the National Medical Association of China in 1932, only 14 had graduated from Japanese medical schools.[15]

In the early Republican period, medical education and the nascent medical administration in China were dominated by Japanese influence, even though European and American medical missionaries provided the majority of modern-style health-care services. Before the Rockefeller Foundation decided to involve itself with medical development in China, it established the famous China Medical Commission in 1914 to survey the state of medicine in China. Concerning the government-funded medical schools, this commission discovered the following situation: "With the exception of the Peiyang Medical School at Tientsin and the Kwong Hwa School at Canton, the Chinese government and private medical schools in China are almost exclusively under Japanese influence, practically all their staff having been trained in Japan or by Japanese teachers in China. Unfortunately most of the Chinese who have studied medicine in Japan have attended second-grade schools."[16] As this report suggests, during the late Qing period, Japanese-trained physicians dominated the newly established governmental institutions of hygiene, especially the medical positions in the army and military schools. When the Republic of China was established in 1912, the organizational framework for hygiene and legal regulations was often simply translated from the Japanese, although many of these regulations had never been put into practice.[17] Since Japan had taken Germany as its model for medical development and sent many students to medical schools there, China developed a distinctive style of Japanese-German medicine.

This situation shifted radically after the Rockefeller Foundation entered the scene in 1917 and began providing generous support to the Peking Union Medical College (PUMC) and other missionary hospitals and colleges. Moreover, independent of the field of medicine, in 1922 the Chinese government launched a general reform of education that transformed its national education system from the Japanese model to the American model.[18] While the Nationalist government strove to strike a balance between these two competing groups in arranging key personnel for the Ministry of Health,[19] the ministry was staffed with many graduates of PUMC and headed by Liu Rueiheng, the first Chinese superintendent of PUMC. It is fair to say that over the course of the Republican period, the Anglo-American faction gradually acquired a dominant position in both medical administration and the medical profession as a whole.

In spite of the tensions among these different factions, Pang advocated the idea of omitting national markers, such as "Japanese medicine" and "German medicine," from professional titles and newspaper advertisements. This attitude is reflected visually in his chart. Within the Shanghai Association of Physicians, Pang demarcated these national divisions with dotted lines because "it is actually a common misunderstanding that the New Medicine can be further divided into English, American, German, and Japanese factions. We should not recognize this division to be true."[20] Nevertheless, many biomedical practitioners regarded the tensions between these national subgroups as a crucial problem for medical development in China.[21]

What united these biomedical practitioners enough to enable them to cooperate in founding the Shanghai Association of Physicians [8]—the circle that transcends these slices—was the lobbying by practitioners of Chinese medicine to assimilate their schools into the national education system in 1925. In response, the most vocal critics of Chinese medicine—Yu Yan, Wang Qizhang, Pang Jingzhou, and Fan Zhouyuan—collaborated to initiate a merger of the established associations for the promotion of Western medicine and thereby founded the association, naming Yu Yan as its first president. When Yu Yan raised his historic proposal to abolish Chinese medicine in the First Public Health Conference in 1929, he did so in his official role as the representative of this association. Since the physicians in this association were the most devoted critics of Chinese medicine, they turned their association and its official journal, *The New Medicine and Society* (*Xinyi yu shehui*), into a public forum for the debates between the two styles of medicine, in the process establishing Shanghai as the central stage for this struggle.

Before we delve into other details of the chart that is the focus of the present chapter, I would like to make two general points. First of all, the medical environment of Shanghai was by no means representative of the general situation in China. According to an often-cited 1935 survey of the distribution of biomedical physicians, 22 percent of the biomedical physicians in China were practicing in Shanghai, while more than half of China's provinces had less than 1 percent of them.[22] We can easily see that it was almost meaningless to talk about the struggle between two styles of medical practitioners in these provinces. Secondly, Pang's medical world seems to have been dominated exclusively by men. While we find a few terms that imply the male gender, such as "old master Chen" and "the third old master," it seems that no counterpart for female practitioners existed, even though there were nine midwifery training schools in operation in Shanghai at this time.[23] The only possible exception is his category of "nurses" (*kanhusheng*) [13], but at that time many pioneering nurses were male, because people were concerned about physical contact across gender lines. As a legacy of this history of male nurses, the Chinese term for nurse is *hushi*, which is a gender-neutral term. By the 1940s, however, there was not a single man left in the training schools for nursing; they had been completely supplanted by women.[24]

Chinese Medicine: Fragmentation and Disintegration

Let us now turn to the practitioners of Chinese medicine, a summary designation for a wide array of medical providers that is rather difficult to define for the period under consideration here. To begin with, it is highly revealing that we are unable to find the term *Chinese medicine* (*zhongyi*) anywhere in this chart. This intriguing phenomenon is the result of two forces. On the one hand, as I pointed out in chapter 3, the practitioners of Chinese medicine had voted during their second national conference to change the name of Chinese medicine to National

Medicine. On the other hand, what we now call Chinese medicine was at that time not yet a distinct entity but was only in the process of being formed. The so-called practitioners of Chinese medicine were busy building institutions (associations, schools, hospitals, and journals), "putting in order" and systematizing doctrines, differentiating themselves from other competing practices of folk medicine, and thereby creating for themselves a recognizable profession.

Against these efforts to construct a systematized entity of Chinese medicine, Pang painted a picture of dispersion and incoherence for Chinese medicine that was the opposite of the picture that he had painted for Western medicine. With the exception of the Association of Western Medicine [12], which Pang did not consider as belonging to Western medicine, Pang put all segments representing Western medicine side-by-side with each other, outside of the shaded sections of "chaotic medicine." In contrast to this relatively unified image of Western medicine, the visual image of Chinese medicine represented in Pang's chart was remarkably fragmentary, chaotic, and even dispersive.

To be sure, a discernible tradition of the scholar-physician (*ruyi*) had existed in Imperial China. From at least the Ming dynasty (1368–1644) on, many well-educated scholars, often after repeatedly failing the civil service examination, decided to give up the aspiration of becoming state bureaucrats and sought to make a living in medicine. To distance themselves from illiterate folk healers and hereditary physicians, these medical newcomers strove to create a new and idealized identity of scholar-physician for themselves by way of emphasizing both textual erudition and clinical experience.[25] Although these scholar-physicians represented a salient increase in the participation of social elites in medicine, as Nathan Sivin has pointed out, "They were not organized, did not think of themselves as a group, and could not set or enforce common standards of medical education, skill, or compensation."[26] By 1921, the missionary doctor Harold Balme still claimed that "in China there was no medical profession, as we understand the term."[27] In Pang's chart, the circumference of about two-thirds of the largest circle is delineated with a dotted line, expressing the fact that people without any medical background, such as some "old master Chen," "gangsters," [21] and the sarcastically added "anybody" [24], could freely enter the field of medicine.

As Pang emphasized in this chart, health-care services in 1930s Shanghai included many religious, even magical, practices: shamans and exorcists [15], Daoist temples [17], divination [23], altars for spirit mediums [40], talismans from Chenzhou [18], and specialists of *zhuyou* exorcism [14]. Clearly, even the seemingly neutral concept of "medicine" had the unfortunate effect of obscuring many important aspects of Chinese health care in the 1930s.[28] As Yu Xinzhong points out, religious and magical practices were still widely used during the Qing dynastry for health maintance in the Jiangnan area. Some of these practices went so far as to change their focus from fortune-telling to the treatment of diseases.[29] In light of this history, it is no surprise that Pang found so many religious and magical practices still active in Shanghai's health-care services. For quite a while, biomedical practitioners and progressive intellectuals had been critical of this lack of differentiation between medical and religious practices.[30] On the eve of the 1929 confrontation, the affinity between medical and religious practices had emerged as one of the most salient weaknesses of Chinese medicine, because the Nationalist government had launched a vigorous antisuperstition campaign in the lower Yangtze provinces in 1928 and 1929.[31] Not incidentally, this government campaign against folk religion was cited by Yu Yan as one of his four rationales for abolishing Chinese medicine.

While the traditional physicians who were active in the political struggle held very different, even conflicting, visions about how to develop Chinese medicine, the vast majority of them were opposed to "superstition" and strove to purge their medical profession of religious practices. For many scholar-physicians, this differentiation had been their objective since the late Imperial period. For example, beginning in the eighteenth century, some scholar-physicians reinterpreted *zhuyou* exorcism [14] as a kind of secular therapy to treat emotional disorders.[32] Partially because of this legacy, almost none of the traditional physicians ever proposed to use this affinity to defend Chinese medicine as a religious practice.

Against this background, it is noteworthy that Pang identified at least five sectors in which traditional practitioners had a certain amount of control over the boundaries of their communities, as represented by solid lines. In comparison to the nine hundred qualified practitioners of Western medicine, the number of licensed practitioners of traditional medicine in Shanghai in 1933 amounted to 5,477.[33] It is reasonable to assume that the majority of these more than five thousand licensed practitioners came basically from the following five sectors, going counterclockwise: the "faction of awakened young practitioners of the Old Medicine" [26], the "conservative faction" [28], especially the Shenzhou Medical Association [29], the Institute of National Medicine [Shanghai Branch] [34], "neo-Confucian medicine" [39], and the "faction of Chinese-Western medical schools" [25].

Given what I have just said about the history of the scholar-physicians, the reader may be surprised to learn that Pang was highly critical of what he characterized as neo-Confucian medicine. I purposely translate Pang's term as "neo-Confucian medicine" rather than as "scholarly medicine" because his compound term *ruli* highlights the Chinese word *li* (meaning "pattern" or "theory," but specifically, since the Song dynasty, "neo-Confucian philosophy" in the compound *lixue*) and therefore foregrounds the connection between this style of medicine and neo-Confucian philosophy. Pang's critical attitude toward neo-Confucian medicine was widely shared by many reform-minded practitioners of Chinese medicine. For example, the traditional physician Yang Zemin, who was admired for pioneering a Marxist understanding of Chinese medicine, pointed out that traditional practitioners had imitated neo-Confucian philosophers since the Jin and Yuan dynasties (1115–1368 CE) in reorganizing experience-based Chinese medicine with philosophical concepts such as *xing* (nature), *li*, and *qi*. These theoretical concepts had penetrated so deeply into the fabric of clinical practice that medical practitioners had started "differentiating the nature of medicinals on the basis of five colors and five tastes, explaining the use of medicinals on the basis of their shapes, and making sense of pulse diagnosis and medicinal formulas with neo-Confucian theories."[34] Just like Yu Yan and Zhang Taiyan, Yang concluded that the empiricism of Chinese medicine had become irreversibly corrupted by neo-Confucian philosophy.

From Pang's point of view, the contemporary practitioners of neo-Confucian medicine were often literati who had neither medical training nor clinical experience. Since they had turned to medicine late in their careers, they tended to read the medical canons as philosophical works and therefore base their medical credibility mainly on the mastery of neo-Confucian doctrines. After the late Qing government had abolished the Imperial examination system in 1904, which had eliminated the need for traditional scholarship, many scholars had moved to Shanghai and turned to medicine for a living. While the entrance of the literati into the medical profession had been valued historically as a welcome development that had raised the social status of medicine, Pang saw the current flood of literati as a degrading force, because these self-taught physicians had little connection to the clinical practice of traditional medicine.[35]

In contrast to neo-Confucian medicine, Pang apparently had more respect and sympathy for the members of the Shenzhou Medical Association [29]. Established in 1912, its precursor, the Shanghai Medicine Association (*Shanghai yihui*), had been the first organized association of

Chinese medicine. While Shanghai had at least four associations of Chinese medicine by this time, Pang highlighted this association partially because it was founded by Li Pingshu (1854–1927),[36] who was widely admired for initiating many aspects of Shanghai's modernization, including education, banking, water supply, electrical utilities, insurance, self-government, and the anti-opium campaign.[37] Since both his father and grandfather had been practitioners of Chinese medicine, Li also excelled in medicine and became famous for treating a few high-ranking late Qing officials, including the acclaimed modernizer Shen Xuanhuai (1844–1916).[38] Cooperating with the Imperial physician Chen Lianfang (1839–1916) and other practitioners of Chinese medicine, Li had organized the Shanghai Medicine Association in 1903.[39] Moreover, in cooperation with the female biomedical physician Zhang Zhujun, Li had founded the first modern medical school for women (*Shanghai nüzi zhongxiyi xuetang*) in 1904, where he taught students Chinese medicine while Zhang taught Western medicine.[40] In 1908, they had again cooperated to build the Shanghai Hospital, which provided both Chinese and Western treatments.[41] Instead of viewing Chinese medicine and modernity as antithetical, the visionary modernizer Li was personally committed to the modernization of traditional Chinese medicine, even though quite a few of his innovative efforts did not survive when he moved to Japan in 1912. This is a story worthy of a more comprehensive and in-depth study.

The early origins of the Shenzhou Medical Association were consolidated when it initiated the first nationwide campaign of practitioners of Chinese medicine against the so-called "omission of Chinese medicine from the school system" in 1912 (discussed in chapter 5).[42] This successful demonstration had a great impact on one of its leaders—Ding Ganren (1865–1926). The anthropologist and historian Volker Scheid has devoted an entire book to this legendary physician and to tracing the allegedly four-hundred-year-old lineage of the Menghe medical current, which was retrospectively given shape and epitomized by Ding and his disciples during the Republican period in Shanghai. In tracing the evolution of this important medical current, Scheid has revealed a densely connected network of practitioners who were related to each other as teachers, students, family members, colleagues, and friends.

Famous as one of the four great physicians of the Republican period, Ding helped found this association and served as its vice chairman. In addition, he established the Shanghai Technical College of Chinese Medicine (*Shanghai zhongyi zhuanmen xuexiao*) in 1916, organized the Shanghai Chinese Medicine Association (*Shanghai zhongyi xuehui*) in 1921, published the *Journal of Chinese Medicine* (*Zhongyi zazhi*), and managed hospitals with the help of family members, colleagues, and students.[43] These institutions became a close-knit community and served as a center for traditional practitioners when they launched the March 17 demonstration. Both of its earliest initiators, Chen Cunren and Zhang Zanchen, were alumni of Ding's Shanghai Technical College of Chinese Medicine, and Xie Guan, the leader of the five-man delegation, was its first dean. Moreover, Ding Zhongying (1886–1978), Ding Ganren's second son and his helpful assistant in managing the college, was elected as the general secretary of the National Federation of Medical and Pharmaceutical Associations during the 1929 demonstration. Their leadership in the National Medicine Movement was just one of the numerous ways in which the Menghe current contributed to the modern development of Chinese medicine.

As Scheid has correctly pointed out, Ding Ganren separated the task of modernizing Chinese medicine into two distinct dimensions—institutional format and conceptual content—that were to be treated with fundamentally different strategies. While Ding and his disciples pioneered in building modern institutions for Chinese medicine,[44] Ding regarded the "fundamental rejection of Western medicine as being of no value to the development of Chinese medical practice."[45] Ding passed away in 1926 and therefore could not take part in the 1929 campaign. Two of his successors—his second son, Ding Zhongying, and his grandson and officially appointed heir, Ding Jiwan (1903–63)—offered their leadership in organizing the March demonstration. Nevertheless, Ding and most of his senior colleagues were not supportive of the project of scientizing Chinese medicine. As Scheid has pointed out, the majority of Menghe physicians, including Ding Ganren, Yun Tieqiao, and Xie Guan, "belonged to the group of conservatives who sought to preserve Chinese medicine."[46] Nevertheless, the conservative position was by no means the exclusive choice for the numerous Menghe physicians in Shanghai, and "many of their friends, colleagues, and even students, however, made other alliances."[47] As Pang was keenly aware, serious tensions existed between the senior practitioners affiliated with the Shenzhou Association and the young reformers who dominated the Institute of National Medicine.[48] After the state established the Institute of National Medicine, those traditional practitioners who were unwilling to "scientize" Chinese medicine were pushed to the periphery, as shown by "the receding frontier of the Old Medicine" [24] in Pang Jingzhou's chart.

In Pang's view, the only sector of Chinese medicine worthy of hope was what he called the "faction of awakened young practitioners of the Old Medicine" [26]. To visually express his approval of the scholarly orientation of this faction, Pang made the radius of its slice in the pie chart extend beyond the circumference of the largest circle—the only sector of Chinese medicine to have this feature. According to Pang, although the members of this group had been trained as traditional practitioners, they recognized the absurdity of Chinese medicine and whole-heartedly embraced the truth of modern biomedicine. Unfortunately, because they lacked the formal educational qualifications, they had to be considered as traditional practitioners and were therefore faced with a very difficult situation in terms of business competition. This difficult situation is visually expressed by their spatial position—the slice representing this group [26] is sandwiched between and rigorously segregated from the "conservative faction" [28] and the "foreign doctors" [1]. Pang was very sympathetic to this group, noting that "they were in the most painful position since they did not dare to practice modern medicine. . . . [As a result], they confined themselves to either theoretical discussions or to conducting historical and textual research on the Old Medicine."[49]

While Pang did not give specific names, he must surely have had in mind people like Fan Xingzhun, who later became a widely respected pioneering historian of Chinese medicine and a beloved teacher, working until his nineties at the Beijing Research Institute of Chinese Medicine.[50] Fan became interested in medical history partially because of the structured position depicted by Pang. More importantly, however, Fan, and Yu Yan as well, found medical history to be an effective tool in accelerating their vision of a Chinese medical revolution. This political function, in combination with cultural nationalism, generated early interest in the study of medical history, attracted talented and committed students to this field, and thereby helped to turn medical history into an academic discipline.[51] At the same time, though, it left a long-lasting legacy of confinement in the historiography of traditional Chinese medicine. Meanwhile, this group represented hope for people like Pang, because its members recognized the truth of biomedicine but, unlike the "communication faction within the Old Medicine" (*goutong pai zhi jiuyi*) [35], they had no intention of combining biomedicine with traditional medicine. In fact, this group soon became the most vigorous critic of the project of "scientizing Chinese medicine." For Pang, these young practitioners embodied the real meaning of awakening—not just seeing the truth, but also relentlessly rejecting falsehood.

Systematizing Chinese Medicine

Among all five of the organized sectors of Chinese medicine, the faction of Chinese-Western medical schools [25] was the only one that Pang placed in the shaded section of chaotic medicine. This unique positioning demonstrates how critical Pang and his colleagues felt of the booming development of such hybrid medical schools. While traditional practitioners had established schools of Chinese medicine since the end of the Qing dynasty and campaigned for their legalization in the 1920s, there was a breakthrough both in terms of the quantity and quality of these schools around the time of the 1929 confrontation. An article by Longxibuyi reviewing the curricula and histories of the seven schools of Chinese medicine in Shanghai in 1928 provides a testimony from that time. Although the author enthusiastically praised Ding Ganren for his pathbreaking efforts to establish the Technical College of Chinese Medicine, he criticized its five-year curriculum, which included almost no courses on modern biomedicine, as "an obstacle to progress."[52] Moreover, even though Longxibuyi also highly praised Ding's equally impressive efforts to build the Technical College of Chinese Medicine for Women in 1926, he could not help commenting that this school was "so focused on interpreting the old [medical] canons and [so opposed to adopting] Western methods, that it was not compatible with the trends of modern society."[53] The author's judgment was captured by Pang with an arrow extending from "traditional practitioners of the Old Medicine" [33] almost half-way across the chart to "outmoded practitioners of the Old Medicine" [42], a phrase that sits smack atop the circle representing the faction of Chinese-Western medical schools [25]. By this time those physicians who adhered strictly to the traditional doctrines of Chinese medicine were often criticized as being outmoded by advocates of the Chinese medical schools.[54] Although Longxibuyi was highly critical of Ding's colleges, he nevertheless concluded that the only college that stood a chance of surviving in the metropolitan area of Shanghai was the most tradition-oriented one among them, namely the Technical College of Chinese Medicine.[55]

Compared with the two colleges founded by Ding, the other five colleges described in the article included more courses on Western medicine in their curricula. Another author compiled a table to compare Ding's college point by point with the Shanghai Institute of Chinese Medicine (*Shanghai zhongguo yixue yuan*), which was founded in 1927 and had the famous scholar Zhang Taiyan as its honorary president. According to this author, the Shanghai Institute was much more accommodating of Western medicine,[56] and it prepared brand-new "lecture notes" for each course, while Ding's College relied completely on traditional books of Chinese medicine.[57] Among the 172 teaching materials now collected in the library of the Beijing Institute of Chinese Medicine,[58] at least 75 contain in their titles the term *lecture notes* (*jiangyi*), which is a modern Chinese loan word, borrowed from the Japanese, and designates a modern lecturing style.[59] The compilation of lecture notes sent a clear signal that the teaching style had started shifting from traditional apprenticeship and canon interpretation toward class lectures.

In 1932, Pang claimed that the Chinese people did not trust the graduates of these schools who had not gone through a traditional apprenticeship.[60] The truth was, however, that the number of such schools had increased dramatically after the 1929 confrontation. The number of Chinese medical schools that were established in the period from 1928 to 1937 was more than triple the number of schools that had been established before 1928.[61] After the policy of equal treatment for Chinese and Western medicine was promulgated by the state in 1935, the number of schools of Chinese medicine skyrocketed.[62]

Conclusion

From Pang's point of view, these medical schools represented a degrading process; the students thus produced were just horrible, according to Pang: "They prescribe Western drugs and give injections. Almost none of them use the traditional therapeutics exclusively; many of these graduates have adopted Western therapeutics and joined the Association of Western Medicine."[63] Against the fact that practitioners of Western medicine repeatedly made this accusation in public,[64] it is noteworthy that hardly any traditional practitioners ever confronted these serious accusations. While practitioners of Chinese medicine appeared devoted to "integrating" Western medicine with Chinese medicine through establishing medical schools, practitioners of Western medicine saw their efforts as nothing more than giving birth to a degraded "mongrel medicine" (*zayi*).

Pang felt appalled by the emergence of this new species of medicine:

> Ten years ago, there were only two styles of medicine: Old Medicine and New Medicine. While the practitioners of the Old Medicine had their own old-style experiences and technique, they did not dare to claim to know anything about the New

> Medicine. Conversely, practitioners of the New Medicine did not dare to appropriate the Old Medicine. The distinction was clear-cut. . . . In the past ten years, opportunistic "integrators of the New and Old Medicine" have emerged who have asserted all kinds of nonsense, such as the theory that "wind strike" (*zhongfeng*) is identical with cerebral hemorrhage." Then this has really become a catastrophe for the people.[65]

The establishment of the Association of Western Medicine, the founding of schools of Chinese-Western medicine, and the emergence of mongrel medicine, from Pang's point of view, had all led to more chaos in the medical environment in Shanghai. The intended general impression of Pang's chart was precisely this increasing chaos—almost every arrow indicating the flow of personnel led to a more chaotic medical environment. Inside a small circle at the very center of the chart and of the largest circle, we find the phrase "the center in charge" [43] surrounding a prominent question mark.

For Pang, the problems facing the profession of Western medicine were not unlike those facing the Nationalist state:

> While the Nationalist revolution has temporarily succeeded, there are many tasks in need of continued attention: overthrowing the warlords (just like abolishing the Old Medicine), wiping out the Communists and purging the Nationalist party (just like outlawing the motley medical practices), committing to construction (just like promoting medical education and research), and defending against military invasion (just like defending against cultural invasion from abroad). . . . Nevertheless, none of the above-mentioned problems facing our medical profession can be solved without the full assistance of politics.[66]

As represented by Pang, practitioners of Western medicine generally believed that "politics" would provide the ultimate solution to the problem of Chinese medicine and the chaos of health care in China. Pang apparently failed to notice that the current situation had been caused directly by the attempts of the biomedical profession to regulate medicine through the state. Even if some of Pang's fellow physicians finally recognized that their strategy had in fact generated this unintended group dynamic, it was too late by then to retreat. Having successfully consolidated their national association and forced the state to give in to their demands, practitioners of Chinese medicine were no longer interested in merely "resisting" the state. They had realized as well that Chinese medicine was in a rather chaotic situation and had accommodated too many unqualified practitioners. Most importantly, they could not agree more with Pang that political authority was the ultimate tool for solving this problem. In response to the question mark at the center of Pang's chart, related to the statement "center in charge," traditional practitioners swore to turn the semi-governmental Institute of National Medicine [34] into the real administrative center of Chinese medicine.

7 Science as a Verb: Scientizing Chinese Medicine and the Rise of Mongrel Medicine

While the noun *science* was valorized and personified as "Mr. Science" (*sai xiansheng*) by the progressive intellectuals of the May Fourth Movement, the Nationalist state pioneered the use of *science* as a verb, most notably while popularizing the project of "scientizing Chinese medicine" (*zhongyi kexuehua*). Before I delve into the birth of this controversial project, I would like to call attention to a little-recognized but crucial linguistic phenomenon. That is, while Chinese-speaking people normally assume the Chinese term *kexuehua* to be a Chinese translation of a Western word, its English counterpart, *scientize*, is actually not a proper English term. If you do not trust your computer's spell check, which will underline every instance of *scientize* in your writing, you will find in the *Oxford English Dictionary* that the word *scientize* is rarely used in English. In the handful of quotations listed in the OED, almost half enclose the term in quotation marks. Apparently, although *scientize* looks like a direct and natural extension of the concept of science, the European countries that jointly created the modern concept of science did not feel the need to turn it into a verb. In sharp contrast, *scientize* is a word used daily in modern Chinese, Japanese, and Korean; many native speakers would find it puzzling that Westerners can do without it.

Given its local origin, there is no way that the discourse of scientization could be simply a local duplication of the globally circulated discourse of European modernity. It is a pity that, as far as I know, scholars have yet to recognize the local and regional nature of the concept of scientization, let alone to document its genealogy and circulation in East Asia. Instead of providing such a regional history, however, this chapter focuses only on China and argues that the project of "scientizing Chinese medicine" played a pioneering role in popularizing the concept of scientization in Republican China. Given that the concept of scientization served as a tool that some Nationalist ideologues used to reconcile the conflict between scientism and cultural nationalism, the project of "scientizing Chinese medicine" served as a crucial innovation in negotiating with the universalist conception of modernity.

In addition to serving as an ideological tool, the project of "scientizing Chinese medicine" signals a crucial break in the way in which leading figures imagined the relationship between Chinese and Western medicine, which was the central theme of chapter 4. The historical discovery discussed in the present chapter is the fact that the project of Scientizing Chinese Medicine, which was later endorsed by Mao Zedong in 1956,[1] and is still actively pursued in today's People's Republic of China,[2] was first proposed and made popular by the Nationalist state when it decided to establish the Institute of National Medicine (*guoyiguan*) in 1931. As advocates of the National Medicine Movement strove to assimilate Chinese medicine into the national medical system, many prominent leaders openly embraced Yu Yan's tripartite categorization of Chinese medicine into theory, drugs, and experience and admitted that it needed to be thoroughly "scientized." This chapter traces the rise of this project, its ideological context, and the three representative positions developed within the debate over the scientization of Chinese medicine. Most importantly, in order to document and to understand the nature of this new species of Chinese medicine—one that was described as being "neither donkey nor horse"—we have to understand how the locally invented project of scientization motivated, shaped, and questioned the very existence of what was known as "mongrel medicine."

The Institute of National Medicine

Almost immediately after the confrontation in March of 1929, science emerged as a possible middle ground for the competing parties of medical practitioners and state bureaucrats. Among the four rationales on which Yu Yan had founded his proposal to abolish Chinese medicine, two concerned its perceived lack of science—namely, those dealing with theory and diagnosis. When traditional practitioners published their petition in all the major national newspapers, they openly admitted that "because the theories [of Chinese medicine] are incompatible with science, the international scholarly community does not believe in Chinese medicine."[3] Furthermore, in response to the numerous telegrams protesting its decision to abolish Chinese medicine, the Ministry of Health repeatedly emphasized that "with regard to Chinese drugs, our ministry strongly supports their development. Concerning Chinese medicine, we plan to improve it so it can be scientized."[4] From this time on, the Ministries of Health and Education apparently subscribed to the idea of improving Chinese medicine with scientific methods. A traditional practitioner put it vividly, "as the result of such a vehement confrontation, a truce was declared: Chinese medicine should be scientized."[5]

This truce was formally declared by the Nationalist government when it decided to support the funding of the Institute of National Medicine and announced its constitution. On March 17, 1930, about one year after the 1929 confrontation, seven high-ranking members of the Central Executive Committee of the Nationalist Party, including the current (also the first) prime minister, Tan Yankai (1880–1930), and head of the Legislative Yuan Hu Hanmin (1879–1936), proposed to establish the Institute of National Medicine on the model of the Institute of National Martial Arts. According to this proposal, the primary purpose of this institute was to "put in order Chinese medical learning and conduct research on Chinese drugs with scientific methods."[6] Since the leaders of both the executive and legislative branches of the government were behind this proposal, the Central Executive Committee passed the resolution and then passed its decision on to the Nationalist government. Keenly aware of this origin, practitioners of Western medicine later ridiculed the notion that the institute had been created by the "great men" from the Nationalist Party.

As noted in chapter 5, it was traditional practitioners who first came up with the idea of the Institute of National Medicine, and their newly established National Federation of Medical and Pharmaceutical Associations proposed it to the government. Because their objective was centered on creating an administrative authority, controlled by them, to take charge of regulating the education and practice of Chinese medicine, the three tasks specified in their original proposal do not mention science at all. As soon as the Nationalist Party approved the proposal in April 1930, the core members of the federation completely lost control over the project that they had initiated. For eight months, they were left in the dark about the preparation of this institute within the Nanjing government. On December 20, 1930, the standing committee of the federation decided to send a delegation to Nanjing to find out about its development.[7] To pursue their agenda, the delegation carried with them a memorandum that aimed at turning the Institute of National Medicine into an administrative agency.[8] The delegation was welcomed warmly in Nanjing, but when they returned to Shanghai their worries deepened that the institute might not be entrusted with the requested administrative authority.[9]

It was highly symbolic that the Institute of National Medicine was established on March 17, 1931, precisely two years after the historic demonstration by the advocates of Chinese medicine in Shanghai. The choice of this date portrayed the institute as the hard-won result of their collective struggle. On its inauguration day, Chen Yu, the chairman of the ceremony and a Nationalist politician, made the purpose of the institute crystal clear: "The most important thing that I have to report for today is the foremost purpose for which the central government has decided to establish the Institute of National Medicine. What the [Central Executive] committee members proposed in the political conference was to put in order (*zhengli*) national medicine and national pharmacy with scientific methods."[10]

By repeatedly emphasizing "the foremost purpose," Chen Yu clearly expressed that he had delivered a deal from the government. If traditional practitioners wanted to have a state-sponsored Institute of National Medicine, they had to accept the purpose of that institute. As a result, the first article of its constitution reads, "This institute has the objective of choosing scientific methods to put in order Chinese medicine and pharmacy and improve the treatment of disease, and the manufacture of pharmaceuticals."[11] Instead of preserving Chinese medicine as an aspect of a distinctly Chinese national essence, the objective of this institute, as expressed in its official manifesto, was "to transform Chinese medicine, as a representative of Eastern culture, into a 'world medicine' [*shijie yixue*]."[12]

Despite the fact that the Institute of National Medicine was not entrusted with administrative authority, traditional practitioners were thrilled to know that finally they had their own official organ within the state, instead of the semi-official organization that they proposed originally.[13] In the journals of Chinese medicine published during this period, practitioners enthusiastically congratulated each other for this breakthrough. As a part of the deal, they began embracing the idea that Chinese medicine needed to be "put in order" and scientized.

So far we have seen two senses in which science was to be connected with Chinese medicine. The first sense was to "put in order" (*zhengli*) Chinese medicine with scientific methods. In this context, "put in order" was not as mundane a phrase as it might appear. Rather it had a well-established meaning created by those who promoted the idea of "putting in order" the "nation's antiquities" (*guogu*). Invented by Zhang Taiyan, the notion of the "nation's antiquities" implied that these cultural artifacts, thoughts, and institutions were all things of the past. While this looks like a scholarly endeavor, as John Fitzgerald insightfully pointed out, the study of the "nation's antiquities" represented a political movement concerned with what he calls the "museumification" of China. In the name of national studies (*guoxue*), the advocates of modernity placed still-vibrant aspects of Chinese culture in museums, to make viewers see these cultural elements as things of the past, or at least as destined to pass away in the near future.[14] In fact, interest in Chinese medical history and the construction of a Museum of Chinese Medicine were inspired by this larger movement of national studies.[15] While emphasizing the role played by scientific methods, the first formulation of scientization as "putting in order" appears to have been modeled on this established idea of putting in order China's past and thereby making Chinese medicine a part of the old China.

In contrast to the first formulation, which relegated Chinese medicine to China's past, the second formulation of *kexuehua* (scientization) carried with it the implication of turning its object into an element of science, and thereby into a contributing and functional aspect of modernized China. In other words, by creating the term *kexuehua*, the Nationalist project of "scientizing Chinese medicine" opened the crucial possibility that the relationship between Chinese medicine and modern science did not need to be antithetical. Since *kexuehua* was a newly created word with uncertain meanings, these two formulations often got mixed up until their different implications manifested as the polemics of scientization intensified in the mid 1930s. For the time being, let us follow our actors in regarding these two formulations as interchangeable. The important point is that although Ding Fubao allegedly invented the slogan of "scientizing Chinese medicine" in the late 1910s or early 1920s, this project did not become popular until the funding of the Institute of National Medicine. This constitutes a sharp break in the attitude of Chinese medical practitioners toward the project of Scientizing Chinese Medicine.

There is a significant time lag between the May Fourth Movement (1919), which turned scientism into one of the most dominant ideologies in the history of twentieth-century China, and the rise of this scientizing project in the 1930s.[16] Because intellectual historians have taken the prevalence of scientism for granted in intellectual circles after 1919, it is easy to overlook the fact that the majority of traditional practitioners at that time did not follow the path of progressive intellectuals in valorizing science. After the founding of the Institute of National Medicine, the sudden popularity of this project made the competing camps of practitioners acutely aware that their struggle now had advanced to a new stage.

Among practitioners of Western medicine, Yu Yan was most excited to see that traditional practitioners were finally accepting the project of Scientizing Chinese Medicine. While we might expect Yu to still feel frustrated about seeing his proposal pigeonholed in March of 1929, only two years later he enthusiastically celebrated the victory of "the theoretical aspect of the Medical Revolution" because "those who identify themselves as practitioners of Chinese medicine now also openly support the idea that Chinese medicine should be put in order with scientific methods."[17] As Yu recalled later, when he heard the news that the Institute of National Medicine had been established to scientize Chinese medicine, he became "too excited to get a good sleep for days."[18] In addition, when Yu wrote the preface for the second edition of his *Collected Essays on Medical Revolution*, published in 1932, he singled out the institute's scientizing project as his main achievement in promoting the Chinese medical revolution.[19] As exemplified by Yu, practitioners of Western medicine saw a clear break in the stance of traditional practitioners toward scientizing Chinese medicine.

While some traditional practitioners advocated this project in the early 1920s,[20] the majority of them became keenly aware of it only after its sudden popularity. For example, Chen Peizhi wrote,

> Just five or six years ago [1929–30], only a very few enlightened people advocated the idea that Chinese medicine needed to be put in order with science. This project was too idealistic to be popular. However, since Tan Yankai and Hu Hanming proposed establishing the Institute of National Medicine to put in order traditional medicine, in the past several years almost everyone in the medical profession has agreed that Chinese medicine should be reevaluated and put in order on the basis of scientific principles.[21]

Many other practitioners of Chinese medicine testified to the fact that the project of scientizing Chinese medicine had become extremely popular almost overnight after the institute had been established to promote the project in 1931.[22] Instead of being associated with any individual thinker or physician, the concept of scientizing Chinese medicine was made popular by the Institute of National Medicine, the first semi-official organ for Chinese medicine. For traditional practitioners, the scientific modernity of Chinese medicine grew directly from the political modernity of the Nationalist state.

The China Scientization Movement

The project of Scientizing Chinese Medicine was accepted by the competing parties as the basis for a truce following the 1929 confrontation, but the curious thing about this government-sponsored truce was that it said almost nothing directly about the medical policy under debate. The project did not clarify whether the government should accept Chinese medicine as part of the national educational and health-care systems or outlaw its practice; it simply put off the decision, waiting for a national policy on Chinese medicine in these two areas. The project was nevertheless accepted by the contesting parties because what concerned the majority of nonmedical participants was not really health-care policy but the nature of China's modernity. As I emphasized in chapter 1, it is very important to bear in mind this coexistence of two separate but related struggles—one about the role of Chinese medicine in national health-care policy, and the other about science as modernity. The project of Scientizing Chinese Medicine was proposed by the state and accepted by the competing parties because, while leaving open the policy issue, it nevertheless provided a middle ground for the related and larger ideological struggle over the cultural authority of science.

The most important point about this project was the fact that it was the Nationalist state that declared it to mark a truce in this struggle. The state did so because it was an acceptable position within the larger ideological struggle. As Prasenjit Duara points out in his analysis of modernity in Republican China, while scholars have thoroughly investigated the ascendancy of scientism among the intelligentsia in the 1920s, "less well documented and analyzed is the commitment of the twentieth-century state to modernity."[23] Duara further points out that "since the establishment of the Republic in 1911, the Chinese state has been caught up in a logic of 'modernizing legitimation'—where its *raison d'être* has increasingly become the fulfillment of modern ideas."[24] For the sake of maintaining its "modernizing legitimation," when the Nationalist state stepped in to support the founding of the semi-governmental Institute of

National Medicine, it simultaneously demanded that the institute take on the project of Scientizing Chinese Medicine as its official objective.

The project of Scientizing Chinese Medicine was acceptable both to the progressive intellectuals concerned about the ideological authority of science and to the Nationalist Party, which wanted to fashion itself as the inheritor of the iconoclastic May Fourth Movement. This was because the project implied that Chinese medicine was unscientific as it stood and could continue to exist in modernizing China only after it had been transformed into a medicine that was compatible with science. In this sense, the project of Scientizing Chinese Medicine, as vague and complicated as its meaning was, resolved the tension between modern science and traditional medicine, even if only in appearance and for the moment. No wonder that some of the modernizing intellectuals felt, as Yu Yan did, that they had won a historic victory in the ideological struggle.

In light of its ideological function, the project of Scientizing Chinese Medicine was not an isolated case but a pioneering exploration of the larger China Scientization Movement (*zhonguo kexuehua yundong*). Launched by some Nationalist elites and natural scientists in November 1932, the objectives of this movement were to popularize science and to reconcile the relationship between science and Chinese culture.[25] In addition to sharing the use of the term *scientization*, the project of Scientizing Chinese Medicine and this larger movement were launched at almost the same time and by the same brothers, Chen Guofu (1892–1951) and Chen Lifu (1900–2001).

By every standard, the Chen brothers were crucial figures for understanding the medical-ideological struggle under consideration. As cadres of the Nationalist Party and Chiang Kai-shek's long-term political allies, the Chen brothers provided strong and steadfast support for Chinese medicine. Chen Lifu, one of the Nationalists' main ideologues, served as the first chairman of the board of the Institute of National Medicine, and his elder brother, Guofu, was on that board of directors. While their interest in medical matters was simultaneously medical, political, and personal, they had no formal medical education at all. Chen Lifu possessed a master's degree in mining engineering from the University of Pittsburgh, and Chen Guofu's formal education ended with military high school. Therefore, even though they were both lifelong supporters of Chinese medicine, their support was generally perceived as motivated purely by the ideological function of cultural nationalism.[26] Partially for the sake of challenging this conventional understanding of the Chen brothers' support for Chinese medicine, I document in chapter 9 how Chen Guofu helped—by means of both his political position and "personal experimentation"—to scientifically validate the Chinese herb *changshan* (*Dichroa febrifuga*) as an effective antimalaria drug in the 1940s.

As Chen Shou has meticulously documented in his doctoral dissertation, the Chen brothers and their powerful faction within the Nationalist Party were the main movers behind the semi-official China Scientization Movement.[27] In addition to being the first nationwide movement to popularize science in China, this movement is important because in his efforts to develop the concept of "scientization" and thereby negotiate the notion of science, Chen Lifu cooperated with one of the most original thinkers about science in Republican China: Gu Yuxiu (1902–2002), the first Chinese to receive a PhD from MIT (1928), founder of the School of Engineering at Tsinghua University, and an internationally renowned scientist and poet.[28] Moreover, as the main ideologue of the Nationalist party, Chen's central objective in promoting the movement was to reconcile the tense relationship between science and "traditional culture" so that the Nationalist party could ally itself with scientism and cultural nationalism all at once. Because of this political orientation, the Chen brothers' conception of scientization stood in sharp contrast to the universalist conception of science promoted by the May Fourth Movement.

Given this background, it is noteworthy that the Chen brothers assigned the mission of scientizing Chinese medicine to the Institute of National Medicine almost twenty months prior to the time when they actually launched the China Scientization Movement in November 1932. In light of this time lag, their strong interest in the notion of scientization might have grown out of their previous endeavors in support of Chinese medicine since the 1929 confrontation. If that is indeed the case, the Chen brothers' support for Chinese medicine might have helped the Nationalist party to develop the novel notion of scientization to reconcile the tension between science and Chinese culture. Despite this intended objective, as Chen Shou points out, most of the movement's activities actually focused on popularizing concrete scientific knowledge rather than on realizing Chen's high-minded and controversial objective of scientizing Chinese culture.[29] The discrepancy between the movement's slogans and historical reality further supports the possibility that Chinese medicine was the main source of leverage in the Chen brothers' efforts to promote their idea of scientizing Chinese culture.

Historical developments after the creation of the Institute of National Medicine in 1931 seem to support such an understanding. In the following year, Chen Lifu gave a talk emphasizing the relationship between the China Scientization Movement and China's cultural renaissance, and in June 1933, Chen Guofu gave a keynote lecture at the Nationalists' Central Committee entitled "The Necessity of Scientizing Chinese Medicine."[30] Both of these speeches were published in the movement's official magazine, which was modeled on the American monthly magazine *Popular Science*. Instead of viewing science and traditional Chinese culture, especially Chinese medicine, as antithetical, the Chen brothers strove to create a mutually beneficial relationship between them through the concept of scientization. This novel use of the concept was prominent in the China Scientization Movement; one of its three central slogans was "to put in order our traditional culture with scientific methods."[31] In addition, the movement made clear that it opposed "those who dismiss or discredit China's traditional culture simply on the grounds that [it is] 'non-scientific.'"[32] As this movement led the way in popularizing the concept of scientization in the 1930s,[33] the notion of scientization represented the Nationalist elites' exploratory effort to reconcile science with cultural nationalism.

As self-proclaimed heirs of the May Fourth Movement, the Nationalists never entertained the idea that Chinese medicine represented a form of Chinese cultural particularism, or a radical alternative to modern science. Instead, they created the novel concept of "scientization" to open up the possibility that the relationship between Chinese medicine and science was not necessarily as antithetical as was normally assumed. With the Nationalist state providing the breakthrough *in discourse* and ideology, the project of scientizing Chinese medicine motivated advocates of Chinese medicine to transform the relationship between the two *in practice*, namely, in medical theory, pedagogy, research design, clinical diagnosis, therapeutic treatment, epistemology, and ontology—to name a few areas that are documented in the second half of the book. In other words, while this project represented a settlement of the ideological struggle, the real-world development of Chinese medicine would prove henceforth to be profoundly conditioned by its engagement with this ideological middle ground. In return, these endeavors and achievements in reforming Chinese medicine gave substance to the project of Scientizing Chinese Medicine as an ideological middle ground that the Nationalist elites desired. In these two interrelated levels of ideological discourse and medical reform, the project of Scientizing Chinese Medicine represented the most salient and thus the most controversial experiment in negotiating the concept of science and modernity.

The Polemic of Scientizing Chinese Medicine: Three Positions

Although the Institute of National Medicine encountered great difficulty putting into practice its two concrete proposals for scientizing Chinese medicine, it elevated them to the level of the central theme of the post-1929 struggle. In 1936, the Medical Research Society of China (*zhongxi yiyao yanjiushe*) retrospectively devoted two special issues of its journal to the topic "Polemics of Scientizing Chinese Medicine" (*Zhongyi kexuehua lunzheng*).[34] Among the twenty-three articles collected in these two special issues, except for one whose date of publication I could not ascertain, only Yu Yan's "Construction and Destruction in Our National Medical Revolution" was published before the 1929 confrontation.[35] Other than that one, all the collected articles were produced after 1929.[36] The dating of these important articles supports my argument that the project of scientizing Chinese medicine did not exist before the 1929 confrontation.

In the opening remarks of the first special issue, the editor classified the twenty-three articles according to three positions:[37] (1) Chinese medicine does not need to be scientized (this group includes those who hold the opinion that Chinese medicine is compatible with science); (2) Chinese medicine can be scientized; and (3) Chinese medicine cannot be scientized.

Instead of being regarded merely as three different scholarly positions, these three stances on the relation between Chinese medicine and science were widely seen as representing the new field of struggle after the Institute of National Medicine was established in 1931. In the call for papers, the editor had specifically pointed out that "the problem of whether we should abolish Chinese medicine actually boils down to the problem of Scientizing Chinese Medicine. If Chinese medicine is compatible with science, we should surely preserve and improve it. Otherwise, we should just abolish it."[38] As this statement indicates, the fate of Chinese medicine had become closely connected with the project of Scientizing Chinese Medicine.

During the period from 1931 to 1937, the main debates focused on two proposals for scientizing Chinese medicine that the Institute of National Medicine issued in 1932 and 1933 respectively. By tracing the primary debates associated with the first proposal, "Proposal for Putting in Order Chinese Medicine and Pharmaceutics" (*Guoyiyao xueshu zhengli dagang*), I examine below how the three positions took shape in and through the polemics of scientizing Chinese medicine from roughly

1929 to 1932. Because the second debate gave birth to the idea and practice of "pattern differentiation and treatment determination" (*bianzheng lunzhi*) as the defining features of so-called Traditional Chinese Medicine (TCM), it receives a detailed analysis in chapter 8.

Embracing Scientization and Abandoning Qi-*Transformation*

We may begin our discussion with the second group, those who held the opinion that Chinese medicine both needed to be and could be scientized. This group was represented by the writings of Chen Guofu, Tan Cizhong, Ye Guhong, and, of course, Lu Yuanlei.[39] Since the first group believed that Chinese medicine did not need scientization, and the third group thought Chinese medicine could not be scientized, the second group had to be the group that actively proposed detailed plans for putting in order Chinese medicine.

On the basis of the first article of its constitution, in 1932 the Institute of National Medicine entrusted Lu Yuanlei with the task of drafting the "Proposal for Putting in Order Chinese Medicine and Pharmaceutics."[40] As a radical advocate of scientizing Chinese medicine, Lu was adamantly opposed to any cultural particularism of Chinese medicine. To insure his vision of scientizing Chinese medicine, Lu insisted on establishing five preconditions for the proposal. One of them required the following: "Among theories about one specific entity, only one can be true. . . . It is not acceptable that more than one theory be considered true."[41] While this statement might have sounded self-evident or even naive to many modernizers, this was the first time that practitioners of Chinese medicine accepted the zero-sum game of scientific realism.

As documented in chapter 4, less than one decade earlier, traditional practitioners had developed a popular strategy, which Yu Yan had characterized at the time as a "move to avoid the place of confrontation," to resist the hegemony of scientific realism. When Yu Jianquan and Yun Tieqiao responded to Yu Yan's criticism of Chinese medicine in the early 1920s, their central argument was that certain entities of the Chinese medical body—internal organs and meridian channels in particular—belonged to the realm of *qi*-transformation and therefore were beyond the horizon of Western science. In their hands, the concept of *qi*-transformation ceased to be a tool of communication, as Tang Zonghai had conceived it in the 1890s; instead, it became a tool for creating an incommensurable dichotomy between Chinese medicine and Western anatomy. Being a passionate follower of Yu Yan and his idea of a "Chinese medical revolution," Lu Yuanlei saw this popular strategy as the crucial obstacle in any serious endeavor to scientize Chinese medicine. For Lu, the project of Scientizing Chinese Medicine meant above all else, as suggested by the title of his famous article, "to overthrow *qi*-transformation at its very foundation."[42]

Ironically, incommensurability can easily be turned into incompatibility when the balance of power shifts. Once Lu Yuanlei was entrusted by the Institute of National Medicine with the task of putting Chinese medicine in order, he was determined to force his fellow practitioners to make a painful choice between the two mutually exclusive options of science or *qi*-transformation.[43] Listing this point as one of the five preconditions for his proposal, Lu asserted,

> While Chinese and Western understandings are different and have their own drawbacks and strengths, there is no way that both are simultaneously true. Some people in the Chinese medical profession have no idea about either the drawbacks of Western medicine or the strengths of Chinese medicine; they suggest a compromise formula, saying that Western medicine is good at anatomy, but Chinese medicine is good at *qi*-transformation, or that Western medicine is a scientific medicine, but Chinese medicine is a philosophical medicine. We should know that one specific disease is a specific thing. Therefore, there is only one understanding that can be true (*zhenshi*). There is no way that both understandings can be simultaneously true. Unless one can thoroughly prove anatomy to be untruthful, *qi*-transformation cannot coexist with anatomy.[44]

Denouncing the established way to compare the two styles of medicine, Lu and other reformers embraced as their own a radically new way to compare them, namely, in ontological terms. As Lu made clear in this paragraph, once this ontological framework was accepted, comparison related to the relative therapeutic efficacy of the two styles of medicine was pushed into the periphery, if not completely out of sight. Moreover, assuming the existence of a single ontological reality with distinct disease entities, the comparison between the two styles of medicine was transformed into a zero-sum game in which "there is only one understanding that can be true."

While the form of the zero-sum game was clearly inherited from the earlier debate between Yu Yan and his contenders, the content nevertheless shifted from the ontology of anatomy to that of disease entities. This shift was the result of the recent medical revolution of specificity,[45] which had made Lu's notion that "a specific disease is a specific thing" almost self-evident. More importantly, instead of adopting the earlier strategy of "avoiding the place of confrontation," practitioners of Chinese medicine became obsessed with debating ontological issues such as whether some entities of Chinese medicine really existed and whether a specific understanding of that entity was true or false. As a result, ontology became the key to deciding China's health-care policies and the nature of its modernity. From now on, Chinese medicine would formally take its place in the strange ontological space of modernity, a space that was deemed too crowded to allow Chinese medicine to coexist side by side with biomedicine.

The impact of this ontological turn became most pronounced in the extreme case of Tan Cizhong (1887–1955).[46] Just like Lu Yuanlei, Tan was an admirer of Yu Yan and published several articles in journals edited by Yu. Two years after the 1929 confrontation, Tan published two books in dialogue with Yu: *My Opinion on Scientizing Chinese Medicine* (*Zhoungyi kexuehua zhi wojian*) and *Preliminary Steps to Scientifically Transform Chinese Drugs* (*Zhongyao kexue gaizhao zhi chubu*).[47] According to Tan, these two books were just one fifth of his more ambitious research project entitled "Chinese Medicine and Science." Tan also published several lengthy articles whose titles included the key word *scientization*.[48] Accepting the dichotomy between the visible materialism of science and the invisible *qi*-transformation of Chinese medicine as a fact, Tan thought that the core task of scientization lay precisely in finding out the material bases for certain selective aspects of *qi*-transformation.[49] Tan summarized as follows: "Unless there is no way to scientize Chinese medicine, as long as certain parts of Chinese medicine do have scientific value, there is no doubt that those parts are the ones that have physical form and a material foundation."[50]

On the basis of this conviction, Tan spelled out the "material foundation" of ten key concepts in Chinese medicine: *yin* and *yang*, *blood* and *qi*, *water* and *fire*, *depletion* and *repletion*, *wind* and *dampness*. While Tan confidently asserted that these ten interpretations of Chinese medical concepts would become idiomatic for the new generation of traditional practitioners, many of his "discoveries" now sound absurd and have never been accepted by his peers. For example, he equated *yin* and *yang* with the heart and *wind* with the brain. Most interestingly, though, the first of his ten "discoveries" was the identity of the traditional conception of *xue* (*blood*) with the biomedical notion of blood. For many critics of Chinese medicine, this assertion was nothing more than an obvious statement of identity, that is, *X* is *X*.[51] For Tan, this statement was far from trivial, because, according to the traditional doctrine, *blood* was much more than blood. For people such as Tang Zhonghai and Yu Jianquan, although the Chinese conception of *blood* did overlap with the biomedical notion of blood, the two could never be considered as identical. For one thing, *blood* was inseparable from *qi*; it was considered the yin aspect of *qi*. In light of this understanding, Tan's apparently trivial definition (*blood* is blood) is most revealing; he was determined to delete any "excessive meaning" for *blood*, anything that could not be mapped onto the material landscape of biomedicine. In spite of their very different interpretations of *qi*-transformation, both Tang Zhonghai in the 1880s and Yu Jianquan in the 1920s had denied the identity of meridian channels with the blood vessels on very similar grounds. Since Tan Cizhong now equated *blood* with blood, he consequently also concluded that the meridian channels were identical with the blood vessels, no more and no less.[52]

In sharp contrast to Yu Jianquan, who appropriated the invisible realm of meridian channels to circumvent the material world of science, Tan Cizhong strove to turn *qi*-transformation into a subset of the material world of biomedical sciences. Although Tan's move looks like a total reversal of Yu Jianquan's argument, on a more fundamental level, it was the structured position inherited from Yu Jianquan's earlier defensive strategy. To elaborate on this counterintuitive point, let me first stress that the very move to "avoid the place of confrontation" presupposes and reinforces the idea that the battle between the two styles of medicine was about the proper territory of each within the ontological space. By way of promoting the invisible and immaterial realm of *qi*-transformation, the proponents of Chinese medicine agreed with their critics that the final justification for medicine was found in a truthful ontology.

Assuming the fundamental importance of ontology, when Tan Cizhong and like-minded reformers were forced to abandon the ontology of *qi*-transformation, they felt the urgent need to reestablish the ontological basis, or "material foundation" in Tan's words, for the key concepts of Chinese medicine. By this time, they were left with no other options than to look for it among the ontological entities that had been constituted anew by the language and conceptions of modern science and biomedicine. When practitioners of Chinese medicine encountered difficulties in associating this new world with such deeply entrenched traditional concepts as *qi*, *yin* and *yang*, *blood*, *wind*, and so on, they often found fault with their theoretical concepts. Throughout this period, many traditional practitioners freely suggested that their deeply entrenched concepts were just like mathematical symbols or pronouns; that is, they represented something other than what they had previously been assumed to symbolize. By repeating this frequently, practitioners of Chinese medicine created a critical distance between the words they used and their referents. As a result, they had the bizarre and appalling feeling that the concepts of traditional medicine on which their predecessors had relied for thousands of years had all of a sudden lost their referents. Even worse, since it had become unavoidable that the remapping of Chinese medical terms onto their new referents would be mediated and obstructed by the language of modern science and biomedicine, from now on, at least for the radical reformers, the most fundamental concepts of Chinese medicine became codes that required scientists to decipher them.

Rejecting Scientization

According to the editor of the special issues of the *Journal of the Medical Research Society of China*, six out of twenty-three authors argued that there was simply no need to scientize Chinese medicine. This faction tended to be very argumentative and prolific, including hard-liners like Zeng Juesou, Chen Wujiu, and He Peiyu. When we examine these six articles, we find a clear continuity of argumentation tracing back to Yu Jianquan's strategic interpretation of Tang Zhonghai's views. Zeng Juesou, for example, argued that inasmuch as Chinese medicine transcended the visible, material realm of science, it did not matter that the two were incompatible.[53] Moreover, Zeng considered it to be intellectual suicide for traditional practitioners to embrace Yu Yan's tripartite characterization of Chinese medicine, especially the idea that medical theories, Chinese drugs, and clinical experience constituted three independent categories, completely unrelated to each other. He was appalled by the fact that many leading practitioners after the historic campaign in March 1929 came to embrace what he considered to be such a faulty and dangerous characterization. In his public letter to Lu Yuanlei, Zeng elaborated his sense of frustration:

> With regard to the idea that the strength of Chinese medicine lies in its verified drug formulas (*yanfang*), Yu Yan formulated this notion because he realized that the experience of Chinese medicine is widely trusted by our society. It is very difficult to overthrow this trust. Therefore, instead of attacking our experience, [Yu Yan] attacked the theories of Chinese medicine. Once

> the theories are overthrown, though, despite the abundance of our experience, we will not be able to compete with him. Therefore, Yu Yan's ideas are a strategy for attacking Chinese medicine. Now, you also hold the same opinion. I really do not understand [how you can share Yu Yan's way of thinking]. Isn't it the case that verified drug formulas are based on experience? Isn't it also the case that [what you call] experience is based on theory and confirmed in practice? Therefore, it is called experience, and the formulas that are demonstrated to be effective in experience are called verified drug formulas. If there are no correct theories, how could there be verified drug formulas?[54]

This is a very revealing statement. In Zeng's eyes, when traditional practitioners advocated the position that Chinese medicine was based on experience and that its greatest contributions were Chinese drugs and drug formulas, they were embracing Yu Yan's strategy against Chinese medicine. It did not take a visionary to see Zeng's points; many traditional practitioners held similar positions up until the 1929 confrontation. Zeng was remarkable merely because of his consistent efforts to turn back the clock to the pre-1929 situation.

Upon examination, it becomes clearer why this group of practitioners has fallen into obscurity in twentieth-century historical accounts of Chinese medicine.[55] They were the last generation of traditional practitioners who could afford to reject the project of scientizing Chinese medicine and to uphold the invisible and immaterial world of *qi*-transformation. What they were adamantly resisting in the 1930s did not take long to become common sense for the new generation of practitioners of Chinese medicine. The popularity of their position declined steeply after the establishment of the Institute of National Medicine in 1931.[56] Even opponents of Chinese medicine openly declared that they no longer considered this group of traditional practitioners, in their eyes the most dogmatic advocates of Chinese medicine, as their main competitors.[57]

In spite of the fact that their general opposition to the project of scientizing Chinese medicine was no longer popular, these traditionalists still constituted a formidable force against any concrete proposal on reforming Chinese medicine. From the beginning, Lu Yuanlei knew that the fate of his proposal would be closely associated with his rejection of *qi*-transformation. When Xie Guan (1880–1950),[58] the senior leader of the delegation discussed in chapter 5, joined the committee for putting in order Chinese medicine as its sixth member, Lu realized that the Institute of National Medicine would probably not accept his proposal and therefore decided to publish it in a few journals of Chinese medicine. In a personal statement attached to his proposal, Lu specifically pointed out that if the task of the Institute of National Medicine degenerated into "putting in order *qi*-transformation," he would have to withdraw from this project.

Lu's hunch was shown to be correct. When the Institute of National Medicine published the official version of *The Schematic Standard for Putting in Order National Medicine and Pharmaceutics* on October 29, 1932,[59] its content was very different from Lu's version. While the official draft still emphasized its objective to put Chinese medicine in order with scientific methods, it deleted almost all the preconditions that had been stated in Lu's proposal. When Yu Yan, who was sympathetic to Lu's proposal, read the official draft, he could only conclude that "the good intention of adopting scientific methods [to put in order Chinese medicine] has been lost for good."[60] In spite of Yu Yan's pessimistic judgment, however, the Institute of National Medicine maintained its reformist agenda and made a number of important decisions in reassembling modern Chinese medicine after the first run of debates ended in 1932.

Reassembling Chinese Medicine: Acupuncture and Zhuyou *Exorcism*

When the Institute of National Medicine strove to put in order Chinese medicine, traditional practitioners had to make many crucial and difficult decisions that gave shape to modern Chinese medicine as a nationally unified system. Nothing illustrates the significance of this decision-making process better than the opposite ways in which the Institute of National Medicine treated acupuncture and *zhuyou* exorcism, which were both located in the "chaotic medicine sector" of Pang Jingzhou's chart of the medical environment in Shanghai (see figure 6.1).

Within that diagram, acupuncture [36] stands alone in its own slice of "chaotic medicine," separated from all the established institutions of Chinese medicine. Pang's negative characterization of acupuncture might look strange to modern readers, because they take acupuncture as the representative practice of Chinese medicine. But the truth is that acupuncture had become marginalized in Chinese medicine as the Chinese increasingly abhorred the idea of intrusive intervention during the Late Imperial period. By the mid-eighteenth century, the famous elite physician Xu Lingtai (1693–1771) complained that he was not able to find anyone learned in acupuncture to teach him this practice.[61] It was only as the result of Japan's inspirational influence that traditional practitioners began to view acupuncture in a very different light in Republican China.

Although the Meiji government in Japan had implemented a one-time registration for practitioners of traditional herbal medicine and had thereby effectively brought it to an end, the Japanese government nevertheless had kept issuing licenses to practitioners of moxibustion and acupuncture, occupations which had long been reserved for the blind population in Japan.[62] Thanks to the inspiration from Japan,[63] when the Nationalist politicians proposed to create the Institute of National Medicine, they listed moxibustion and acupuncture as one of the four major areas of reform, placing it on equal footing with theory, diagnosis, and pharmacy.[64] The institute's approved proposal for sorting out Chinese medicine lists acupuncture as one of eleven applied subjects of Chinese medicine and emphasizes that "with regard to the location of the channels and acupuncture points, [we] should strive to put them in relation to modern anatomical and physiological [structures]. . . . Moreover, [we] should pay special attention to the issue of disinfection during the procedure of needling."[65] As one traditional practitioner pointed out, this modern knowledge was essential because lack of it could cause either serious damage to the neuromuscular system or bacterial infection.[66] Another traditional practitioner contested this approach by stating that, "since modern scientists have difficulty comprehending these acupuncture points and channels and tend to distrust them, we should use [acupuncture] to serve as the representative for the supreme refinement of Chinese medicine."[67] For the purpose of making this point, he suggested limiting the role of anatomy to that of a useful reference but not the ultimate foundation for acupuncture.

In the end, the Institute of National Medicine had it both ways. On the one hand, it incorporated the once marginal practice of acupuncture into its modernized version of Chinese medicine as a core representative and "unique invention" of Chinese medicine vis-à-vis the world of global medicine.[68] On the other hand, the institute proudly produced the newly published *Verified Charts of Acupuncture Channels and Points* (*Zhenjiu jingxue tukao*), which did not include any traditional illustrations, since "they were incompatible with modern anatomical illustrations."[69] Within this supportive atmosphere, the most influential innovation in acupuncture came from Cheng Dan'an (1899–1957), who founded the first school of acupuncture in modern China. In 1932, he produced large, four-color posters that depict the acupuncture channels and points in relation to anatomical structures.[70] As medical historian Bridie Andrews has pointed out, "Cheng Dan'an not only rescued Chinese acupuncture from superstitious oblivion, paving the way for a 'scientific' acupuncture to rescue the status of Chinese medicine as a whole, as it did during the Communist era; he also created anew the Chinese-medical body."[71] With Cheng, "for the first time, a Chinese-medical physiology [was] superimposed upon an unambiguously Western anatomy."[72]

In contrast to their efforts to revive acupuncture, practitioners of Chinese medicine were highly cautious about the practice of *zhuyou* ([14] in Pang's chart, figure 6.1), a form of exorcism that had been a formal subject of the Imperial medical examination since the Sui dynasty (681 BCE). While its exact meaning is still under debate, scholars agree that originally *zhuyou* referred to a form of healing practice that involved neither herbal medicine nor acupuncture.[73] Against the early history in which *zhuyou* had been perceived simply as a magical practice, during the eighteenth century some scholar-physicians reinterpreted *zhuyou* as a kind of secular therapy to treat emotional disorders.[74] Nevertheless, outside medical circles, *zhuyou* prospered as a form of exorcism that was closely associated with the locality of Chenzhou,[75] which is why Pang put the description of "Talismans from Chenzhou" [18] underneath the term *zhuyou* in his chart. To distance themselves from these magico-religious practices, almost none of the reformers of Chinese medicine considered the possibility of allying this formal subject of Chinese medicine with the newly introduced psychotherapy.[76] The salient exception was Chen Guofu, who repeatedly promoted the idea that *zhuyou* should be considered as a kind of psychotherapy in modern Chinese medicine.[77] As an enthusiastic promoter of "mental hygiene," he saw it as one of a few key therapeutic skills, along with acupuncture and massage, that were worthy of scientific investigation.[78] Nevertheless, the majority of Chinese medicine practitioners did not embrace the idea of reviving *zhuyou*, and the approved proposal for reforming Chinese medicine therefore did not include the subject of psychotherapy.[79]

The contrasting fates of acupuncture and *zhuyou* exorcism reveal the crucial importance of the project of "putting in order" Chinese medicine for determining the structure and content of modern Chinese medicine. In both of these cases, neither history nor science could serve as an unambiguous guide to determine the direction for reforming Chinese medicine. When practitioners of Chinese medicine decided to revive acupuncture and to further distance Chinese medicine from *zhuyou* (which had once been a formal subject of Chinese medicine), their decisions could not have been logically derived from the historical reality. Likewise, in these two cases, science alone could not have provided sufficiently unambiguous criteria for traditional practitioners to make these decisions. While I have emphasized the crucial role played by science in patterning the emergence of modern Chinese medicine, it is equally important to remember that some aspects of the reform efforts had less to do with the external authority of modern science than with the difficulty of making choices among the various internal elements of Chinese medicine so as to transform it into a standardized coherent system.

The Challenge of "Mongrel Medicine"

Against this predictable, almost mechanical deployment of three positions concerning the project of Scientizing Chinese Medicine, it is surprising to find that the editor of the special issue of the *Journal of The Medical Research Society of China* "Polemics of Scientizing Chinese Medicine" (1936) was ambivalent about Yu Yan's real position. In his eyes, this ambiguity was partially caused by Yu's cordial relationship with Tan Cizhong and Lu Yuanlei, two of the main advocates of Scientizing Chinese Medicine. Yu published Tan's papers in the journal he edited and made sympathetic comments on Lu's proposal. The editor suggested, however, that Yu Yan "was the first person to raise the banner of [the] Chinese Medical Revolution and has struggled strenuously against traditional practitioners for decades. Therefore, it should not be wrong to classify him into the third group,"[80] that is, the "group of individuals who denied that Chinese medicine could ever be scientized at all." The editor's ambivalent feeling about Yu demonstrates the radical reshuffling that had taken place since the rise of this project as the official objective of the Institute of National Medicine.

By no means was the editor the only person who felt ambivalent about Yu Yan. When Yu published the second edition of his *Collected Essays on Medical Revolution* in 1932, he opened his new preface with his friends' bitter complaints. They said,

> Recently, the half-old half-new, "neither donkey nor horse" [*feilü feima*] medicine has run wild. They have learned this trick from you. If you had not attacked them, they would not have metamorphosed into this. The old medicine will remain old-fashioned, and the new medicine will stay purely new-

> fashioned. The boundary is clear-cut, and the problem will be easy to solve. . . . Because of your attack on the old medicine, now they have learned the trick of survival.[81]

After the 1929 confrontation, many opponents of Chinese medicine witnessed what they regarded as a terrible phenomenon: the sudden emergence of a mongrel form of medicine, one that was recognizably "neither donkey nor horse." Within a very short period of time, the hybrid medicine achieved tremendous popularity in scholarly medical circles, catching up with the earlier development in the business sector of pharmaceutics.[82] Although the first and the third groups held opposite opinions about Scientizing Chinese Medicine, they both attacked Lu Yuanlei's and Tan Cizhong's project as being "neither donkey nor horse."[83]

Let me make explicit my reasons for translating the idea of "neither donkey nor horse medicine" (*feilü feima yi*) as "mongrel medicine" (*zazhongyi*) rather than as the more positive-sounding "hybrid medicine." First of all, unlike the concept of hybrid medicine, mongrel medicine is an actors' category. As critics of Chinese medicine used "neither donkey nor horse" medicine and "mongrel medicine" interchangeably, these expressions were meant to indicate a double betrayal of two pure parental traditions.

This strong negative connotation leads to my second point: As an actors' category, "neither donkey nor horse" was not a term of self-identity for traditional practitioners but a derogatory device imposed by critics of Chinese medicine. In contrast to this strong negative connotation, the postcolonial concept of hybridity was meant to emphasize "the hybridized nature of post-colonial culture as a strength rather than a weakness."[84] As a result, the related concept of hybrid medicine cannot serve my purpose of conveying the humiliating constraints imposed upon the advocates of integrating the two systems of medicine by their critics. It is for the sake of conveying the derogatory sense of the term "neither donkey nor horse" that I have decided to paraphrase it as mongrel medicine.

In response to attacks from both sides, Lu emphasized the acceptance of mongrel medicine as one of the five preconditions in his controversial proposal to scientize Chinese medicine that he had prepared for the Institute of National Medicine: "As [scholars] endeavor to put in order Chinese medicine and pharmaceutics, they will have to take advantage of scientific principles. Therefore, they should not be blamed for damaging our national essence."[85] Clearly, Lu distanced himself in his proposal from the need to preserve the authenticity of traditional medicine as part of the "national essence." This is telling evidence that, at least for Lu, the National Medicine Movement should not be reduced to a movement of cultural nationalism. Citing as an example the successful integration of Confucianism and Buddhism that had taken place during the Song dynasty (960–1278), Lu saw his project of Scientizing Chinese Medicine to be the same kind of audacious, creative blending of Chinese and foreign cultures.[86] In this sense, people like Lu launched the project and willingly endured the derogatory labels imposed on them, because they were not committed to preserving Chinese medicine as it was, but to developing a new, hybrid medicine, as announced by the Institute of National Medicine.[87]

Yu Yan's friend was right to blame him for helping to create this mongrel medicine. The rise of mongrel medicine was a response to the threat of medical revolution that had been proposed by Yu Yan and other critics of Chinese medicine. This is not to say that no one had ever tried to integrate the two styles of medicine before Yu's attack—Tang Zonghai serves as a salient counterexample.[88] The important point is that the desirability, the derogatory connotation, and the dangerous potential of mongrel medicine were all inseparable from the idea of putting in order Chinese medicine with scientific methods, or in other words, the project of Scientizing Chinese Medicine. For the first time ever, practitioners of Chinese medicine were forced to cope collectively with the concept of science when they imagined the relationship between Chinese and Western medicine.

The notion of mongrel medicine and the project of scientizing Chinese medicine can be said to have existed in a dialectic of constitution and suppression. The relationship between them was mutually constitutive, because the strange creation of mongrel medicine would never have become a desirable objective for traditional practitioners if the Scientization of Chinese Medicine had not been promoted by the Nationalist state and thereby accepted by competing parties as a truce in the battle over Chinese medicine's future in the first place. It was this goal of scientization that forced traditional practitioners to take seriously the concept of science and its related discourse of modernity, such as Yu Yan's tripartite characterization of Chinese medicine, in their endeavors to reform Chinese medicine. In this sense, their reform efforts were modernist by nature and thereby distinctively different from the earlier, premodernist syncretism represented by Tang Zonghai.

On the other hand, the relationship between them was also suppressive because it was the very idea of science (of scientific biomedicine) that made it difficult to imagine a productive crossbreeding between Chinese medicine and biomedicine. The simple act of mixing the two styles of medicine might not produce a monster, but the idea of crossbreeding science with its alien other was seen as amounting to blasphemy.[89] This assumed impossibility justifies the characterization of this new medicine as "neither donkey nor horse." Just like the mule, which is not capable of reproducing itself, mongrel medicine, while vigorous in appearance, would never be able to carry on by itself as a valuable living tradition. Its popularity called for the derogatory use of the concept of mongrel medicine to emotionalize the potential danger of transgression, a danger that actually originated in the project of scientizing Chinese medicine. To summarize, it was because the two contending parties embraced the project of Scientizing Chinese Medicine as a way to bring a truce to their struggle that mongrel medicine had become a popular, desirable, though seemingly impossible project for traditional practitioners and a potent danger for biomedical practitioners.

Conclusion

It was in the context of guarding against the formation of mongrel medicine that we can see most clearly why practitioners of Western medicine strongly preferred the slogan of Scientizing Chinese Medicine to the more general idea of putting in order Chinese medicine with scientific methods. As this slogan included the locally invented concept of *kexuehua*, it brings us back to the question that I proposed at the beginning of this chapter, namely, what kind of functions did the project of Scientizing Chinese Medicine serve as a crucial historical force in the struggle between practitioners of Chinese medicine and practitioners of Western medicine in the early 1930s? To put it bluntly, the very act of turning science into a verb was the most effective way of suggesting the existence of a homogeneous entity in itself that was called science.[90] Unless science could be understood in this context as such a homogeneous, single entity, it would be hard to comprehend what was meant by the act of "scientizing" something. Moreover, the routine and unproblematic use of this verb presupposed and reinforced the idea that both science and its opposite—traditional medicine in this context—were identifiable entities, as real as tangible things and distinct from each other.

Given that *science* is understood as such a homogeneous, singular entity, its verb form inevitably implies the dissolution of the object in the process. To elaborate a bit, the term *scientize* suggests a set of rigorous procedures that segment the object into pieces and then selectively assimilate some elements into parts of a new homogeneous entity. That is why the true topic of discussion in so-called polemics about Scientizing Chinese Medicine was the correct process for reforming Chinese medicine, even though in appearance the point of debate seemed to be about the possibility or impossibility of this project. It might be surprising to find that the authors who advocated the absolute incompatibility of Chinese medicine and science concluded that "all theories of Chinese medicine should be demolished, but its experience can be scientized. When this experience has been scientized, Chinese medicine will meet its doom."[91] This is not an opinion that one would expect to hear from those who considered it impossible to scientize Chinese medicine. Rather, it sounds like they were saying that Chinese medicine, at least some of its elements, such as experience, could be scientized, but that the process of scientization should be conducted with great rigor so that it would not lead to further development in Chinese medicine. As asserted by Fan Xingzhun (1906–98), a pioneering historian of Chinese medicine, "If Chinese medicine is thoroughly scientized, the term 'Chinese medicine' will no longer make sense. Therefore, the so-called scientization of Chinese medicine actually means using science to demolish Chinese medicine."[92] If the verb form of *science* guaranteed the understanding of science as a single entity, the accompanying concept of experience ensured the dissolution of Chinese medicine into atomized units, readily absorbable into cosmopolitan medicine.

Given that the Nationalist state had established the Institute of National Medicine to overhaul Chinese medicine by means of science, the opponents changed their tactic to focus on enforcing a particular kind of research program, which they characterized as Scientizing Chinese Medicine. While the project of Scientizing Chinese Medicine was carefully crafted to ensure the disintegration of Chinese medicine,[93] the concept of mongrel medicine expressed the deeply felt danger of failing to impose this kind of "destructive scientization" on Chinese medicine. Nevertheless, at the very moment that practitioners of Chinese medicine were forced to embrace this project as their own agenda, they strove to appropriate the disunity and heterogeneity of science, to renegotiate the locally invented concept of *kexuehua*, and to take on the historic challenge of developing the seemingly impossible and stigmatized project of mongrel medicine. To appreciate their efforts to transcend the received limit of imagination and to create a viable species of mongrel medicine, we must examine the concrete and material ways in which they put into practice and thereby negotiated the project of Scientizing Chinese Medicine. For this purpose, chapter 8 below examines the incorporation of the germ theories of disease into Chinese medicine, while chapter 9 explores laboratory and clinical research on Chinese drugs. As this project was considered an oxymoron by its opponents, the development of mongrel medicine would in turn challenge the cultural authority of science as modernity and thereby constitute the most provocative force for China's exploration of its own modernity.

8 The Germ Theory and the Prehistory of "Pattern Differentiation and Treatment Determination"

Do You Recognize the Existence of Infectious Diseases?

Nothing reveals the challenges faced by Chinese medicine in the 1930s better than the following fictitious exchange between a critic of Chinese medicine and a reform-minded practitioner of Chinese medicine:

> CRITIC: Do you admit the existence of infectious diseases?
>
> PRACTITIONER: Yes.
>
> CRITIC: Do you also admit that the cause of infectious diseases is germs?
>
> PRACTITIONER: Yes.
>
> CRITIC: Do you admit that the symptoms of infectious diseases are caused by the toxin of the germs?
>
> PRACTITIONER: Yes.
>
> CRITIC: In treating infectious diseases, do you have any newly invented drugs that can either destroy germs or counteract their toxin?
>
> PRACTITIONER: No such things. I have only the old formulas of National Medicine.
>
> CRITIC, WITH AN ANGRY LAUGH: The old formulas of National Medicine recognize

> nothing but concepts such as *wind*, *cold*, and the six *qi*. If you have succeeded in using these formulas to treat infectious diseases, you are either bluffing, or the patients recovered by way of self-healing, and you have taken the credit for this natural process. Otherwise, how could a formula that neither destroys germs nor counteracts toxins succeed in treating infectious diseases?
>
> PRACTITIONER: Neither is the case.[1]

This imaginary conversation was composed in 1935 by Ye Guhong (1876–1940s), who represented his own views as those of the traditional practitioner. Ye trained at Japan's Kyoto Imperial University and became an active member in literary and artistic circles after he returned to China. Although he did not adopt Yu Yan's position on abolishing Chinese medicine, he also advocated the idea of a "Chinese medical revolution." As a noteworthy exception, his articles were published in the journals of both competing camps and praised by leaders on both sides.[2] Ye composed this fictitious exchange because the dialogue format helped to highlight the crucial questions that he thought practitioners of Chinese medicine had to address: (1) Did they admit the existence of infectious diseases? (2) Did they accept the germ theory of infectious diseases? (3) How could Chinese medicine be effective in treating infectious diseases, given that it did not recognize the concept of germs? This chapter traces the historical process by which practitioners of Chinese medicine developed competing answers to these three crucial questions and thereby created ways of integrating Chinese medicine with germ theories of disease. It was through this contentious process of striving to accommodate and negotiate the germ theory that practitioners of Chinese medicine developed the incipient form of what later became the defining feature of so-called Traditional Chinese Medicine (TCM), namely "pattern differentiation and treatment determination" (*bianzheng lunzhi*).

The Critic's first question reveals the surprising fact that not very long ago it had still been possible for practitioners of Chinese medicine to not recognize the existence of infectious diseases. In other words, "infectious diseases" did not constitute a traditional disease category of Chinese medicine; rather, I would argue, it was the most neglected, invisible categorical development of modern Chinese medicine.

As discussed in chapter 2, the category of infectious diseases, translated into Chinese as *chuanranbing*, did not exist in premodern Chinese medicine as a separate etiological category. The very term *chuanranbing* had only recently been borrowed from the Japanese,[3] and it was formally introduced into China when the Republican government promulgated the first Regulation Concerning the Prevention of Infectious Diseases in 1916, six years after the end of the Manchurian plague. The first item of this regulation stated that only the following diseases were included in the government's definition of infectious diseases: cholera (*huleila*), dysentery (*chili*), typhoid fever (*changzhi fusi*), smallpox (*tianran dou*), typhus exanthemata (*fazhen zhifusi*), scarlet fever (*xinghongre*), diphtheria (*shifu dili*), and plague (*baisituo*).[4] These eight diseases were exactly the same as the set of diseases regulated by the Japanese colonial government in Taiwan in 1896 and by the government in Japan in 1897.[5] It is noteworthy that each of these eight names is a loan word created from Japanese translations, and at least five of them are simply transliterations of the Western biomedical terms. Clearly, Chinese and Japanese medical modernizers agreed on a crucial point: Both individual infectious diseases and their collective category should be marked as radical novelties in the East Asian medical universe.

Given that this regulation was the first Chinese legal document concerning infectious diseases (with the exception of items from the seaport Customs Service), it effectively defined what counted as *chuanranbing* for the government. Instead of presenting a formal definition, the first article of this regulation offered an exclusive list of the eight infectious diseases named above. Following the first article, the regulation made notification of cases of these diseases compulsory for all registered medical practitioners. This list of infectious diseases is by no means exhaustive; both tuberculosis and syphilis are missing, for example. The main concern of this regulation was to legalize responsibilities, both of the state and of medical practitioners, regarding acute epidemics that might cause social upheaval. Given that the concept of *chuanranbing* was introduced to China mainly through official regulation and public health measures, *chuanranbing* thus became a legal and medical category of the state, used interchangeably with "eight main infectious diseases" (*ba da chuanranbing*), which was the official Chinese term for "notifiable infectious diseases." For the Chinese population during the Republican period, the newly introduced medical concept of *chuanranbing* was hence used very often in a restricted sense identified with the legal concept of the eight notifiable infectious diseases.

Notifiable Infectious Diseases

The government promulgated the above-mentioned regulation concerning notifiable infectious diseases in 1916, although *chuanranbing* had already become part of the legal-medical network of governmental control that had emerged locally and regionally over the past quarter century. In addition to elevating *chuanranbing* to the status of an official governmental category, the state effectively redefined this concept by making microscopic testing and the germ theory central to its official method of identification. As revealed by their awkward names (which were updated and improved when the Republican government reissued the regulation in 1928), the eight notifiable infectious diseases listed in the regulation and their unequivocal identification by means of the germ theory signal a break within the history of Chinese medicine. As the result of this radical reconstruction as the first generation of Chinese biomedical practitioners were promoting the control of *chuanranbing* as their new vision for Western medicine (see figure 3.1) and the key project for the construction of the modern state, infectious disease also became the greatest weakness of Chinese medicine.

That is why the control of *chuanranbing* became the key issue during the 1929 confrontation. In his famous proposal, Yu Yan specifically pointed out that Chinese medicine, on account of being an "individualistic medicine," was useless for preventing epidemics.[6] Moreover, in his original proposal, senior practitioners of Chinese medicine would automatically have been granted a special license to practice by the government, but they would have been "prohibited from treating notifiable infectious diseases and from issuing death certificates."[7] During the three-day campaign in March 1929, the Nanjing Association of Chinese Medicine proposed to the assembly that in order to defend itself against the threat from outside, Chinese medicine should build up its internal defenses by doing research, especially regarding pestilential diseases (i.e., infectious diseases).[8] The proposal explained,

> What Chinese medicine considers as pestilential disease is identical with the concept of infectious disease in Western medicine. With regard to pestilential disease, Chinese medicine is strong in therapeutics but weak in prevention. That is why Western medicine has taken this weakness as an excuse [to outlaw Chinese medicine] and proposed to set up intensive courses to train practitioners in how to handle cases of infectious disease. We should train ourselves and sort out [related teachings in Chinese medicine], so they will have one less excuse [to outlaw Chinese medicine].[9]

The difficulty of dealing with infectious diseases became an urgent practical problem when the National Medicine Movement started pushing the Nationalist government to draft legal regulations for practitioners of Chinese medicine after the 1929 confrontation. In 1934, the Jiangzhou provincial government led the way by inaugurating a tentative regulation that demanded practitioners of Chinese medicine to acquire knowledge of hygiene and infectious diseases. Among the fourteen subjects on examinations, hygiene was listed as one of the four required subjects. Candidates who failed the subject of hygiene would have to attend specially designed intensive training courses.[10] In addition, the regulation specified that if practitioners encountered any case of infectious disease, they were required to report it to the police within twelve hours.[11] In response to this initiative, the Union of Chinese Medicine at Wu County offered a public criticism, aimed specifically at the regulations concerning infectious diseases. They argued that

> infectious diseases as defined by the Department of Hygiene include disorders such as typhoid,[12] influenza,[13] malaria, dysentery, cholera, and smallpox. These are the diseases that practitioners of Chinese medicine treat most often in their daily practice. Normally we treat more than a few dozens of such cases in one day. It is impossible for us to report every one of these cases. Besides, the scope covered by infectious diseases is quite wide. In the beginning, when the symptoms have not manifested yet, it is impossible to report them right away. If practitioners then report such patients later when the symptoms have fully manifested, should they be blamed for the crime of belated reporting? . . . [In such situations], the practice of Chinese medicine becomes a very dangerous business.[14]

This resolution reveals several important points about the difficult position held by traditional practitioners with regard to *chuanranbing*. First of all, contrary to conventional wisdom, Chinese medicine was not only helpful in treating chronic diseases, but was moreover widely consulted as late as the 1930s for the treatment of acute infectious diseases.[15] Second, even just the reporting of notifiable infectious diseases presupposed the germ theory of diseases and its associated diagnostic methods. To adhere strictly to the proposed regulation would have caused tremendous difficulties for practitioners of Chinese medicine. If practitioners used the conventional way of diagnosis in Chinese medicine, which was based on manifested symptoms, they would run the risk of breaking the law, as the union clearly described. On the other hand, if a practitioner of Chinese medicine tried to obey the law and reported a case as a type of infectious disease on the basis of traditional Chinese diagnostic methods, he or she would then be at risk of a malpractice suit later if the bacterial test turned out negative. In light of these difficulties, the union opposed the proposal to list hygiene as one of the required subjects. It was for the sake of resolving these practical, even administrative, difficulties posed by infectious diseases that the Institute of National Medicine had considered this field as its first priority in the project of Scientizing Chinese Medicine and publicized the much-debated, controversial proposal for "unifying [Chinese] nosological nomenclature" (*tongyi bingming*) in June 1933.

Unifying Nosological Nomenclature and Translating Typhoid Fever

This controversial proposal was written by Shi Jinmo (1881–1969),[16] vice president of the Institute of National Medicine. Before addressing the specific content of this document, I would like to point out that, for Shi and other leaders of the institute, this was far from a proposal with only scholarly interest. On the contrary, they saw it as the first step toward issuing compulsory national regulations for practitioners of Chinese medicine. In Shi's own strong language, "Once the disease names have been settled, The institute will promulgate the regulation, inform practitioners all over the country, and demand that they use this set of standardized [disease] names within a designated time. Those who disobey the regulation will be properly punished. If they again disobey the regulation after being punished, they will be prohibited from practicing medicine."[17]

Shi openly linked the scholarly proposal for unifying nosological nomenclature to the administrative issue of compulsory regulation because the latter was the key objective that he and the other leaders of the Institute of National Medicine had in mind. As discussed in chapter 5, these leaders had a vested interest in using the project of scientizing Chinese medicine as a tool to help turn the institute into a full-blown administrative branch of the state. I would argue that it is impossible to make sense of this seemingly academic proposal without understanding its instrumental role in realizing the political objective of promulgating the Regulation of National Medicine.

In the opening paragraph of his proposal, Shi provided a detailed, step-by-step procedure for unifying the nomenclature of diseases. The first step was "to list the [biomedical] disease names that have been translated into Chinese by practitioners of Western medicine in a table." The second step consisted of "consulting the prominent books of National Medicine . . . and then investigating which diseases of Western medicine the diseases listed [in these books] correspond to."[18] The final step that Shi proposed was to consider the traditional disease names listed below and decide which ones were best suited to serve as the standardized names. Shi added, "If there is no name listed as the Chinese translation of the Western names, or if the traditional names listed below are not compatible with scientific principles, we will just adopt the name of Western medicine."[19]

Shi's three-step procedure made clear that the proposal was not designed to unify the nosological nomenclature within the framework of Chinese medicine; rather, it aimed at providing a unified system of standardized disease names for both Chinese and Western medicine. More importantly, the organizing rubric was the nomenclature of Western medicine. According to his stated procedure, there was absolutely no room for any disease name of Chinese medicine if no counterpart could be found in Western medicine. To defend this radical design, Shi provided an answer to the question, "Why should we depend upon the disease names of Western medicine?" He explained his position as follows:

> Isn't it the case that the Institute of National Medicine itself claims to have adopted scientific methods [in putting in order Chinese medicine]? The traditional disease nomenclature of our national medicine has always been incongruent with science. If we wanted to incorporate it into the scientific paradigm, it is not feasible that the few members of the academic committee could have achieved this objective within such a short time. Even if this project were not entirely infeasible, we have to note that *there is only one truth concerning the things in the world* [author's emphasis]. Since Western nosology is deeply rooted in science, there would be no way for our newly created nosological nomenclature to be different from the Western one. If it were indeed different, by no means would it be compatible with science. Otherwise, science would have to include ambiguous "truths."[20]

Here again, we find that Shi subscribed to the modernist discourse when he stated that "there is only one truth concerning the things in the world." This is one of the five preconditions that Lu Yuanlei had demanded in his proposal for putting in order Chinese medicine. We see here how this kind of abstract epistemo-ontological position penetrated into crucial but seemingly technical decisions about reorganizing the nosological structure of Chinese medicine. As a result, the project of scientizing Chinese medicine, for which this proposal was the most high-profile and concrete plan, involved no less than a remapping of Chinese nomenclature onto the world of diseases, which had just been constituted anew by means of Western medical science.

Such remapping obviously would have caused devastating destruction and sacrifice of Chinese medical knowledge. It is remarkable that Shi had no desire at all to circumvent this issue; instead, Shi again volunteered an answer to a second question: "Why can the unified nomenclature not build upon the tradition of Chinese medicine [instead of destroying] the traditional system?"[21] To answer this crucial question, Shi took the two most prominent schools of traditional medicine—Cold Damage (*shanghan*) and Warm Disease (*wenbing*)—as examples to show how internally contradictory and confusing the division was between these two kinds of disorders and the schools that had been named after them.

While Warm Disease originated in the tradition of Cold Damage, Marta Hanson points out that, "during the second half of the seventeenth and through the nineteenth century, the disease concepts Warm Disease and febrile epidemics became a fulcrum around which a new discourse on epidemics developed and a distinct current of medical learning formed."[22] As a result, traditional practitioners hotly debated the proper boundary and relationship between Cold Damage and Warm Disease, a debate that reflected the larger challenge surrounding the universal cosmology within which Cold Damage was conceptualized.[23]

On the basis of these arguments, which had started in the seventeenth century, Shi summarized three ways in which traditional practitioners differentiated between these two types of disorders. First, all forms of fever should be called Cold Damage because they were caused by cold in winter. Depending on the specific season in which the illness manifested, these fevers were given specific names. If the disease manifested right away in the winter, it was called Cold Damage. If it did not manifest until the spring, it was called Warm Disease. The second method of differentiation was based on the patients' subjective sensations: patients with a fear of cold suffered from Cold Damage, while patients with a fear of heat suffered from Warm Disease. As a third alternative, the two kinds of disorders could be differentiated on the basis of their etiology, that is, the fact that they were caused by different unseasonal *qi*. For example, one could become ill by unseasonably cold weather in the spring and summer. No matter which criterion was used, contemporary practitioners in the Jiangnan area adamantly insisted on the crucial importance of differentiating between Cold Damage and Warm Disease. From their point of view, these two kinds of diseases "should be treated with distinctively different therapies, with no borrowing from each other."[24]

After delineating the three ways of differentiating these two kinds of disorders, Shi confidently concluded, on the contrary, that they were actually the same kind of disease in Western medicine: "What people conventionally call 'warmth and dampness' (one major form of Warm Disease) is actually identical with typhoid fever in Western medicine. On the other hand, because the symptoms [of typhoid fever] match those of Cold Damage one-by-one, Japanese scholars translated typhoid fever as Cold Damage and Chinese practitioners of Western medicine followed suit. Clearly, 'warmth and dampness' is Cold Damage and Cold Damage is 'warmth and dampness.'"[25] From this argument, Shi concluded that Chinese medicine is simply too self-contradictory to serve as the foundation for a unified, standardized nosological nomenclature.

Nothing can demonstrate more vividly than this argument how the seemingly mechanical process of remapping was bound to result in a complete refutation of the traditional system of nosology. Even if Cold Damage and Warm Disease were considered as simply two clusters of symptoms rather than two diseases, they were far from identical, as long as one took into account the difference in sensitivity towards cold and heat experienced subjectively by the individual patient. Nevertheless, since the proposed procedure rigorously demanded that these two Chinese diseases be mapped onto the disease entities of Western nosology, it was unavoidable that they were listed below under the same biomedical name of *typhoid fever*. Two distinct disorders in the framework of Chinese medicine were thus forced to be conflated in the framework of biomedicine. This remapping resulted in a confusing conclusion: Cold Damage and Warm Disease actually amounted to the same disease. Shi took this devastating result as his strongest evidence that there was nothing valuable about Chinese nosology; the standardized nosological nomenclature had no choice but to "destroy the traditional system" of Chinese medicine.

Wanting to demonstrate his determination to "destroy the system of Chinese medicine," Shi offered the case of Cold Damage as a concrete example for the necessity of unifying disease names. In presenting this example—in fact, the only example found in his proposal—Shi was careful to specify the structural position of Cold Damage within the newly unified nosology. The proposal classifies Cold Damage as one of the infectious diseases, which constitute the first category for conditions of internal medicine. Since there was no such category as "infectious dsisease" in traditional Chinese medicine,[26] Shi's approach of creating that category and treating it as the first branch of internal medicine reveals a revolutionary, although largely invisible,[27] effort to reorganize the nosology of Chinese medicine. Moreover, Cold Damage, the name of the most important traditional disease category since the Song dynasty, was chosen to exemplify this revolutionary reorganization of Chinese medicine. The proposal explained, "This disease was originally called Cold Damage because the ancient people took wind and cold to be its cause. Because we now know that the real cause is the bacteria *Salmonella* Typhi (*zhifusijun*), the old name of Cold Damage is no longer appropriate. We should adopt the name of the bacterium as the name for the disease."[28] Clearly, Shi took as his departure point that "the disease [which] was originally called Cold Damage" was in fact identical with typhoid fever.

Shi's apparent intention in this paragraph was to introduce the disease of typhoid fever, as defined by the germ theory, into the nosological system of Chinese medicine. If Shi had been satisfied with this limited objective, however, the chosen name should have become something like *zhifusi*, which is indeed the name used by the Japanese to refer to typhoid fever. The transliterated name of *zhifusi* signaled something new but did not challenge the traditional nosological system. Having said so, however, Shi continued by pointing out that this transliterated name of the disease-causing germ is very difficult to remember and pronounce in Chinese. Therefore, he concluded, "Since the majority of practitioners of Western medicine have adopted the name Cold Damage [to refer to typhoid fever], we [had] better translate bacteria of *Salmonella* Typhi as 'germs of Cold Damage' and then still call this disease Cold Damage."[29]

Traditional practitioners rightly recognized that Shi's translation would have horrific consequences for Chinese medicine, even though his suggestion sounds like a reasonable compromise for the sake of convenience. If the term *Cold Damage* was used to translate typhoid fever into Chinese, it would then become the official name for one of the eight notifiable infectious diseases that were defined by means of the germ theory and whose diagnosis, prevention, and notification supposedly were all under the control of the government and biomedical practitioners. This was the legal-medical context in which practitioners of Western medicine were most capable of controlling the definition of Cold Damage, to make sure that it was understood, managed, and treated exactly as typhoid fever. Shi's translation had profound practical and political implications for practitioners of Chinese medicine. If it had been accepted, it would have popularized the revolutionary idea that Cold Damage was an infectious disease caused by germs.

Incorporating the Germ Theory into Chinese Medicine

Among the critiques that rained down on this proposal, one of the most prominent responses was drafted by Yun Tieqiao.[30] By no means was Yun opposed to the project of unifying disease names in general; his specific concern related to the notion of infectitious disease, because it unavoidably involved the thorny issue of the germ theory. While Yun had focused on anatomy in his debates with Yu Yan in the early 1920s, since the late 1920s Yun had begun to consider the germ theory to be the pivotal challenge facing Chinese medicine. Two factors made this challenge particularly serious. First, Yun pointed out, among the traditional categories of Chinese medicine, the least unified was that of febrile diseases (*rebing*). At the same time, however, this category was also the very home base of Chinese medicine.[31] Second, since Western medicine took germs to be the cause of febrile diseases, this type of disease caused an unavoidable head-on collision between Chinese and Western etiologies. Seeing no way to reconcile the traditional Chinese etiology with the germ theory, in his public response to Shi's proposal, Yun proposed postponing the efforts to assimilate the germ theory into Chinese medicine and setting aside the naming of febrile diseases for future discussion.[32] For the time being, Yu suggested to just keep using the disease names from *The Treatise on Cold Damage* as the unified names for febrile diseases in Chinese medicine.

Yun's response reveals that the controversy over unifying nosological nomenclatures was not concerned with disease names in general, but only with febrile conditions or, more precisely, with the introduction into Chinese medicine of a new category of disease, namely, infectious diseases (*chuanranbing*). For proponents of this controversial proposal, incorporating the category of infectious diseases into Chinese medicine at once posed the most daunting challenge and promised the most crucial breakthrough for Chinese medicine. More importantly, this proposal provoked a heated public debate over the crucial issue of whether or not to incorporate the germ theory of diseases into Chinese medicine. Unlike previous debates over the relationship between the germ theory and Chinese medicine, which had taken place between individual actors only and therefore had little influence, the debate launched by the Institute of National Medicine promised to result in a governmental regulation that would shape the future development of Chinese medicine for the years to come. In this sense, this proposal was a public debate about incorporating the germ theory into Chinese medicine.[33]

Against both Shi Jingmo's proposal to remap Chinese diseases onto Western nosology and Yun Tieqiao's suggestion to postpone this discussion, Shi Yiren (1896–1966)[34] provided a third response and a more positive way for Chinese medicine to cope with the germ theory. From his point of view, the fundamental problem was how to accommodate the new category of infectious diseases within the framework of Chinese medicine while maintaining its integrity. To achieve this objective, Shi Yiren claimed that in ancient texts "epidemics" (*yi*) and Cold Damage were often placed in opposition to each other. The forefathers of Chinese medicine thus had already known that infectious diseases and Cold Damage should be treated as two distinct categories of disease. Based on the belief that they had always been considered as two separate kinds of disorders, Shi published separate books on each of them: *Chinese Study of Seasonal Diseases* (*Zhongguo shilingbing xue*) in 1930 and *Chinese Study of Acute Infectious Disease* (*Zhongguo jixing chuanranbing xue*) in 1933. As Shi pointed out in the preface to the former, he separated the discussion into two books to make sure that people would not confuse Cold Damage with infectious diseases.[35] He asserted that it was very inappropriate to translate *typhoid fever* as *Cold Damage*.

Given his objective, it is not surprising that Shi's strategy was applauded by quite a few practitioners of Chinese medicine. What is surprising is the virtues that they now attributed to Shi's endeavor. In his review of Shi's book, Tian Erkang argued that "Chinese medicine is capable of treating acute infectious diseases but has several weaknesses. In contrast, Western medicine is incapable of treating acute infectious diseases but has several strengths in preventing them."[36] Among the weaknesses of Chinese medicine, the first one was its "imprecision of disease names"; therefore it was a popular saying that "if a person consulted Chinese medicine but died nevertheless, he or she had no idea which disease had been the cause of death."[37] To address this weakness, Tian first suggested compiling a book specifically devoted to acute infectious diseases and adopting the bacterial tests of Western medicine for diagnostic purposes. That was exactly the objective that Shi had tried to achieve with his book on acute infectious disease. Second, according to Tian, Chinese medicine lacked both the concept and practice of disinfection, which was crucial for curtailing the spread of infectious disease among the population. Third, practitioners of Chinese medicine needed to learn to protect themselves and their patients by wearing masks and isolating patients who were suffering from acute infectious diseases. From Tian's point of view, the crucial contribution of Shi's book was his analysis of the relative strengths and weaknesses of the two styles of medicine in coping with infectious diseases.

Again it is by the example of typhoid fever that we can best appreciate how Shi planned to buttress the perceived weaknesses of Chinese medicine. The first few paragraphs in the chapter on typhoid fever, which cover etiology, pathology, and symptomatology, were all borrowed from contemporary biomedical literature. This information is followed by a special appendix on "the way of diagnosing this disease with Chinese medicine by means of the pulse and tongue fur."[38] It is important to point out here that Shi never tried to suggest a counterpart disease for typhoid fever within Chinese medicine. Rather than creating direct one-to-one correspondences between Western and Chinese disease names, which had been the publicized objective of Shi Jingmo's controversial proposal, Shi Yiren endeavored to develop from the traditional skills of pulse and tongue diagnosis a set of *new* patterns for typhoid fever, which was treated as a new kind of disease foreign to Chinese medicine. Furthermore, even within traditional pulse diagnosis, Shi strove to show how the patterns of typhoid fever were distinctively different from those of noninfectious diseases discussed in *The Treatise on Cold Damage*.[39]

Shi's discussion of typhoid fever continues with a section on prevention that includes advice on isolating the patient, handling the patient's excretions, washing hands, and controlling flies. Only after he had effectively employed all the strengths of biomedicine in understanding, preventing, and controlling typhoid fever, did Shi move on to the discussion of treatment methods offered by Chinese medicine, consisting of dozens of formulas and corresponding sets of symptoms. To conclude, Shi considered typhoid fever a newly defined disease that was caused by bacteria and was preventable and controllable with Western hygienic measures but could be diagnosed and treated with Chinese medicine.

To highlight his vision that the Chinese concept of Cold Damage did not belong to the category of infectious diseases, Shi devoted another book to the argument that both Cold Damage and Warm Disease were caused by seasonal changes in the weather and therefore required a separate category, which he called "seasonal diseases" (*shilingbing*). In spite of his innovative efforts, Shi was fully aware that the idea of creating a new disease category specially for Chinese medicine would be a highly controversial project for the world's medical community. Although the Shanxi Association for Reforming Chinese Medicine, which

Shi had helped found, had considered adopting this new category of seasonal diseases, as Shi himself pointed out, he refrained from insisting on this agenda because, "for the purpose of following the global trend of medical development, one had better be cautious."[40]

While Shi's strategy never became mainstream, it reveals a fundamental shift in the ways that traditional practitioners imagined the relationship between Chinese medicine and the modern medical sciences, especially the germ theory. I have pointed out in chapter 4 that when Yun Tieqiao and Yu Jianquan developed the strategy of using the concept of *qi*-transformation to "avoid the place of confrontation" in the early 1920s, they had agreed to accept ontology as the foundation of their debate with Western medicine. Instead of debating the relative ontological status of germs and the Six *Qi* of Chinese medicine, however, from this point on, people like Shi switched to a new endeavor with dual objectives. They strove to integrate what they perceived as the practical strengths of the germ theory—its ability to diagnose, prevent, and control infectious diseases—into Chinese medicine, while preserving, even developing, the valuable strengths of Chinese medicine in treating these diseases.

Shi's approach clearly signaled a shift away from ontological debates to efforts to synthesize the respective strengths of both styles of medicine for practical purposes. This integration of two styles of medicine was intended to be practical in two senses: (1) fulfilling the governmental regulations concerning notifiable infectious diseases, on the one hand, and (2) protecting patients from contracting infectious diseases by measures of disinfection and so forth, on the other. In the final analysis, however, the project had clear limits. Since the disease categories of seasonal diseases and infectious diseases were construed as mutually exclusive and separated into two unrelated books, the two styles of medicine were "united but divided." In this sense, Shi's approach was too similar to the earlier strategy of "avoiding the place of confrontation" to be a real solution.

Last but not least, I would like to point out that the capacity of biomedicine to control and prevent infectious diseases was indeed an unprecedented and unique strength of the modern germ theory and its clinical applications. As the philosopher K. Codell Carter has pointed out, when Louis Pasteur first formulated his germ theory of human disease around 1876, he was misunderstood and hence ruthlessly criticized by his contemporaries for supposedly asserting that germs were the "sufficient cause" of disease. In other words, Pasteur was misinterpreted as having claimed that the presence of a pathogenic organism invariably meant its necessary manifestation in the related disease. In fact, however, the practical orientation of his research led Pasteur to be concerned rather with the "necessary cause," a kind of cause without which certain diseases would not be able to occur. To use Pasteur's language, he discovered the necessary cause of disease because he could "give and prevent it at will."[41] Equipped with the means to manipulate the necessary cause of disease, modern germ theory then developed effective ways to prevent the spreading and even formation of certain life-threatening diseases. The power of prevention was such a unique strength of modern germ theory that practitioners of Chinese medicine like Shi Yiren and Tian Erkang could not help appreciating its obvious practical benefits and therefore strove to integrate it into their own practice.

Pattern versus Disease

The published "Proposal for Unifying Nosological Nomenclature" soon came to be regarded as an aborted project, if not a total failure. Just five months after its publication, the Institute of National Medicine announced to the organizations of Chinese medicine all over the country its decision to withdraw the proposal. Moreover, to signal a radical policy shift, it dismissed the reform-minded physicians Ye Guohong, Lu Yuanlai, and Guo Shoutian from the committee reponsible for putting in order Chinese medicine.[42] On the surface, this proposal indeed proved to be a great setback for the reformers of Chinese medicine.

Around the time that the reformers were forced to realize how difficult the unification of Chinese and Western disease nomenclatures truly was, some practitioners of Chinese medicine started considering the possibility that the project of integrating Chinese and Western medicine did not need to focus on disease to begin with. As they argued, the strength of Chinese medicine did not lie in treating diseases (*bing*), but in treating *zheng*, understood as "patterns" or "syndromes."

Because this argument later became a fundamental doctrine of Traditional Chinese Medicine, created in the mid 1950s, it is noteworthy that hardly any traditional practitioner had ever characterized Chinese medicine this way before the debate over the "Proposal for Unifying Nosological Nomenclature" broke out in the summer of 1933. In his lead article for the special issue on this proposal, for example, Zhang Zanchen pointed out that one of the central problems of the reform project was "the confusion between disease names and names of 'symptoms' [*zheng*]" in Chinese medicine.[43]

Zhang Zanchen's short essay reveals two important points. First of all, instead of using the term *pattern* (*zheng*), Zhang relied on another word with the same pronounciation (*zheng*) but represented by a different character, with the meaning of "symptoms."[44] Second, precisely because Zhang complained about the confusion between diseases and symptoms, it was clear that, at least from his point of view, Chinese medicine also concerned itself with treating diseases—not just with *zheng*, no matter whether this term was understood as meaning "symptoms" or "patterns." As Volker Scheid points out, "For long periods, especially from Han to the Tang, diseases rather than patterns functioned as the most important diagnostic classifiers."[45] Even after patterns reemerged as a focus of concern for literati physicians in the Song dynasty and thereby became definitive of elite Chinese medicine,[46] the concept of disease kept playing a central role in the theory and practice of Chinese medicine. By the late 1920s, there was no sign that Chinese medicine did not concern itself with disease. Besides, there was no consistent effort to delineate the boundaries between diseases, patterns, and symptoms.

Just like the discourse on experience discussed in chapter 4, this contrast between disease and pattern was also inspired by the writings of Japanese advocates of traditional Chinese medicine, especially via Chinese translations of works authored by the Japanese doctor Hiroshi Watanabe. According to Watanabe's own confession, he had decided to switch his career from studying biomedicine to becoming an advocate of traditional medicine because he felt deeply disappointed by the former's lack of effective therapeutics. When he had been pursuing his studies in Germany at the turn of the twentieth century, precisely during the height of German scientific medicine, he had found that German medicine "bore no fruit for practical therapeutics and cures."[47] To dramatize his disillusion with modern medicine, Watanabe quoted a statement allegedly made by Joseph Skoda (1805–81), the founder of the modern medical school of Vienna and the champion of therapeutic nihilism in the 1840s: "Whether or not a disease is cured does not concern medicine. Our objective is to satisfy our intellectual curiosity."[48] As Watanabe admitted, this was an imprecise quotation. In fact, it is perhaps a paraphrase of Skoda's famous comment:[49] "Whilst a disease can be described and diagnosed, we can dare not to suspect to cure it by any manner of means."[50]

In his book, *The True Essence of Japanese-Chinese Medicine*, Watanabe devotes an entire section to criticizing "Skoda's erroneous opinions." Following the above-mentioned quotations, Watanabe commented that, "This opinion is what I have hated most in my life, and the very opposite of the spirit of Oriental medicine."[51] After Watanabe's articles and book were translated into Chinese in the late 1920s and in 1931, this pointed criticism of therapeutic nihilism was cited by quite a few leading figures of Chinese medicine, including Zhang Taiyan,[52] Lu Yuanlei, Shi Yiren,[53] and the extremely conservative Wu Hanqian.[54] In various ways, they each found inspiration in this article, specifically in the criticism of Skoda's purported statement.

Placing himself in sharp disagreement with Skoda's original position, Zhang Taiyan paraphrased Watanabe's purported translation of Skoda's statement and turned it into the opposite conclusion: "What doctors take as their responsibility is the cure of diseases. As long as they can have the therapeutic methods, there is no need to insist on knowing the cause of disease (*bingyin*)."[55] It goes without saying that Skoda's therapeutic nihilism as it was being criticized by both Watanabe and Zhang was merely an eighty-year-old relic and by no means representative of the contemporary practice of biomedicine in the late 1920s.[56] Still, Skoda's statement was instrumental for advocates of Chinese medicine in their efforts to envision a new division between their practice and that of modern scientific medicine. Because of this strategic purpose, what Zhang emphasized was not really Skoda's original opposition between diagnosis and therapeutics, but the opposition between etiology and therapeutics.

It was in opposition to the biomedical emphasis on etiology, that is, on the "cause of disease," that Zhang developed the idea that the key concern of Chinese therapeutics was the notion of pattern rather than disease. In the preface to Lu Yuanlei's *Contemporary Interpretation of the Shanghanlun* [Treatise on Cold Damage], Zhang Taiyan elaborated this paired opposition. The first of the two key points in studying *The Treatise on Cold Damage*, Zhang emphasized, is to realize that the traditional names of diseases—Cold Damage, Wind Strike, and Warm Disease—were actually based not on the cause but on the "pattern" found in the patient.[57] The differentiation between cause and pattern is very important because, as Zhang immediately pointed out, "when nowadays Westerners discuss febrile conditions, they often take germs as the root cause."[58] In other words, a proper understanding of *The Treatise on Cold Damage*, especially its conception of pattern, provided the key for Chinese medicine to circumvent the threat of the germ theory of diseases.

The context of Lu's book reveals a great deal about the historical origin of this opposition between pattern and disease. First of all, Zhang had originally articulated this opposition not as a general statement concerning the nature of Chinese medicine, but for the purpose of elaborating on the nature of *The Treatise on Cold Damage*. In this sense, the concept of pattern is closely associated with this ancient text. Second, Lu's book is actually an elaboration of Yumoto Kyushin's *Japanese Style of Chinese Medicine*. As discussed in chapter 4, Zhang Taiyan had used Yumoto's book to promote the idea that *The Treatise on Cold Damage* constituted an "empirical tradition" within Chinese medicine, elevating it above the *Inner Canon of the Yellow Emperor* as the seminal canon of Chinese medicine.[59] For Zhang Taiyan and Lu Yuanlei, the "empiricism" of *The Treatise on Cold Damage* went hand in hand with its focus on treating patterns and symptoms; these two features lent support to each other and were both inspired by the Japanese scholarship on the *Treatise*. In the process of engaging with the biomedical conception of disease, the *Treatise on Cold Damage* came to be seen as the origin of the definitive feature of Chinese medicine, namely, "pattern differentiation" (*bian zheng*).

Finally, this notion of pattern was placed in opposition not to the concept of disease in general but to the newly developed ontological conception of disease as defined by its cause, that is, its etiology. It is actually two sides of the same coin that, on the one hand, some practitioners of Chinese medicine realized all of a sudden that Chinese medicine had "confused" the concepts of symptom and disease and, on the other hand, that both Hiroshi Watanabe and Zhang Taiyan felt the need to defend Chinese medicine against the primary importance of etiology by emphasizing the concept of pattern. Both were reactions against a recent development in Western medicine that had sharpened the boundary between diseases and symptoms. As the medical historian Michael Worboys has pointed out, "germ theories, and then bacteriology, have been seen as important factors in a major shift in the dominant conception of disease in Western medicine: moving from defining diseases by their symptoms and results to defining them in terms of processes and causes."[60] As a result of this historical shift, the ontological notion of "disease specificity" emerged—the notion "that diseases can and should be thought of as entities existing outside the unique manifestations of illness in particular men and women."[61] While we have come to take this idea of disease specificity for granted, the cultural impact of this "specificity revolution," according to Charles E. Rosenberg, is comparable to "the effects of the Newtonian, Darwinian, or Freudian revolution."[62] It was in response to this "specificity revolution" in Western medicine that the opposition between pattern and disease arose in East Asian traditional medicine.

A Prehistory of "Pattern Differentiation and Treatment Determination"

It was only when the reform-minded practitioners of Chinese medicine were forced to take the germ theory of disease seriously in the 1930s that they started to realize that this time their challenge was not just the incorporation of a new theory of disease, such as pathological anatomy, into Chinese etiology. Rather, the accommodation of the germ theory involved a fundamental reconceptualization of the nature of disease, that is, the definition of disease in terms of its cause, germs. Instead of arguing that Chinese medicine also defined diseases by their causes, albeit with different kinds of causes, such as cold, wind, and so forth,[63] Zhang Taiyan proposed the revolutionary idea that Chinese medicine defined them in a fundamentally different fashion, namely by means of "patterns."[64]

With this revolutionary idea, Zhang articulated a strategy with long-lasting consequences. While he refused to accept the modernist remapping proposed by other modernizers like Shi Jingmo,[65] he also did not want Chinese medicine to keep fighting against the germ theory in the latter's favored battlefield. Zhang was willing to give up that battle and concede that Chinese medicine did not know the cause of disease, because, just like Watanabe, he wanted to emphasize that the real objective for medicine, and also the real strength of Chinese medicine, lay in its therapeutics.[66] Although its theories might be incorrect in certain aspects, Chinese medicine was nevertheless effective in treating many diseases, including many life-threatening infectious diseases. The point was not which medical theory could identify the real cause of diseases, but which medicine could provide effective treatments, regardless of how it conceptualized health problems. Therefore he advocated that the value of any conception of disease should be judged in terms of its clinical function or, in other words, its efficacy in treating disease.

Zhang's emphasis on therapeutics as opposed to etiology came at a particularly advantageous historical moment. While disease specificity was a well-established concept in the late nineteenth century, the closely related promise of a specific cure (namely, a treatment targeted at the specific cause of disease), which had been enthusiastically awaited ever since Robert Koch (1843–1910) developed tuberculin in 1890, remained yet to be fulfilled. By the 1930s, only two celebrated achievements of specific cures existed: the development of diphtheria antitoxin in 1913 by the German physiologist Emil von Behring (1854–1917), who received the first Nobel Prize in medicine and physiology in 1901, and the development of "compound 606," also known as Salvarsan, for treating syphilis by Paul Ehrlich (1854–1915) and his Japanese student Sahachiro Hata (1873–1938) in 1909. Salvarsan came very close to Paul Ehrlich's conception of a "magic bullet," since it allowed physicians to stop trying to treat the body in its entirety. It is no surprise that Pang Jingzhou mentioned both treatments in his account of Shanghai's health-care services and highlighted the fact that some practitioners of Chinese medicine had also appropriated these popular biomedical treatments. These exemplary achievements of Western medicine substantiated the notion of a specific cure—the idea that since disease had a specific cause, the most rational therapy must be aimed at what was causing the disease itself.

While calling Ehrlich's chemotherapy "one of the most significant achievements of the present [twentieth] century," the medical historian Erwin H. Ackerknecht points out, "Nonetheless, during the twenty years following Ehrlich's death, the field lay barren and was almost abandoned as hopeless. . . . It seemed impossible to find effective anti-bacterial drugs."[67] This gloomy situation continued until the late 1930s, when scientists developed the sulfa drugs and later penicillin, thereby giving birth to the era of antibiotics. In light of this historical background, when the debates about the unification of Chinese nosological nomenclature took place in the early 1930s, practitioners of Chinese medicine had good reasons to argue that disease specificity had failed to bring forth effective therapy and that Chinese medicine had a lot to offer in terms of therapeutics.

It was at this historical juncture that the reform-minded Ye Guhong wrote the pointed article entitled "The Therapy of National Medicine in Treating Infectious Disease," with which I began this chapter. Following Zhang's strategy of emphasizing the therapeutic efficacy of Chinese medicine, Ye went on to explain why Chinese medicine, without any knowledge of germs and therefore of the genuine cause of disease, was still capable of providing effective treatments. Being a student of Japan's Imperial Medical College, Ye was keenly aware of the gloomy situation surrounding the search for specific cures. Ye pointed out that, with the exception of the effective syphilis treatments, compounds 606 and 914, Western medicine had discovered not a single synthetic chemical compound that "could be used directly to kill germs or counteract toxins."[68] All the other kinds of Western therapies, such as vaccine therapy and antitoxin therapy, worked by injecting toxins or germs into the human body in order to induce its "natural power of anti-toxin resistance."

Ye elaborated as follows: "In this sense, in terms of killing germs and counteracting toxins, even Western medicine relies on the natural power of the human and animal body; drugs do not play any unmediated and substantial role in this process. Similarly, when we use ancient formulas from our National Medicine to treat infectious disease, we merely help to strengthen the natural resistance of the human body; by no means do we try to kill germs with formulas for Wind and Cold."[69]

As indicated by this revealing paragraph, people like Ye appropriated the concept of resistance from Western medicine to explain the perceived efficacy of traditional Chinese therapy. While Ye did not mention a specific person, he was probably referring to the idea first developed by Hans Buchner (1850–1902) that the human body was equipped with a "natural resistance" to diseases and natural antibodies.[70] On the basis of the biomedical idea, Ye argued that Chinese medicine had always focused on boosting the body's natural resistance, as represented by the concept of "proper *qi*" (*zhengqi*) in Chinese.[71]

Instead of arguing for the incommensurability of the germ theory with Chinese medicine, Ye strove to identify potential allies for Chinese medicine from among the heterogeneous and disunified fields of the medical sciences. By means of the concept of resistance, Ye was able to have it both ways: he could embrace the germ theory for identifying the cause of disease while insisting that Chinese medicine could offer effective and valuable treatments, even though it did not address this narrow cause of disease specifically. To the practitioners of Chinese medicine, the collective and official acceptance of the germ theory was conditioned on their acceptance of the biomedical concept of disease resistance.

It was precisely for the purpose of preserving their valuable therapeutics that reform-minded practitioners felt the need to articulate the conception of pattern from among the conventional usages, which had not clearly differentiated between disease, symptom, and pattern. This new conception no doubt has historical roots in traditional practices, most notably in *The Treatise on Cold Damage*. The therapeutic practice advocated in this text consists of several integrated steps: After identifying the types of diseases, "the diseases are then broken down to more specific patterns characterized by a particular constellation of symptoms and signs. . . . They are usually named after the prescriptions suggested for their treatment (e.g., *Xiao chaihu tang* or Minor Bupleurum Decoction pattern)."[72] It is no accident that Watanabe also emphasized that Chinese medicine often takes the names of prescriptions to stand in for the names of diseases.[73] To say, for example, that a patient is suffering from a "minor bupleurum decoction pattern" means that he or she is afflicted by a condition in which Minor Bupleurum Decoction will be effective.

To appreciate the function of this conception of pattern in preserving and valorizing treatment methods, let us again take Cold Damage as a concrete example. If practitioners of Chinese medicine had accepted the definition of Cold Damage in terms of its biomedical "cause," that is, as identical to typhoid fever, in accordance with Shi Jingmo's controversial proposal, they would immediately have been confronted with the irreconcilable conflict between Cold Damage and Warm Disease in Chinese medicine. On the contrary, though, if they had focused on therapeutics instead and thereby been able to value the different ways of achieving therapeutic efficacy, they would have had good reason to preserve both schools of Chinese medicine, at least for the time being, since they both had proven useful in treating different patterns. It is important to note here that the logic of reasoning had just gone through a complete reversal. These practitioners of Chinese medicine proposed to preserve the notion of pattern not because patterns served as truthful representations of disease entities but because they were a necessary tool for utilizing supposedly valuable therapeutics.

This conception of pattern and the related emphasis on therapy is rather close to "pattern differentiation and treatment determination" (*bianzheng lunzhi*), a phrase that is repeated in virtually every modern textbook on Traditional Chinese Medicine to explain the direct link between diagnosis and therapy. As Volker Scheid documented, it was only in the 1950s that this four-character phrase began to be used to characterize the essence of the clinical practice of Chinese medicine.[74] As Eric I. Krachmer points out, this phrase became further substantiated in practice in the 1960s, when the textbook *Chinese Medicine Diagnosis* started providing a comprehensive list of patterns and their associated symptoms.[75] As far as I know, the phrase was never used in this exact fashion during the Republican period, even though much of its conceptual content and linguistic fragments were already present and had been correlated with each other. In light of the discussion in the 1930s, "pattern differentiation" (*bian zheng*) was based upon and applied for the purpose of "determining treatment" (*lun zhi*): the former was the means to the latter. To give a clinical example of this approach, which is still the key to Traditional Chinese Medicine as taught in Chinese institutions today, it is important to diagnose a patient as suffering from a Minor Bupleurum Decoction pattern so that traditional practitioners can treat him or her with Minor Bupleurum Decoction.

To use Watanabe's similar language, the therapeutic method of

Japanese-Chinese medicine was summarized as "prescribing medicinals by matching the pattern" (*duizheng touyao*).[76] In both formulations, what was valuable and therefore worthy of being preserved as a top priority was the practical therapeutics rather than the "theoretical foundation" of traditional medicine. In this sense, "pattern differentiation and treatment determination" was not merely the expression of an alternative theory of disease; more importantly, it represented the rise of a philosophical alternative to the representationist conception of realism.

Since the advocates of Chinese medicine at that time considered pattern differentiation and treatment determination a defensive strategy, they never thought about valorizing it as a competing philosophy against Yu Yan's representationist conception of realism. By "representationist conception of realism," I mean to refer to a specific, and dominant, kind of philosophy of science that considers the art of the experimentalist to be about making representations that reliably imitate the act of unmediated seeing. From this point of view, the function of theoretical concepts is to represent the material entities that already exist in the real and eternal world. Since the sole purpose of these concepts is to represent these entities as precisely as possible, there is absolutely no value to concepts for which we cannot find corresponding referents in the real world. Yu Yan's seminal critique of Chinese medicine was built squarely upon this philosophical tradition, regardless of whether his specific target was yin/yang, *qi*-transformation, or meridian channels. When Western medicine developed the ontological conception of disease as part of the "specificity revolution" in the late nineteenth century, disease became one of such ontological entitites, for which Chinese medicine was then accused of failing to provide correct representation.

As a matter of fact, the idea of pattern differentiation and treatment determination was designed to transcend the confinement of this representationist conception of realism. For the advocates of this idea, the emphasis was on what they could do with the help of their theoretical constructs—what kind of treatment they could offer if they diagnosed diseases by means of the patterns of Chinese medicine. In this sense, their relation to theoretical concepts is quite close to philosopher Ian Hacking's idea of the interventionist conception of realism. In Hacking's language, "If you can use it as a tool to create material effects, it is real."[77] The point of realism is not to represent the reality out there but to intervene in the world right here. With the development of pattern differentiation and treatment determination, the struggle between the two styles of medicine thus shifted from a competition over which medicine better represented the world of material reality to one over which provided more useful tools for the practical purpose of treating illness in the here and now. The difference between these two forms of competition is huge, because the latter did not have to be a zero-sum game.

Conclusion

Rather than having always been the defining feature of Chinese medicine, the idea of pattern differentiation and treatment determination emerged out of a very specific context of historical struggle. It was shaped by the following four historical developments: first, the threat posed by the ontological conception of disease, especially the idea of a specific cause of disease, as developed in Western medicine; second, the Japanese-inspired strategy to emphasize the practical efficacy of therapeutics instead of engaging in an ontological struggle with the proponents of Western medicine; third, the political need to incorporate the new notion of notifiable infectious diseases, defined in terms of the germ theory, into Chinese medical doctrine; and, fourth, the practical benefits of incorporating the biomedical practices of prevention, disinfection, control, and therapeutics of infectious diseases into clinical applications in the framework of Chinese medicine. Because of this complicated combination of threats, interests (political as well as clinical), and innovative efforts, practitioners of Chinese medicine developed the incipient form of the novel approach of pattern differentiation and treatment determination. Nevertheless, since they did not yet express this idea with the exact four-character phrase in Chinese, as it is known to practitioners of Traditional Chinese Medicine today and did not yet have the authority to accord it the status of a national standard, I call this stage the prehistory of pattern differentiation and treatment determination.[78]

Three salient features of this methodology are actually the result of the historical process that took place in the 1930s. First of all, its inventors deliberately avoided a fundamentalist war over ontology and refrained from fashioning Chinese medicine into a system incommensurable with the modern germ theory.[79] In fact, the community of Chinese medicine rejected the two proposals of either excluding the germ theory from Chinese medicine (Yun Tieqiao) or establishing Cold Damage as a new category of "seasonal disease" independent of the system of biomedicine (Shi Yiren). As most reform-minded practitioners appreciated the strength of the germ theory in diagnosing, controlling, and preventing infectious disease, they embraced, if selectively and thus partially, the germ theory of disease for the practical purpose of making Chinese medicine a constitutive part of the national health-care system. Instead of resisting the germ theory on the grounds of ontology, they developed this strategy to borrow from and negotiate with the germ theory so that they could maximize their political and clinical interests. In other words, pattern differentiation and treatment determination constitutes a modern development of Chinese medicine partially because its formulation draws on the germ theory of disease and the "specificity revolution" of modern medicine.

Second, as pattern differentiation and treatment determination emphasized the value of therapeutic practice (in opposition to knowledge and representation), it offered the potential of transcending the modernist confinement of representationist realism. Rather than becoming trapped in the modernist framework that forced them to engage in ontological wars, this approach allowed practitioners to keep preserving and developing the valuable aspects of their therapeutics, even though the ontological status of related patterns was far from certain. Differing drastically from the conventional understanding of this phrase today, the key feature of the phrase at that time was not perceived to lie in its opposition between pattern and disease, an opposition that could be easily but wrongly interpreted as another kind of constructed incommensurability, but in its foregrounding of clinical practice and of the practical value of therapy. In other words, pattern differentiation was not carried out for the sake of representing disease entities or phenomena, but for the purpose of taking action to transform the disease by offering therapy. Moreover, because the notion of pattern, when conceived in this way (as a practice-based methodology), was not part of the world as defined by the representationist conception of realism, it was no longer locked in a zero-sum game with the ontological conception of disease. These two medical paradigms were consequently able to coexist without contradicting each other. In contrast to the previous period, in which the two styles of medicine had been locked in the modernist space of representation, pattern differentiation and treatment determination opened up a nonmodernist space for coexistence, mutual learning, and exchange between them, at the level of clinical practice.

Third, pattern differentiation and treatment determination demonstrated that the accommodation of the germ theory did not necessitate the complete destruction of Chinese medicine. Instead, the function of this approach was to preserve what Chinese practitioners considered to be the crucial strength of Chinese medicine. People could be wrong about their specific evaluations, and their view about the most valuable elements of Chinese medicine could keep evolving as the situation changed. For example, as trends in the leading causes of death shifted along with a decline in the death rate from acute infectious diseases in the 1800s to the chronic diseases in the 2000s, people would value very different aspects of Chinese medicine. Of undeniable importance was the presumption that there was something valuable to be found in Chinese medicine that was worthy of careful preservation and innovative development. In a nutshell, the core of the much-debated project of Scientizing Chinese Medicine lay in these problems: How valuable was Chinese medicine? Where did these valuable elements reside within Chinese medicine? And in terms of realizing these values of Chinese medicine, what were the most appropriate ways of either conducting scientific research or incorporating it into a national health-care system? The following two chapters investigate these questions, and the concluding chapter, "Thinking with Modern Chinese Medicine," offers a general reflection on the crucial question of value.

9 Research Design as Political Strategy: The Birth of the New Antimalaria Drug *Changshan*

Changshan *as a Research Anomaly*

Among all the discourses of modernity that the advocates of Chinese medicine were faced with from the 1920s on, the one that they resisted least, if at all, was the idea that Chinese drugs were just "grass roots and tree bark" (*caogen shupi*), that is, raw material from nature. While this expression was meant to denigrate Chinese medicine as primitive, it nevertheless implied that there were valuable elements to be found within it, elements that had not been corrupted by its allegedly erroneous theories. At a time when many viewed Chinese medicine as nothing but a "laughing stock," a representative of China's backwardness, even such a derogative notion helped to remind people that it was not completely without value. Thanks to the world-famous discovery of ephedrine from the Chinese drug *mahuang* (*Ephedra sinica*) discussed in chapter 4, the program that came to be known as Scientific Research on Nationally Produced Drugs (*guochan yaowu kexue yanjiu*) emerged as a national consensus, enthusiastically supported by both sides in the struggle over China's medical future. As a testimony to the support that this research program received during both the Republican and Communist periods, the American Association for the Advancement of Science observed in 1960 that "pharmacology occupies an exalted position in Communist China, where the emphasis given to pharmacology is probably greater than in any other country."[1] Moreover, in 2011 a study of Chinese drugs won the most prestigious medical research award ever bestowed on Chinese scientists: the Lasker Award, nicknamed "America's Nobel Prize," was granted for the discovery of artemisinin from the Chinese drug *qinghao* for the treatment of malaria.

For many critics of Chinese medicine at that time, this research program enjoyed a unique status, representing the only acceptable approach to the controversial project of Scientizing Chinese Medicine. Unlike other research programs, which carried the obvious risk of fostering what was conceived as "mongrel medicine," scientists believed it to be capable of validating the effective elements of Chinese medicine while keeping its corruptive elements at bay. In this sense, this research program served the political function that critics of Chinese medicine intended for the so-called project of Scientizing Chinese Medicine, a function that, as one critic put it bluntly, could be characterized as "using science to demolish Chinese medicine."[2] This strong statement reminds us that, after the founding of the Institute of National Medicine in 1931, both sides of the medical struggle agreed to determine the fate of Chinese medicine by the outcomes of the research effort to scientize Chinese medicine. Within this context, the program called Scientific Research on Nationally Produced Drugs could not avoid serving two roles at the same time: that of a research program for the study of Chinese drugs and that of a political strategy in the struggle between two styles of medicine.

In light of the dual role of this research program, the following questions arise: How could a political strategy designed to contain Chinese medicine in turn serve as a productive research program for the study of Chinese drugs? More specifically, how did this research program bring about its two major achievements during the Republican period, namely, the discovery of ephedrine and the validation of the antimalarial efficacy of *changshan* (dichroa root)? Finally, how did the outcomes of this popular research program transform the political landscape of the medical struggle, influencing the public perception of the proper relationship between Chinese medicine and biomedicine? As these questions require us to go beyond the world of words into the world of acts, this chapter reveals how experimentation and the scientific laboratory brought unpredictable and thus creative forces into the modern history of Chinese medicine.

To answer the above-mentioned questions, this chapter investigates the second major success of this research program during the Republican period, namely, the study of *changshan* in the 1940s. Contrary to common belief at that time, this research demonstrated that Chinese drugs were in fact effective for treating infectious diseases, even though Chinese medicine had no knowledge that various kinds of micro-organisms were their causes (see the introduction to chapter 8). Moreover, the validated antimalarial efficacy of *changshan* had paved the way for the governmental project to search for antimalarial drugs from Chinese herbs in the 1960s,[3] which culminated with the world-renowned discovery of artemisinin. In addition to its historical significance, I have chosen to focus on the case of *changshan* because it occupied an anomalous status within the tradition of this research program.[4] Although the research on *changshan* was celebrated as one of its two most important achievements, it was far from representative of this research program. To put it bluntly, it was only because participating researchers violated some key procedures in this research protocol that they were able to demonstrate the antimalarial efficacy of *changshan* as quickly and efficiently as they did. The "anomalous case" of *changshan* thus makes visible an often neglected debate over the design of research protocols in studies on Chinese medicine, a debate that continues to be of central significance in the scientific evaluation of traditional medicines to this day.

To turn the case of *changshan* into a critical reflection on this popular research program, especially on the design and political effect of its research protocol, I have drawn on the scholarship of science and technology studies. In particular, I have employed the concept of "translation" and the related actor-network theory developed by Bruno Latour, Michel Callon, and John Law.[5] Latour uses the concept of "translation" to question the preexisting distinction between the "content" of science and its so-called "context." Instead of passively allowing their research to be determined by the context, Latour argues, scientists must first actively build their own socio-technical network and recruit allies. Second, because in the network-building process, the so-called context gets continuously displaced and transformed, we can never understand the "content" of science in terms of its context. Third, to emphasize the role played by nonhuman objects in "associating" people, Latour and his colleagues call this framework of analysis the "socio-technical network." This emphasis on nonhuman objects reflects the crucial contribution of science and technology studies to the fields of humanities and social sciences; it highlights the fact that people can always re-engineer their socio-technical network by bringing new objects into the network. As Latour put it aptly, "We live in communities whose social bond comes from objects fabricated in laboratories."[6] To go beyond the conventional understanding of Chinese medicine as a symbolic or cultural system, this chapter focuses on a laboratory study that foregrounded the materiality of Chinese medicine, especially the agency of Chinese drugs.

With the help of this analytic framework, I show that the key to the success of these studies lay in the fact that the researchers treated traditional drugs such as *changshan* not as "raw material from nature," but as a practice-based, fabricated object sustained by the socio-technical network of Chinese medicine. More specifically, they treated *changshan* as a time-tested, carefully studied, and trustworthy pharmaceutical agent, designing their research accordingly and innovatively. By highlighting the crucial—but so far neglected—role of research design in shaping the modern history of Chinese medicine, the case of *changshan* on the one hand reveals the political role played by this research program in reinscribing and negotiating in practice the modernist divide between Chinese medicine and Western medicine since its emergence in the 1920s. At the same time, though, this case also demonstrates that the relationship between Chinese medicine and science was by no means bound to be antithetical, or a zero-sum game. Instead, their relationship depended on the specific way that people "related" them to each other in practice, such as in the design of research protocols for the scientific investigation of Chinese drugs, and for the project of Scientizing Chinese Medicine, by extension. To highlight the theoretical implications of this case study, I have given this chapter the thematic title "Research Design as Political Strategy."

Scientific Research on Nationally Produced Drugs

Independent of the breakthrough in the scientific research on ephedrine, Chinese drugs acquired an important role in the context of economic nationalism. As discussed in chapter 5, when practitioners of Chinese medicine gathered on March 17, 1929, to inaugurate the National Medicine Movement, a pair of giant posters spelled out their main goals: "Promote Chinese medicine to prevent cultural invasion!" and "Promote Chinese drugs to prevent economic invasion!"[7] To recruit allies, traditional practitioners adopted not only the rhetoric of cultural nationalism but also that of economic nationalism. By characterizing Chinese drugs as "national products" (*guohuo*), traditional practitioners were able to recruit not only people involved in the industry of traditional Chinese pharmacy, but also supporters of the National Products Movement, who would otherwise have had little interest in this medical struggle. Emerging in the early twentieth century, the National Products Movement urged people to resist imperialism and express nationalist sentiments by purchasing "national products" as opposed to "foreign products."[8] Since this movement intensified after the May 30th Incident in 1925 and started receiving endorsements from the Nationalist government in 1928, it is no wonder that advocates of Chinese medicine at this juncture sought to build an alliance between themselves and this popular movement. As a result, the newspaper campaign by traditional practitioners focused on the economic consequences of abolishing the "national products" of Chinese medicine. Once Chinese medicine had been wiped out, the propaganda by traditional practitioners suggested, Western pharmaceutical companies would take over the market and thereby substantially increase China's trade deficit.

While practitioners of Western medicine paid little attention to the argument of cultural nationalism, they took the issue of economic nationalism very seriously, caring greatly about the accusation that they were traitorous salesmen of Western medicine.[9] To emphasize their different stances toward these two kinds of nationalism, they emphasized the distinction between "Chinese drugs" (*zhongyao*) and "nationally produced drugs" (*guochan yaowu*). From their point of view, the term *Chinese drugs* by itself was highly misleading. Being "grass roots and tree bark," the so-called Chinese drugs were raw materials from nature, having nothing to do with either Chinese culture or Chinese medical theories. While they adamantly rejected any association between Chinese drugs and cultural nationalism, they did recognize the importance of at least some Chinese drugs to economic nationalism. I emphasize *some* here because, as the critics of Chinese medicine pointed out, many so-called Chinese drugs had not even originated in China,[10] and some of them were actually no longer being produced in China but were imported from foreign countries.[11] Even worse, according to a report from the Chinese Customs Service, the amount of money spent on imported "Chinese" medicinals—including Western ginseng, Japanese ginseng, rhinoceros horn, and the like—actually exceeded the amount spent on "Western pharmaceuticals" by more than one million dollars.[12] Therefore the issue of whether Chinese drugs qualified as national goods was highly dubious at best, and traditional practitioners seemed to be at least as guilty of advertising "foreign products" as their biomedical colleagues.

To emphasize the fact that only a subsection of the traditional Chinese materia medica could correctly be viewed as national products, Yu Yan coined the term *nationally produced drugs*. Instead of shouting "Chinese people should consume Chinese products" (the slogan of the National Products Movement), practitioners of Western medicine promised that they could do better than that by selling nationally produced drugs to foreigners around the world—as long as they succeeded in creating modern pharmaceuticals out of these nationally produced drugs. Practitioners of Western medicine hoped that if they succeeded in this scientific project, many people might switch camp to support Western medicine including workers in the traditional pharmaceutical industry,[13] advocates of the National Products Movement, and even cultural nationalists.

Luckily for them, the successes achieved by the scientific research on ephedrine provided timely support for this best-case scenario. According to K. K. Chen, *mahuang*, the plant source of ephedrine, had become a commodity in the global market right after the publication of their findings in 1924. Chen created a table in his paper to show the remarkable growth of *mahuang* exports to the United States since then and went so far as to include a photograph of a happy Chinese worker with a shovel in his hand, standing on top of a pile of *mahuang*. As Chen explained, "Many men in China have found a new calling in the annual collection of *mahuang*. The crude drug is shipped in bales from China to all parts of the world, for the manufacture of ephedrine."[14] Since the case of ephedrine demonstrated in concrete terms how domestically produced drugs could contribute to the political cause of economic nationalism, it provided the archetype for the program Scientific Research on Nationally Produced Drugs.

Stage One: Overcoming the Barrier to Entry

The discovery of the therapeutic efficacy of ephedrine created great enthusiasm for the investigation of Chinese drugs in China and elsewhere in the mid 1920s, but no other naturally occurring product of similar therapeutic value was found in the following dozen or so years. In an article entitled the "Scientific Research on Chinese Drugs in the Past 30 Years" in the leading Chinese science journal in 1949, the author praised the research on *changshan* in the 1940s as the second most important achievement of this research tradition during the Republican period.[15]

In addition to being separated by two decades, these two research projects were conducted in very different environments. In sharp contrast to the *mahuang* research, which was started at the prestigious Peking Union Medical College, research on *changshan* originated in a humble school clinic in the mountains of southwestern China. While the latter research never attracted similar international attention, its success led to the establishment of the government-funded Research Institute for Chinese Drugs with Specific Efficacy (*zhongguo texiaoyao yanjiusuo*), and the Jingfo Mountain Research Farm for Chinese Drugs.

Unlike most other research on Chinese drugs, the research on *changshan* was initiated, recorded, and officially supported by an enthusiastic advocate of Chinese medicine, Chen Guofu. As pointed out in chapter 7, Chen Guofu and his younger brother Chen Lifu had played a crucial role in establishing the Institute of National Medicine and in identifying the scientization of Chinese medicine as its official objective. Although not a single account of the history of *changshan* research has recognized Chen Guofu's role in "discovering" the antimalarial efficacy of *changshan*, Chen Guofu saw himself as having played a groundbreaking role. For him, the history of the scientific discovery of *changshan* could be roughly divided into two stages. The critical first stage occurred in 1940, when Chen sent a traditional Chinese prescription for malaria treatment to the clinic at the Central School of Politics (*zhongyang zhengzhi xuexiao*). Once *changshan*, one of the seven drugs in this prescription, had entered the school clinic, it was handled almost exclusively by biomedical scientists and physicians. The crucial question to ask in this context now becomes whether or not any significant barrier existed that would have made it difficult for *changshan* to get into the school clinic.

Fortunately for historians, Chen Guofu was so proud of his contribution that he wrote a script for an educational film on the discovery of *changshan*.[16] His script described in detail how this anonymous prescription became the inspiration for scientific research. Reading this script closely, we can understand what role Chen Guofu saw himself as having played in this history and in particular what he saw as the main barrier for the introduction of *changshan* into the network of biomedicine.

The first stage in the scientific discovery of *changshan* appears clear and simple. According to Chen Guofu's recollection, the story began when one of the school guards found a prescription for treating malaria in the local newspaper of Chongqing. Before long, this guard started distributing copies of this prescription to the staff members of the whole school, including Chen Guofu. Chen Guofu described what followed next:

> At that time, the director of our school clinic, Doctor Cheng Peizhen, happened to be in Chen Guofu's office. Involved in his work, Doctor Cheng mentioned to Chen Guofu that the price of quinine had risen steeply, and he was therefore very worried about a potential shortage. Chen Guofu then asked Doctor Cheng why he did not use Chinese drugs instead. Doctor Cheng responded that he could not take advantage of them since he had no idea which Chinese drugs were effective in treating malaria.[17]

At least two crucial points emerge from this conversation. First of all, the terrible risk that advocates of Chinese medicine had been warning the Nationalist government about for decades had finally become a real threat. To link Chinese medicine to some important problems worthy of the state's attention, they had predicted that the annihilation of Chinese medicine would put the whole nation in grave danger if certain Western pharmaceuticals came to be in short supply.[18] No matter how far-fetched it might have sounded when they had first raised this argument, this was exactly the situation in China at the end of the 1930s. At that time, the Nationalist central government had retreated to the southwestern provinces of China, an area with various endemic strains of malaria, and many government and military personnel were falling victim to the disease. Meanwhile, the Dutch East Indies had been lost to the Japanese, cutting the Allied forces off from 90 percent of the world's supply of quinine.[19] To cope with this health crisis, a group of biomedical scientists had been searching among Chinese drugs for substitutes for quinine even before Chen Guofu initiated the research on *changshan*.[20] In this context, it is no wonder that Chen Guofu became interested in the problem of malaria. If he succeeded in finding a "drug with specific efficacy" among traditional Chinese drugs, his success would demonstrate that, contrary to the commonly accepted wisdom, Chinese medicine was able to contribute to the solution of "national medical problems." Nevertheless, since Chen Guofu set out specifically to look for a substitute for quinine, his project could result in nothing but the translation of Chinese drugs into another therapeutic tool in the biomedical arsenal of scientific weaponry.

Second, although Dr. Cheng Peizhen claimed his ignorance as the reason why he did not take advantage of Chinese drugs, lack of information was by no means the real problem. Chinese prescriptions for the treatment of malaria were easily accessible to anyone who bothered to look for them. In the opening summary of a brief pamphlet about Chinese drugs, for example, the author Ding Fubao recommended an antimalaria prescription that included *changshan* among its ingredients, emphasizing that it was marvelously effective and inexpensive.[21] In one of the most commonly consulted English studies of Chinese materia medica, F. Porter Smith also recognized that "there is no form of the malarial fever for which *changshan* is not recommended [by Chinese materia medica]."[22] Furthermore, when the *Chinese Medical Journal*, the leading biomedical journal in China to this day, published a special issue on malaria in 1932, it included a historical article called "A Study of Malaria in China." Its author, Li Tao, a historian of Chinese medicine, pointed out that a certain prescription of Chinese drugs (including *changshan*) was one of four major traditional methods for treating malaria. Nevertheless, Li Tao was quick to add, "no one is sure about the curative efficacy of this prescription."[23] It was thus not difficult at all to find such bits of "information" about traditional Chinese drugs; the problem was to trust them and use them in the same way that they were trusted and used by traditional practitioners. Practitioners of Western medicine were simply unable to trust Chinese drugs unless they had successfully been "translated" into the socio-technical network of biomedicine.

Nothing illustrates my point more clearly than the following episode. Many years after the antimalarial efficacy of *changshan* had been validated by scientists, Xu Zhifang, a chemist at the Chinese Academy of Sciences (Academia Sinica), recalled how he had cured his own malaria with "malaria-blocking pills" (*jienüewan*) that included *changshan* as their key component. After taking the pills, which he had prepared himself according to a prescription in the *Systematic Materia Medica (Bencao Gangmu*, completed in 1590), Xu Zhifang recovered completely. Nevertheless, he wrote, because "there had not been any physiological and pharmacological experiments [on the efficacy of *changshan*], I did not dare to announce my results to the public. Moreover, since I could not figure out the chemical composition of the pill, I could not trust it myself."[24]

As revealed in this case, Xu was not lacking information. He not only knew of the reputed curative efficacy of the "malaria-blocking pill" but went so far as to conduct an experiment on himself that confirmed its efficacy. Unfortunately, the success of his experiment meant nothing to the biomedical community. As long as he could not share it with his colleagues and would not publish a scientific paper on his success, it remained his personal experience. Most importantly, as Xu's confession made clear, before he dared to announce his "personal experience" to the public, the drug had to first go through a series of scientific trials—chemical, physiological, and pharmacological. That is, in order to be recognized publicly as valid treatments, Chinese drugs had to be translated materially into substances that were identifiable in the sociotechnical network of biomedicine. Biomedical practitioners (including Cheng Peizhen and Xu Zhifang) were unable or unwilling to utilize Chinese drugs not because they lacked information about them, but because as a group they would not recognize the efficacy of such "foreign" substances unless they had first been translated into constituents of their biomedical network.

While Xu did not dare to announce his positive experience right away, he began producing the pills for sale four years later, when he found out that so many poor villagers were suffering from malaria and could afford neither Western nor available Chinese medications. Within eight months, Xu sold more than a hundred thousand packages of his "malaria-blocking pills." One might find Xu inconsistent in his actions on these two occasions, but he faced a difficult situation. As a chemist, he could not confirm the efficacy of *changshan* without first translating it into familiar chemical components. As long as Chinese drugs were foreign to the community of biomedical practitioners, his claim of efficacy would not have been regarded by his colleagues as anything more than the claims made by traditional practitioners. Theoretically speaking, Xu should have devoted himself first to the necessary research before producing the pills for sale. The practical problem was that it would have taken a lot of resources and hard work to get *changshan* translated into biomedically recognizable components, not to mention validating its efficacy. Because of these professional and practical constraints, Xu ended up employing *changshan* in its traditional network—compounding the "malaria-blocking pills" in accordance with the formula from the *Systematic Materia Medica* and then selling them to Chinese patients.

Very similar constraints prevented Dr. Cheng Peizhen from recognizing the efficacy of Chinese drugs. His feigned ignorance concerning Chinese antimalaria drugs was far from accidental; his professional membership required him to claim ignorance in situations where the average Chinese person would know what to do, or at least whom to turn to. No matter how many "personal experiences" had confirmed the efficacy of this prescription, Cheng Peizhen was therefore simply unable to openly recognize its efficacy. In other words, before *changshan* had been assimilated into the socio-technical network of Western medicine, any clinical success with it would still have been categorized as nothing but personal experience in need of subsequent experimental verification.

Curing Mrs. Chu

Let us return to Chen Guofu's narrative history of the biomedical discovery of *changshan*: After receiving a copy of the prescription that the school guard had been circulating, Chen Guofu decided to test this prescription on Mrs. Chu, a friend who was suffering from malaria. When she did in fact recover from the disease after taking this prescription, "Director Chen [Guofu] took ease and developed a strong interest in this prescription. A grant project emerged in Director Chen's mind."[25] Chen Guofu excitedly boasted about his "experiment" on Mrs. Chu, but his experiment did not matter as much as he thought. Ultimately, how could Chen Guofu's experiment have been more convincing or trustworthy than that of either the chemist Xu Zhifang or the renowned traditional practitioner Zhang Xichun, who claimed to have cured his own malaria with *changshan* in 1917?[26] If Chen Guofu had not been in a position to promote the prescription in this school clinic, his cure of Mrs. Chu would have just added another anecdote with a highly questionable scientific status.

This is one of the few commonalities between the two successful research projects on *mahuang* and *changshan*: When biomedical practitioners first selected these Chinese drugs for investigation, they were influenced by people with whom they had personal relationships. Chen Guofu was Dr. Cheng Peizhen's superior at the college, and K. K. Chen trusted his uncle's recommendation.[27] In both cases, scientists' preliminary trust in Chinese drugs was based on their trust in their personal acquaintances. Second, as Carl Schmidt put it for the case of ephedrine, the successful outcomes in both studies appeared to be inseparable from "a series of coincidences."[28] K. K. Chen had happened to meet his uncle at a family reunion in Shanghai, and the school guard had happened to find an effective prescription in a local newspaper and be daring enough to spread it around. Both of these successful research projects thus appear to have been helped by extraneous factors and coincidences. While I refer to these factors as extraneous, that does not mean that they necessarily distorted the normal course of scientific research. It simply means that they were systematically positioned outside the network of biomedicine. In light of this understanding, it is no wonder that these two cases were helped by extraneous factors and coincidences; the information flow and social exchange between the outside and inside of a network is unavoidably irregular and uncertain.[29]

> Three days after [the successful conclusion of the experiment on Ms. Chu], Director Chen [Guofu] checked out some books to study the characteristics of the seven medicinal components in the prescription. Then he invited Dr. Cheng Peizhen to his office, where he told him about the prescription and his successful experiment. Director Chen then asked Dr. Cheng to buy these drugs and initiate systematic experimentation on them. Later, Director Chen visited Zhang Jianzhai [a famous practitioner of Chinese medicine] to discuss the properties of this prescription and the procedure for making the remedy. Director Chen learned a lot through this discussion.[30]

Chen Guofu never spelled out what he had learned from Zhang Jianzhai, and neither did any of the biomedical doctors or scientists involved in his research. In fact, none of the other actors gave any credit at all to either Chen Guofu or Zhang Jianzhai; instead, they trace the prescription back to the classical materia medica literature, for which no one but the ancient authors could claim credit. They had good reasons to dismiss the role played by Chen Guofu: he had used Mrs. Chu as a "guinea pig" and facilitated the introduction of *changshan* by means of his position, not to mention that he lacked any medical qualifications whatsoever. Except for Chen's own account, his experiment on Mrs. Chu was systematically omitted from every scientific record of *changshan* research because, from the biomedical perspective, the scientific value of Chen's "personal experiment" amounted to nothing.

Stage Two: Re-networking Changshan

Dr. Cheng Peizhen immediately organized a research team to investigate the effectiveness of the prescription he had received from Chen Guofu.[31] The researchers first recruited a group of fifty students who were reportedly suffering from malaria. Once the presence of malaria parasites in their blood had been confirmed, the students were ordered to take the medication. On the night when the results revealed that the remedy was indeed capable of killing the malaria parasites, "Dr. Cheng became extremely excited and ran through pouring rain to report this result to

Director Chen."[32] Cheng Peizhen thus contained his excitement (or suspicion) about Chen Guofu's success until he had confirmed the efficacy of the prescription himself. Cheng's initial reservation further supports the possibility that he was willing to conduct a clinical experiment on this prescription mainly because the suggestion came from his superior, Chen Guofu.

In terms of assimilating the prescription into the socio-technical network of biomedicine, Cheng Peizhen's clinical experiment was no more than the first step. Nevertheless, it was a step of critical importance. Instead of relying on patients' reports of their physical situation, Dr. Cheng measured the level of malaria-causing plasmodia in the patients' blood, taking the disappearance of those parasites as the criterion to judge the efficacy of the prescription. Although biomedical practitioners at this point still had no knowledge about the chemical composition of the seven drugs that made up this prescription, they at least knew the crucial information: just like quinine, these drugs in combination were able to kill the malaria-causing parasites.

After Cheng Peizhen succeeded in this first clinical experiment, he next discovered that *changshan*, when used alone, was as effective against malaria as when used in combination with the other substances. Chen Guofu now reported his discovery directly to Chiang Kai-shek. According to Chen, Chiang immediately appropriated special funds to establish a Research Laboratory for National Drugs at the Central School of Politics and ordered the National Health Administration to support Chen's project. With Chiang's support, Chen Guofu set up a relatively well-equipped research laboratory and recruited a team of scientists from various fields to join his project.

The results of their research were collected in the *Preliminary Research Report on Changshan as a Malaria Treatment* (*Changshan zhinüe chubu yanjiu baogao*), published in 1944 (cited hereafter as the *Report*). Opening with Chen Guofu's preface, the *Report* was subdivided into four disciplines: (1) pharmacognosy, by Guan Guangdi; (2) chemistry, by Jiang Daqu; (3) pharmacology, by Liu Chengru; and (4) clinical research, by Chen Fangzhi and his colleagues. As the project director, Cheng Xueming, admitted in his introduction, the common departure point for all four research groups was the fact that *changshan* had been clinically confirmed to have an antimalarial effect.[33] Taking Guan Guangdi's pharmacognostic study as an example, I demonstrate below how Dr. Cheng Peizhen's successful clinical experiment crucially shaped the course of *changshan* research.

Identifying Changshan

Before Guan Guangdi conducted modern pharmacognostic research on *changshan*, he had to solve one tricky problem: What exactly was this substance called *changshan*? For Guan Guangdi, the problem of identifying *changshan* consisted of two subproblems. First of all, what was the historical identity of the herbal drug *changshan* in the tradition of Chinese materia medica? Second, in terms of modern pharmacognosy, which plant was the source of the herbal drug *changshan* described in the Chinese materia medica?

By comparing descriptions of *changshan* in the Chinese materia medica literature, Guan Guangdi differentiated three different subtypes: *jigu changshan* (*Alstonia yunnanensis*), *haizhou changshan* (*Clerodendrum trichotomum*) and *tu changshan* (*Hydrangea strigosa*). Then, citing remarks on *changshan* by the famous nineteenth-century scholar of Chinese materia medica Wu Qijun (1789–1847), Guan Guangdi concluded that "at that time there were so many types of *changshan* that Wu Qijun could not differentiate the genuine from the fake."[34] Just like Wu Qijun almost a century before him, Guan Guangdi was thus unable to conclude positively which among these three possibilities was the genuine *changshan*. Throughout his article in the *Report*, Guan therefore referred to his tentative conclusion, namely, that the genuine *changshan* was identical with the subcategory *jigu changshan*, as a "hypothesis."[35]

For the purpose of Guan's research, the historical materia medica literature provided only partial information on identifying *changshan*; the clinically confirmed antimalarial effect of *changshan* turned out to be the most useful guide (or constraint) in determining its identity. For example, Guan Guangdi eliminated the possibility that *tu changshan* could be identified as genuine *changshan*, because it was reported to have no efficacy in curing malaria.[36] This decision, while reasonable, directly contradicted the conventional procedure for conducting research on Chinese drugs as advocated by the program Scientific Research on Nationally Produced Drugs. To conduct chemical and pharmacological experiments on Chinese drugs, biomedical doctors had initially demanded "a complete identification of mineral drugs, animal drugs, and herbal drugs."[37] In this case, the logic of actual practice went in the opposite direction: the clinically confirmed antimalarial effect of *changshan* had played the pivotal role in determining the most likely historical identity of *changshan*. I say "most likely" not only because Guan Guangdi self-consciously viewed his identification of *changshan* with *jigu changshan* as merely hypothetical, but also because he clearly did not care very much whether this drug did in fact correspond to the historical reality of *changshan*—or whether a unique and consistent historical identity of *changshan* existed at all.[38] The ancient practitioners could get confused, just as Wu Qijun had admitted. Among the many types of *changshan* recorded in the literature, Guan's project aimed at identifying a specific variety of *changshan*, the variety that could be judged as "effective" by current scientific standards.[39]

Because of this biomedical focus on efficacy, the already confirmed efficacy of a specific type of *changshan* again played a crucial role in determining the botanical identity of *changshan*. Assuming that *jigu changshan* was in fact identical with the *changshan* described in the historical materia medica literature, Guan Guangdi yet again found two possible plant sources even for this subcategory of *jigu changshan*: *Orixa japonica* Thunb. and *Dichroa febrifuga* Lour.[40] The former was produced mostly in Japan and had been thoroughly studied by Japanese scientists in terms of chemistry, pharmacology, and pharmacognosy. According to their investigation, *Orixa japonica* showed a clear antipyretic effect but had no efficacy in the treatment of malaria.[41] Since the Japanese scientists had generally accepted the identification of *Orixa japonica* as the botanical source of *changshan*, they were suspicious about the antimalarial efficacy of *changshan* as recorded in the Chinese materia medica.[42]

The truth was, however, that the Japanese variety of *changshan* was completely different from the Chinese variety. Although Japanese scholars had traced *changshan* back to *Orixa japonica* in 1827, the root of this plant was substantially different in its morphology from the root of *changshan* as it had been described in Chinese materia medica literature.[43] Because *changshan* was not native to Japan and had not been cultivated there in premodern times, Japanese doctors had historically used a variety of native substitutes, including a plant called *kokusaki* (*Orixa japonica*).[44] Due to the leading position of Japanese scholars in the scientific research on Chinese materia medica, Chinese scholars had up to now just copied the conclusion drawn by Japanese scholars, namely, that *Orixa japonica* was the unequivocal botanical source of the Chinese herb *changshan*.[45] Therefore, if Chen Guofu had been determined to look for *Orixa japonica*, he would not have been able to buy the "right" drug from the local drugstore. Now that this local *changshan* had been proven to be effective in treating malaria, Guan Guangdi felt confident enough to reject the noneffective *Orixa japonica* as the botanical source for *changshan*, and instead identified *Dichroa febrifuga* as the plant source for *changshan*.

The ultimate evidence was presented when Guan juxtaposed a sliced specimen of the root of *Dichroa febrifuga* (grown in the laboratory) and one of *jigu changshan* that had been bought from the local drug market in Chongqing under the microscope and demonstrated their identity. *Jigu changshan* could now serve as the ultimate referent because it was the same as the local drug that Chen Guofu's and Cheng Peizhen's clinical experiments had demonstrated to be effective against malaria. In principle, the proper research protocol for determining the efficacy of *changshan* would have required its unequivocal botanical identification prior to any clinical experimentation. In practice, however, it would have been extremely difficult to sort out the botanical identity of *changshan* if scientists had not known beforehand that the specific substance was effective in treating malaria. I am not suggesting that there was no way out of this paradox; my point is merely that Chen Guofu and his colleagues crucially shortened the time and effort needed to escape this cycle by first conducting a clinical experiment with human subjects that verified the historically transmitted curative efficacy of *changshan*.

After clinical experiments on human subjects had thus proven the medical efficacy of *changshan*, the drug was sent for study to many foreign places. Joseph Needham (1900–1995), at that time the director of the Sino-British Science Co-operation Office and later the leader of the monumental research project (the results of which were published as *Science and Civilization in China*), sent the same local drug to K. K. Chen's research group at Eli Lilly & Company and to an American group for animal testing.[46] When the US experiments confirmed the antimalarial efficacy of *changshan*, samples of the same drug were sent to the University of London for pharmacognostic research.[47] Another Chinese governmental laboratory also requested an aqueous extract of *changshan* from the Research Institute for Chinese Drugs with Specific Efficacy.[48] Due to the clinical success achieved by Chen Guofu and his colleagues, the local variety of *changshan* and its extract became the standard object of further research. In the end, however, Chen Guofu did not wait for the scientists to isolate the active agents or to synthesize the chemical component by artificial means. While scientists were busy performing such experiments and writing papers (efforts, we may add, that never led to the development of a useful synthetic substitute for *changshan* because of the serious side-effects),[49] Chen Guofu devoted himself to the mass cultivation of *changshan* on Jingfo Mountain. Looking at the Jingfo Mountain farm, Chen Guofu said that "this green hill and the laboratory, pharmaceutical factory, hospital, and malaria had jointly formed an inseparable ring."[50] In this way, an international network on *changshan* research and production came to be established; the drug produced in Chongqing henceforth began to be circulated all over the network.

Two Research Protocols: 1–2–3–4–5 versus 5–4–3–2–1

Although Chen Guofu's strategy of "clinical experiment first" had substantially accelerated the results of scientific research on *changshan*, his approach was severely criticized by biomedical practitioners as "5–4–3–2–1," that is, research in reverse order.[51] Being neither a scientist nor a physician, Chen Guofu did not care too much about this criticism. Nevertheless, the scientists collaborating with him on this project responded to this criticism with great concern; they recognized that their approach was generally perceived to be violating an ethical consensus in the scientific community.

To elaborate on the larger context of this criticism, let me first explain the meaning of the expression "5–4–3–2–1."[52] In his 1952 paper, "Now Is the Time to Study Chinese Drugs!" Yu Yan juxtaposed two competing protocols for conducting scientific research on Chinese drugs. These two protocols—the "conventional protocol" and the "reverse-order protocol"—were differentiated mainly by the order of their research procedures, symbolized by the expressions "1–2–3–4–5" versus "5–4–3–2–1." The order of the "conventional protocol," the standard research procedure prescribed by biomedical researchers, was chemical analysis—animal experiment—clinical application—artificial synthesis—structural modification.[53] As applied to research on nationally produced drugs, this meant that scientists first subjected the Chinese drugs in question to a chemical analysis to find the active component that was supposed to be responsible for the action of the drug. Then they conducted pharmacological experiments with the isolated chemical, injecting a solution of the chemical into laboratory animals and observing the effect on blood pressure, respiration, heartbeat, and other vital signs. It was only after the proven success of these animal experiments that they began clinical experiments, testing the extracted chemical on human subjects. Once the clinical efficacy was confirmed on humans, chemists would try to synthesize the chemical by artificial means, forming the basis for production on an industrial scale. Finally, in order to get rid of unwanted side effects or to improve curative efficacy, scientists would try to modify the chemical structure of the extracted active chemical. Predictably, chemical research as the first step in the conventional protocol disproportionally dominated the study of Chinese drugs in the 1930s and 1940s.[54]

What Yu Yan called "conventional protocol" roughly captures the historical procedures that led to the discovery of ephedrine from the Chinese drug *mahuang*.[55] The pioneering research on *mahuang* had been carried out entirely by Japanese scholars in the late nineteenth century. As the first step, G. Yamanashi had isolated the active principle in *mahuang* in 1885; in 1887, the pure alkaloid had been obtained by N. Nagai and Y. Hori, confirmed in 1888 by E. Merck. A physiological investigation by K. Miura not only demonstrated the mydriatic action of the drug but also disclosed the toxic effect of large doses on the circulation. Consequently, ephedrine was initially used only as a new mydriatic drug for the purpose of dilating the pupil and was regarded as a very toxic substance. For many years thereafter, research on ephedrine was limited to investigating its chemistry—analyzing its chemical composition and trying to synthesize it artificially. This situation continued until 1924, when K. K. Chen and C. F. Schmidt discovered the clinical use of ephedrine through animal experiments.

K. K. Chen and C. F. Schmidt demonstrated that ephedrine could be used as a substitute for epinephrine, a hormone that increases heart rate, constricts blood vessels, and dilates air passages. Their discovery justified many traditional uses of the original Chinese drug *mahuang*, including as a circulatory stimulant, diaphoretic, antipyretic, and sedative for coughs.[56] In light of the historical procedures that had led to the discovery of ephedrine, some traditional practitioners later argued, scientists should learn a lesson and take the traditional "experience" of Chinese drugs as their guide.[57] *Mahuang* had, after all, been a key ingredient in many well-respected prescriptions described in classics such as the *Treatise on Cold Damage*; those who really valued Chinese medicine would never have been satisfied with using *mahuang* simply as a mydriatic.

Strictly speaking, even the research on *mahuang* had not followed the "conventional protocol." As discussed in chapter 4, it was Nagai Nagayoshi, the widely admired founder of Japanese pharmacology and the first president of the Pharmacological Society of Japan, who had isolated the pure alkaloid of *mahuang* and given it the name of ephedrine in 1887. According to Chen, however, he and Schmidt had had absolutely no knowledge of this fact when they began their research.[58] Having heard from Schmidt that his investigation on Chinese drugs had been disappointing, Chen consulted his uncle and received from him the recommendation to study *mahuang*, a Chinese drug that was not even on the list that B. E. Read had given to Carl Schmidt. Following this recommendation, Chen initiated the study of *mahuang* on his own accord when he arrived at PUMC in August 1923,[59] purchasing a small sample from a local pharmacy. "We made a decoction of the drug and injected a portion of it into a narcotized dog. . . . At once, we observed a prolonged rise in the dog's blood pressure. This result of our first experiment aroused our enthusiasm and stimulated us to carry out an exhaustive investigation," Chen recalled.[60] Completely unaware of the work of his Japanese predecessors, Chen and Schmidt went through great trouble to isolate the alkaloid and to figure out its chemical structure. They were on the verge of giving a new name to this alkaloid when they found the published results of the Japanese research in their small library in Beijing.[61] I would like to point out here that this ignorance reveals the third common point shared by the studies on *mahuang* and *changshan*; neither research project built upon the research on its subject that had been meticulously conducted previously by Japanese scientists. Once they had become aware of this Japanese research history, K. K. Chen realized an ironic fact: long before biomedicine recognized the need for a substitute for epinephrine, such a substitute had already been made available by Japanese scientists in a pure form as ephedrine.[62] From the perspective of "re-discovering" ephedrine, a perspective that K. K. Chen developed to recognize the Japanese contributions, it took almost half a century to turn the isolated alkaloid into a clinically useful drug.

Reverse-Order Protocol: 5–4–3–2–1

The "reverse-order protocol," referred to by Chen Guofu as "5–4–3–2–1" and by Yu Yan as "acting in the reverse order" (*dao-xing-ni-shi*), proceeded in the following sequence: clinical experiment—animal experiment—chemical analysis—retesting and artificial synthesis—structural modification. As this sequence indicates, in comparison with the conventional protocol (1–2–3–4–5), the "reverse-order protocol" should have been characterized more accurately as 3–2–1–4–5, rather than 5–4–3–2–1. Although the conventional protocol and the reverse-order protocol consisted of two rather different sequences of research steps, the key difference between them lay in just one question: whether or not to place clinical experimentation on human subjects as the first step. The function and position of clinical trials in scientific research was crucial in the 1930s because it involved a highly charged ethical issue. Practitioners of Western medicine at that time were forcefully attacking traditional practitioners for treating their patients as "guinea pigs" by feeding them life-threatening drugs.[63] To emphasize the danger of taking Chinese drugs, Yu Yan repeatedly quoted the famous saying "Medicine is learned by sacrificing human lives" (*xue-yi-fei-ren*).[64] Rather than simply being an alternative research strategy, Chen Guofu's "clinical experiment first" approach therefore invoked serious ethical concerns.

As a matter of fact, the scientists involved in Chen Guofu's project appeared quite uncomfortable with conducting direct clinical experiments on human subjects. Chen Fangzhi, the scientist responsible for clinical experimentation, specifically commented on this issue: "According to the standard procedures for developing new drugs, one should not skip animal testing before proceeding directly to clinical experimentation on humans."[65] Nevertheless, Chen Fangzhi justified the nonstandard progression of research steps in the case of *changshan* by arguing the following: "Our *changshan* is actually not a new drug. The practitioners of the Old Medicine have used it for thousands of years. . . . Therefore, even though we did apply it directly to the human body, our actions do not go against the ethical standards. Consequently, it does not matter that we skipped animal experimentation."[66]

Chen Fangzhi's rationale presupposed that the researchers had trusted the nontoxicity and efficacy of *changshan* before they started conducting clinical trials. Otherwise, how could he have claimed that *changshan* was "not a new drug"? Chen Fangzhi did concede this point earlier in his paper: "In our entire research on *changshan*, the clinical experiment was probably the least rewarding one because the antimalarial efficacy of *changshan* has been well known in our country for more than one thousand years. . . . Everyone agrees that it is effective; no one disagrees on this fact. Therefore, our report of efficacy is no more than a duplication of the established opinion and, consequently, by no means as valuable as the reports from the chemical, pharmacological, and pharmacognostic research."[67]

Chen Fangzhi's claim here that no one disagreed about "the antimalarial efficacy of *changshan*" is puzzling. Before Chen Guofu had handed the antimalaria prescription to Dr. Cheng Peizhen, Dr. Cheng had claimed that he had no idea which Chinese drug was effective for treating malaria. In addition, the chemist Xu Zhifang had not dared to announce the curative efficacy of *changshan*, even with his positive personal experience. With Chen Fangzhi's assertion in the above paragraph, all of a sudden it seemed that everybody had always recognized the antimalarial efficacy of *changshan* and, by implication, that even the biomedical community no longer viewed and treated it as a "new drug."

By switching the status of *changshan* from that of a "new drug" to that of a well-known one, biomedical researchers had done themselves two big favors. First, as Chen Fangzhi argued, "even though we did apply it directly to the human body, our actions did not go against the ethical standard." Second, since "the antimalarial efficacy of *changshan* has been well known in our country for more than one thousand years," it was no longer necessary to carry out clinical experimentation to verify its curative efficacy. Given the fact that all the clinical "reports of its efficacy are no more than a duplication of the established opinion," Chen Fangzhi and his colleagues saw no heuristic value at all in the clinical experiments that had been carried out by Chen Guofu, Cheng Peizhen, and themselves. To justify the reverse-order protocol of the *changshan* research project, Cheng decided to accord *changshan* the status of a well-established drug with a long safety record, instead of that of a "new drug." Nevertheless, when his colleagues later published the research results in the journal *Nature*, the title of the article was "A New Anti-Malarial Drug."[68]

Research Protocol as Political Strategy

The above-mentioned arguments against testing Chinese drugs on human subjects were shaped as much by the political struggle surrounding Chinese medicine as by the concern over research ethics. Practitioners of Western medicine were highly critical of the reverse-order protocol, in part because they viewed it as having dangerous repercussions for their political struggle against Chinese medicine. Most likely, they saw its greatest danger in the fact that it echoed the discourse on "experience" advocated by the practitioners of Chinese medicine, especially their idea of "experience with the human body" (*renti jingyan*).

After the 1929 confrontation, the concept of "experience" arose, along with Chinese drugs, as the last bulwark of Chinese medicine in its struggle against Western medicine. The concept of "experience with the human body," an idea borrowed from Japanese Kampo medicine practitioner Yumoto Kyushin, became particularly important when traditional practitioners began to shy away from ontological debates with biomedical practitioners. Instead of engaging in debates over the ontological status of specific aspects of Chinese medicine, practitioners of Chinese medicine consequently strove to justify its value on the practical grounds that certain therapeutic methods were indeed effective. As a result, they argued, the key to scientizing Chinese medicine lay not in conceptual reasoning but in using the tool of clinical trials to validate the therapeutic efficacy of their "experience with the human body." For the traditional practitioners who valorized "experience with the human body" as the essential strength of Chinese medicine, the "clinical experiment first" approach of the reverse-order protocol was not only a more appropriate protocol for conducting research on Chinese drugs, but also a political strategy for further developing Chinese medicine.

For example, in a comprehensive article published posthumously in 1951, Chen Guofu concluded his thoughts on the issue of Chinese medicine with a series of questions: "When Western scientists investigate drugs, as a norm they conduct animal experimentation first and then test the drugs on human bodies. In contrast, scholars in China have tested drugs directly on 'human bodies' for thousands of years. For what reasons do contemporary medical scientists not trust these tests? For what reasons do these precious experiences that have been accumulated over thousands of years count as nothing?"[69]

More revealingly, in his article entitled "Pharmaceutical Experiments Should Not Ignore *Jingyan*," traditional practitioner Tan Cizhong (1897–1955) argued, "No matter what kind of drug is concerned, if one were to first verify the principle [of its action] scientifically and then evaluate its efficacy clinically, one would find this approach logical in principle but difficult in practice. On the other hand, if one were to first evaluate its curative efficacy clinically and then verify its principle scientifically, one would find this approach illogical in principle but fruitful in practice."[70]

At least in the case of *changshan*, Tan Cizhong was right to assert that the "clinical experiment first" approach was "illogical in principle but fruitful in practice." As mentioned before, the project director, Chang Xueming, openly admitted that "all the other studies [on *changshan*] began with clinical experiments that verified the curative efficacy of *changshan*."[71] Besides, it was precisely because Chen Guofu adopted the "reverse-order" approach that his team had gained the subversive power of demonstrating that Japanese scholars had been wrong for decades in identifying *Orixa japonica* as the botanical source of *changshan*. Without first confirming the curative efficacy of *changshan*, scientists would have encountered much greater difficulties both in sorting out its pharmacognostic identification and in analyzing its chemical composition.

In addition to appreciating its heuristic value, practitioners of Chinese medicine preferred the reverse-order protocol to the conventional protocol for two important political reasons. To begin with, if clinical experimentation was generally accepted as the first step in studying Chinese drugs, the autonomy of Chinese medicine would to a certain degree be automatically maintained, since clinical experimentation would have to be conducted more or less in the traditional fashion. On the other hand, if the research procedure for Chinese drugs were to follow the conventional protocol, after the first step of chemical analysis, the extracted active component would already be a totally foreign substance, out of the reach of traditional practitioners.

Second, the conventional protocol was built on a reductionist framework, presupposing that the efficacy of Chinese drugs could be traced to one active chemical. As long as researchers were operating within this protocol, they could not prove this reductionist framework to be wrong or counterproductive. If the extracted principal component of a Chinese drug turned out not to show the curative efficacy claimed by tradition, this failure would simply be taken as a refutation of traditional pharmaceutical knowledge, just as had been the case in the research on *mahuang* before K. K. Chen's involvement in 1924. On the contrary, though, in studies where the curative efficacy of a substance had been clinically verified from the outset, the failure of chemists to find the active component was bound to call into question the reductionist framework. While scientists had succeeded in isolating the alkaloid that was capable of killing malaria parasites in the case of *changshan*, the reductionist program had still failed to bring about a clinically useful drug. Because of its side effects of vomiting and nausea, the isolated alkaloid never managed to replace the extract of crude *changshan* as an effective treatment for malaria.[72]

Practitioners of Western medicine refused to recognize the fact that, at least in some cases, the reverse-order protocol was "illogical in principle but fruitful in practice," because research productivity was far from their only concern. Just like the practitioners of Chinese medicine, the biomedical practitioners were keenly aware that these two research protocols had been functioning as political strategies. The apparently scholarly debates over research protocols had, and would continue to have, a profound influence on the relationship between the two styles of medicine. At the time when practitioners of Chinese medicine compared their "experience with the human body" to the clinical trials of biomedicine, thereby blurring the boundary between the two styles of medicine, practitioners of Western medicine in response felt the ever more urgent need to police the boundaries of their privileged network. As cogently stated by Zhao Yuhuang (1883–1960), the first and probably most important Chinese pharmacognosist during the Republican period, "If we have no idea about the chemical composition of a traditional drug and still go ahead with clinical testing for its efficacy [on human subjects], we are just like the practitioners of the Old Medicine who treat human beings as animals for experimentation. In that case, there will be no hope for the scientization of Chinese drugs."[73] To discourage the controversial practice of conducting clinical trials of Chinese drugs directly on human subjects, Zhao Yuhuang excluded such trials from the central tasks of the Research Institute for Chinese Drugs to be established at the Academia Sinica.[74]

Because research protocols functioned as political strategies in this medical struggle, it is illuminating to compare the methods employed by practitioners of Western medicine to guard the boundaries of their socio-technical network to the strategies employed to maintain the exclusivity of a privileged club. In such clubs, each member is instituted as a custodian of the club's boundaries, because with each new entry, the definition of the criterion of entry is at stake. In the case of Chinese medicine, however, practitioners of Western medicine had to guard not only against traditional practitioners, but also against the material objects used by them. Therefore, they had to renounce, or at least withhold, the trust and knowledge historically associated with thousands of well-known Chinese drugs and medicinal formulas. Just as strangers are often associated with danger in the popular mind, Chinese drugs, after having been conceptualized as "new drugs," were consequently eyed with unprecedented suspicion. As a result of this boundary-policing, a drug such as *changshan*, which "everyone [had] agreed to be effective for more than a thousand years," became a simple plant-based substance about which it could be said that "no one [is] sure about its antimalarial efficacy."[75] By designating traditional Chinese drugs as "new" and therefore potentially dangerous, biomedical practitioners furthermore made it unethical to conduct clinical experiments on them. As well-known Chinese drugs began to be perceived as untrustworthy and life-threatening, it became difficult to promote the productivity of the reverse-order protocol, a protocol that had been stigmatized as nothing more than "learning medicine by sacrificing human lives."

In the end, the contest between these two research protocols on Chinese drugs helped to shape the relationship between the two styles of medicine for years to come; at the same time, however, it boiled down to a deceptively simple question: Was *changshan* a new drug or not?

Conclusion: The Politics of Knowledge and the Regime of Value

What exactly was this Chinese medicinal called *changshan*? Was it, as many critics of Chinese medicine suggested in contempt, simply "grass roots and tree bark,"[76] that is, a raw material from nature? Was it then also true that, as such, it merely indicated a primitive developmental stage in Chinese medicine, so primitive that it was destined to be nothing more than an object of scientific research? After having traced the anomalous case of *changshan* research through such a long journey, at least one thing has become clear: in terms of research design, it would have been far from the most cost-effective strategy, if not outright counterproductive, to follow the conventional protocol (1–2–3–4–5), that is, to treat Chinese drugs like *changshan* as nothing but "grass roots and tree bark." It was precisely because the researchers had followed the reverse-order protocol (5–4–3–2–1) that they had been able to crucially reduce the time and labor needed to achieve their goal of validating the antimalarial efficacy of *changshan*. Even though Chen Guofu had been forced to accept the demeaning characterization of his research progression as 5–4–3–2–1, his approach was not only justified by the successful results, but ultimately turns out to be consistent with the current policy of the World Health Organization on the evaluation of traditional medicinals.[77] Once we situate the case of *changshan* research in the context of two competing research protocols, the story documented here does not just concern a singular case but offers insights into the structural limitations of the conventional protocol and the associated estimation of the value of Chinese medicine.

Contrary to the valuation on which the conventional protocol was based, *changshan* in particular and Chinese drugs in general, are not categorically different from other objects created by modern science. Instead of being a raw material from nature, *changshan* was articulated and reproduced in a socio-technical network that involved at least the historical materia medica literature, local drugstores and regional markets, reputed prescriptions, disease classifications, traditional practitioners, and their clinical experience. When the average Chinese person used the term *changshan*, bought the authentic and thus effective herb from a local pharmacy, or used "malaria-blocking pills" to treat malaria patients, this did not just happen because *changshan* existed "out there" as a physical substance in nature. Rather, it was because traditional physicians, herbalists, and other practitioners had succeeded in bringing *changshan* "right here," integrating it into their heterogeneous networks by accumulating related knowledge, skills, and material objects. In this practical sense, *changshan* had been transformed over thousands of years into a practice-based, fabricated object.[78]

If we look at *changshan* as a practice-based, fabricated object, the so-called "process of discovery" actually becomes a multilayered "re-networking" process through which biomedical practitioners and scientists decoupled *changshan* from the traditional network and assimilated it into their own socio-technical network. In this re-networking process, the material existence as well as the conceptual characterization of *changshan* underwent dramatic changes, being stabilized interactively within different conceptual and material settings at different stages.[79] And while *changshan* was transformed materially and conceptually, the historical actors associated with it also changed accordingly. The key point of my argument here is that since the conceptual identity and material existence of *changshan* were constructed and reproduced within these heterogeneous networks of practice, it was highly problematic to appropriate *changshan* while simultaneously devaluing and discounting the knowledge, skills, people, institutions, and objects of the network in which *changshan* had acquired its identity and entered circulation. When researchers discounted all of these aspects to treat Chinese drugs as nothing but raw materials from nature, it became unnecessary, if not impossible, to conceive of research on Chinese drugs as a scientific cooperation between two styles of medicine. It is unfortunate that practitioners of Western medicine at this time were so preoccupied with guarding against the rise of "mongrel medicine" that they chose research protocols partly as a political strategy to regulate the relationship between the two styles of medicine. It is no wonder that their preferred research protocol—the so-called conventional protocol—ended up aiming at anything but a cooperative project between two different medical paradigms.

By way of tracing the history of re-networking *changshan*, I have intended to provide a critique of the conventional protocol of the program of Scientific Research on Nationally Produced Drugs. But in light of the fact that many practitioners of Western medicine considered this drug research program as the only acceptable approach to the scientization of Chinese medicine, my critique is actually directed at their entire project of Scientizing Chinese Medicine. My point here becomes easier to see if we interpret their project as a "translation," or re-networking, of certain elements of Chinese medicine into the language of biomedical sciences.

If we compare this translation into the language of science to the process of translating one language into another, we find that the project of Scientizing Chinese Medicine has the following unusual features: First, unlike regular translation, it presupposes a highly asymmetrical relationship between the target, namely, the socio-technical network of science and biomedicine, and the source (the socio-technical network of Chinese medicine).[80] To maintain this asymmetric relationship, rigorous boundary policing is required to control the flow between the two networks. Second, scientists by definition monopolize the means of this translation/re-networking. Third, this translation is strictly one-directional; the fundamental categories of the target language (science and biomedicine) must not be transformed at all in the translation process.

Finally, and most importantly, this translation is assumed to be a complete translation. By "complete translation," I mean to refer to the quasi-tautology that any information considered worthy of being translated at all is, in principle at least, successfully translated because these researchers evaluate their translation by the practical purposes defined by the target language of science. In any other nonscientific field, such as philosophy, literature, or Buddhism, for example, people do not evaluate their translations this way; therefore, they are keenly aware that much valuable information gets lost in translation and that their translations are bound to be incomplete, if not distorted. To return to the specific case of *changshan,* the practical purpose of this translation was to find a magic bullet among Chinese drugs to replace the antimalaria drug quinine. As scientists were preoccupied with isolating from *changshan* the one pure compound that was responsible for killing the malaria-causing parasites, they dismissed the other six drugs in the original formula as irrelevant. What got lost in this translation was the fact that the patients who took the original formula did not suffer from the side effect of vomiting, for example.[81]

The truth is that the authors of traditional Chinese materia medica had always been aware of the serious side effects of *changshan*, characterizing it as a "toxic herb" (*ducao*). In light of this understanding, it is no accident that the original formula of seven ingredients effectively alleviated this side effect of vomiting. If researchers had not been preoccupied with isolating one chemical compound from the seven Chinese drugs, they might have realized that knowledge about other valuable substances could be gained from this formula. To summarize this fourth feature, when scientists assume that a translation into scientific language is a complete translation, they become unconcerned with any other potentially valuable elements of Chinese medicine that are destroyed in the process of translation.

Troubled by the asymmetric nature of this translation project, most scientists found little incentive to devote themselves to such a risky and despised project unless special circumstances, such as a war, demanded such efforts.[82] As Yu Yan sarcastically remarked, "Those who are well-equipped and qualified to study Chinese drugs are not interested; those who are interested in studying Chinese drugs are not qualified."[83] Those who were daring enough to undertake such research were forced to follow a research protocol that was, as Tan Cizhong put it insightfully, "logical in principle but difficult [counter-productive] in practice."[84] Although the conventional protocol did not appear to be a productive one for the study of Chinese drugs, it nevertheless appealed to biomedical practitioners because it served the political function of confining the development of Chinese medicine, especially the rise of "mongrel medicine."

I am not suggesting that this research program was purely a tool of political domination disguised as a protocol for scientific research. Many of its advocates believed in their hearts that it was the best available protocol for scientific research. They did not find this research program counterproductive, because it was at once built upon and reinscribed a particular regime of values and risks, a regime that assumed many traditional uses of Chinese drugs to be ill-founded and even life-threatening. When the conventional protocol turned out to produce only meager fruits, this otherwise disappointing result simply reconfirmed the valuation on which it had been based. The conformity between low valuation and meager results allowed the advocates of this research program to overlook the fact that they were involved in a politics of knowledge against the rise of mongrel medicine. In light of this understanding, the historical struggle over Chinese medicine should never be equated with a politics of knowledge. More fundamentally, the struggle concerned a politics of valuation.

In a nutshell, the core of this much-debated project of Scientizing Chinese Medicine centered (and continues to center) on three key questions: How valuable is Chinese medicine? Where in Chinese medicine do these valuable elements reside? What are the most appropriate methods by which to realize and maximize the potential value of Chinese medicine, in order to conduct scientific research, standardize its practices and nomenclature, incorporate it into the emerging health-care system, and put it to productive use in response to the national concern over epidemic disease or in service to the health needs of China's rural population? These questions do not concern themselves narrowly with the truth or falsehood of Chinese medicine as judged against contemporary scientific knowledge, but with the forward-looking objective of realizing the potential value of Chinese medicine in the future. In other words, value is not an innate and fixed property of Chinese medicine. Instead, the value of Chinese drugs, and of Chinese medicine in general, is created in the above-mentioned process of investigation, standardization, and social appreciation. To transcend the constraints of the project of Scientizing Chinese Medicine, we should view the modern history of Chinese medicine as a historic endeavor in the creation of value in Chinese medicine.

10 State Medicine for Rural China, 1929–49

In the previous chapters, I have documented the process by which proponents of Chinese medicine strove to "scientize" Chinese medicine and thereby assimilate it into the emerging national system of medical administration. Instead of providing a stable historical background for this transformation of Chinese medicine, the relationship between Western medicine and the Nationalist government went through an equally profound transformation in the decades after the establishment of the Ministry of Health in 1928. This transformation culminated in the rise of a so-called State Medicine (*gongyi*, literally "public medicine," but translated as "state medicine" in English-language publications to emphasize its similarity to the system of British State Medicine) as the consensus for China's health-care policy in the 1940s. In sharp contrast to the government's lack of interest in health care in previous decades, the Nationalist government included in its first constitution, promulgated in 1947, a commitment to the policy of State Medicine, a health-care system in which all Chinese citizens were guaranteed equal and free access to health-care services. The administration of this system was controlled, staffed, and subsidized by the state. If we compare the promulgation of this policy with the government's response to the outbreak of the Manchurian plague, with which I began this book, we are bound to notice a remarkable change in the state's commitment to the health of its citizens. Moreover, while the Nationalist government never had the opportunity to realize fully its policy of nationalized medical services in China, this stated intention spelled a real beginning for China. Building upon the infrastructure of State Medicine as it had been developed during the Republican period, by the 1970s the health-care system of Communist China, which had achieved "90% coverage of a vast population, was the envy of the world," said Dr. Margaret Chan, director-general of the World Health Organization.[1]

As a number of scholars have pointed out, there is at once a salient continuity and a discontinuity between the policies of State Medicine developed during the Nationalist period and those of the Communist era. While the Nationalist government provided the leadership, blueprints, and, to a lesser degree, infrastructure for its development in the 1950s, the Communist system decisively departed from the Nationalist plan by drawing upon the human resources of traditional Chinese medicine and thereby assimilating it into the national health-care system. In these two separate but related senses, the rise of State Medicine conditioned the subsequent developments of both traditional medicine and of modern health care in Communist China. Moreover, this stage of Chinese medical history had global implications, because it inspired the concept of primary health care—especially the importance of utilizing indigenous medical resources—that was later promoted by the World Health Organization as an exemplary achievement for other developing countries in its 1978 Declaration of Alma-Ata.[2]

Despite this retrospective historical significance, at the time when the Ministry of Health was established in 1928, it was not at all clear whether a state-centered medical system would emerge as a policy of consensus. Even with the enthusiastic promotion by John B. Grant, the Nationalist government shied away from committing in writing to a "governmental responsibility for people's health," as I have discussed in chapter 3. In light of this earlier history, it becomes puzzling why the Nationalist government first refused to take on the "responsibility of protecting people's health" in the late 1920s but agreed to accept the much more daunting responsibility of providing State Medicine in the early 1940s.

The key to this question, as suggested by the title of this chapter, lies in the emergence of rural China as the crucial arena for the political struggle between the Nationalist and Communist Parties in the 1930s. With this political development, "China's health problem" got re-defined as the question of how to provide modern health care to "the

84% of the polulation who live in the rural areas and are incapable of paying for private medical care."[3] In their attempt to address this seemingly impossible task, various historical actors—the Rural Reconstruction Movement, the China Medical Association, the Nationalist government, and the advocates of Chinese medicine—came to the conclusion that State Medicine represented the "only solution" to China's health problem.[4] In four separate sections devoted to each of these players, the present chapter documents how each of them came to accept the concept of State Medicine and to develop its own specific vision for it.

By documenting this process of invention, experimentation, and debate around the concept of State Medicine, this chapter shows how these various competing visions emerged, evolved, became ever more diversified, and even remained locked in tension against each other in the process of reaching a nominal consensus by the end of the Second World War. Despite such a contested and heterogeneous origin, the idea of State Medicine thus became the ultimate answer to the question that proponents of both Chinese and Western medicine had struggled with since the beginning of the twentieth century, namely, how to connect medicine with the state. To understand this historical process, which gave birth to State Medicine, I first investigate how rural China emerged as the crucial connection between medicine and the state.

Defining China's Medical Problem

As discussed in chapter 3, it was when the Nationalist government established the Ministry of Health in Nanjing in the spring of 1928 that John B. Grant and his colleagues first publicly proposed their ambitious vision of State Medicine.[5] Before I move on to trace the development of this medical vision since that time, two aspects of Grant's concept of State Medicine need clarification. First of all, the standard Chinese term used by its promoters at that time was *gongyi*, literally "public medicine." Nevertheless, when this idea was first proposed by John Grant, the Chinese term he used was *guojiahua de yixue*, which literally meant "nationalized medicine."[6] Since the Chinese term *gongyi* fails to highlight Grant's objective of "full state control over all medical matters,"[7] the English term *State Medicine* better catches the gist of Grant's medical vision. In addition, *State Medicine* was also the exact term used by Grant and other promoters of *gongyi* in English publications, perhaps because their medical vision had been inspired by British State Medicine.[8] Second, this State Medicine was by no means a wholesale importation from the West, even though Grant and his Chinese colleagues explicitly took the British system of State Medicine as representative of the global trend of national health-care services. In the opening sentence of his article "State Medicine: A Logical Policy for China," Grant asserted that "China is one of the last of the world's communities to undertake the utilization of scientific medicine."[9] But Grant did not encourage the Chinese people to retrace the path that more advanced countries had taken, with many unnecessary sidetracks, such as the dominance of privately owned curative medicine. Instead, he urged them to develop their own local innovation that took into account China's concrete situation while at the same time anticipating global trends of medical development.[10] Because this envisioned State Medicine would be a local innovation with full self-consciousness, it would be concerned less with an exact duplication of developments in the West than with the practical conditions in China.

This vision of State Medicine was developed by means of local experimental efforts by Grant and his colleagues at the Department of Hygiene and Public Health at Peking Union Medical College (PUMC). To collect information about specific local conditions, Grant and his colleagues began a highly experimental project in 1925 that involved a "health demonstration station" in Beijing.[11] In close cooperation with the police department, they set up this health demonstration station in the Second Left Inner Ward of Beijing, which had a population of approximately fifty thousand. In a way, this experimental project represented Grant's response to Roger Greene's concern about "the need for a pilot study" (discussed in chapter 3), since its objective, besides furnishing fieldwork opportunities for PUMC students, was "to learn by actual experience what kinds of public health work are most needed in China, what kinds are feasible under present conditions, and what methods are most likely to succeed."[12] Like many other celebrated experiments in the history of science, Grant's experiment virtually defined "China's medical problem."

According to Grant, the key to understanding China's medical problem was to conceptualize it in terms of the major causes of "excess mortality" in China. Grant assumed that the "normal death rate," as commonly found in Europe and the United States, was fifteen deaths per thousand. Taking this mortality rate as his baseline, Grant considered China's medical problem to lie in the fact that it had a death rate of thirty per thousand, which amounted to six million additional deaths per year. Grant never insisted on the reliability of his statistical data; on the contrary, he admitted that his estimation was based on an "analogy with other countries possessing statistics coupled with available fragmentary local figures."[13] Nevertheless, both Grant's analysis of China's medical problem in terms of "excess deaths" and his numerical data were widely followed by public health advocates throughout the Republican period.[14]

Once China's medical problem had been identified as consisting of these six million excess deaths per year, Grant's plan was that efforts should focus on the causes of this excess mortality,[15] namely, "Gastro-intestinal diseases, Tuberculosis, Smallpox and the infectious causes of Infant Mortality," which together caused three-quarters of these additional deaths.[16] Since curative medicine could obviously do very little to alleviate these causes, Grant did not consider the lack of modern medicine to be a fatal weakness for China. On the contrary, "China could possess a more effective balanced application of medical science sooner than many countries which, because their viewpoint is so largely curative, are [reluctant to accord] adequate recognition to the other branches."[17] In order to solve its medical problem, which was defined in this way, China would have to find a shortcut instead of using the cure-centered "normal path" of medical development.

As discussed in chapter 3, John Grant, the architect of this vision, viewed the Ministry of Health as the instrument designed to realize the ambitious project of State Medicine.[18] Nevertheless, just one year after the establishment of the Ministry of Health in 1928, the Nationalist government downsized it, renaming it the National Health Administration (*weishengshu*). In the first few years of its existence, the project of State Medicine was far from well-received. While he was sympathetic to the idea of State Medicine, Huang Zifang, trained at the Harvard School of Hygiene and then serving as director of the Health Department for the municipality of Beijing, considered State Medicine a long-term goal that might be realized in fifty years.[19] In July 1928, Chen Zhiqian (1903–2000), a favorite student of Grant's who had graduated from PUMC and gone on to become a famous pioneer of rural health and an ardent supporter of State Medicine, published a long article to promote Grant's idea of full state control of medical affairs. He admitted there that "even intellectuals do not think that medicine should become a state-owned public enterprise. This idea is viewed as extremely strange."[20] To make the idea of a state-run medicine more acceptable to the public, Chen instead used the term *comprehensive medicine* (*quanyi*) to highlight a balanced application of preventive and curative medicine.[21] As a reflection of this shared reservation, the Chinese translation of Grant's English article carried a different title from the original: Instead of the original "State Medicine: A Logical Policy for China," the literal translation of the Chinese title was "Outline for Constructing Modern Medicine in China";[22] within the article itself, the term *State Medicine* was demoted from its prominent status in the original article. In the official plan for a Chinese Ministry of Health, Liu Ruiheng, vice minister of health since 1928, mentioned the notion of State Medicine in passing and suggested that it was "being studied by the Ministry. It is hoped that a small program will be worked out for trial in a selected demonstration area."[23] A year later, the National Board of Health reached the following conclusion: "The National Government shall be petitioned to set aside a special fund for state medicine [*sic*] and the prevention of epidemics. State medicine [*sic*] should be introduced by health organizations whenever conditions are favorable."[24] Even within the Ministry of Health, therefore, no serious efforts were undertaken to launch State Medicine as envisioned by its early proponents.

Discovering Rural China

Because Chen Zhiqian (1903–2000; also known as C. C. Chen) was closely associated with the idea of State Medicine both before and after his famous Ding County experiment (1931–37), it is easy to assume an undisrupted continuity starting from Chen's student years in PUMC under the mentorship of John Grant to the later development of his own vision of State Medicine that was based on his experiences in Ding County. Nevertheless, this conventional view unfortunately obscures the crucial innovations that resulted from Chen's Ding County experiment. Without such innovations, the Nationalist state might not have embraced the idea of State Medicine. To shed light on both the development of Chen's conception of State Medicine and the government's response and eventual acceptance of this idea, the following two sections show how Chen's Ding County experiment constituted a breakthrough in the evolution of State Medicine between 1929 and 1940.

As late as 1928, when Chen Zhiqian published the above-mentioned article on what he called "comprehensive medicine" to support Grant's vision of State Medicine, he did not mention the importance of rural health.[25] He became interested in the problem of rural health only after he had assumed the position of director for a health demonstration station at Xiaozhuang, a rural village near the nation's capital, Nanjing, in the summer of 1929. Nevertheless, a key career-defining move occurred only several years later, right after his return from studying at Harvard University's School of Public Health in January 1932. By way of his connections to Grant, Chen was invited to serve as director of the Department of Rural Health in the Mass Education Movement organized by Yan Yangchu (also known as James Yen, 1890–1990). Before Chen's participation, the movement had already devoted itself to rural health, considered one of the integral aspects of rural reconstruction and supported by the Milbank Foundation since 1929.[26] By the time Chen arrived in Ding County in 1932, the rural health project of the Mass Education Movement had already been in operation for three years. While Chen might have been interested in rural public health beforehand, the undivided devotion of the Mass Education Movement to rural China was instrumental in facilitating his great contribution to rural health care. In this sense, a crucial influence on the history of modern health care in China—the emphasis on rural China—came from outside of medicine.

The lack of concern for rural public health was by no means a unique Chinese phenomenon. When modern public health emerged in the early nineteenth century in England, it responded mainly to problems caused by industrialization and urbanization,[27] and therefore was slow to take rural health needs into account. This relative lack of interest in rural areas struck Hu Ding'an, who would soon become the commissioner of health for the nation's capital, Nanjing, as a big disappointment when he studied public health at Berlin University in the early 1920s. Under the guidance of Alfred Grotjahn (1869–1931), the great advocate of social hygiene and eugenics,[28] Hu drafted a comprehensive and detailed report entitled *Plan for Public Health Administration in China*.[29] Nevertheless, Hu noted with disappointment that "rural public health was lacking in Europe; I have to study it in the future."

Although it is self-evident that China at that time was a country consisting overwhelmingly of village dwellers, historians can identify the specific time—that is, the late 1920s—when the country as a whole finally started to pay attention to its rural population. For very different reasons, the young Chairman Mao, left-leaning intellectuals, experts on social surveys, literary figures, and the Rockefeller Foundation all became intensely interested in rural China. More importantly, while advocates of the New Culture Movement in the 1920s had often blamed the peasants for China's weakness, from the late 1920s on these same peasants were pitied as poor victims of social and economic oppression and were eventually even seen as potential agents for reviving China. According to Charles Hayford's insightful study of Chinese peasants in the Republican period, this public concern over rural China can be associated with a subtle but important shift from a view of China as a country of farmers to seeing it as a nation of peasants who lived under a medieval feudal system and were therefore destined to revolt against it. As a result, Hayford concluded, "This 'peasant' construction of the countryside gave cultural and ideological legitimacy for Mao Zedong's political and organizational revolution."[30] Without going into the details of this ideologically important conceptual shift from farmer to peasant, it suffices here to point out that the problem of rural China consequently became the center of political concern in the late 1920s.[31] To provide a salient example, the widely acclaimed *Eastern Miscellany* (*Dongfang zazhi*) contained an average of one article per year on rural questions in the 1920s, but included more than eighty per year by 1935.[32]

One of the crucial efforts to find a solution for China's rural problems was led by the Yale-educated Yan Yangchu. Before moving its national headquarters from the capital, Beijing, to Ding County in 1929, the movement organized by Yan had been called the Mass Education Movement and had focused not on rural issues but on the nationwide problem of illiteracy. In addition, many of its members had never lived in a country town before the relocation, and over the next few years, only about one-third could endure the difficult rural environment well enough to stay on for more than a year.[33] The relocation of the Mass Education Movement to Ding County caused it to transform from a nationwide federation of literacy movements to an agency for the Rural Reconstruction Movement (*xiangcun jianshe yundong*).

Precisely because its primary objective was not to improve health, let alone to popularize scientific medicine, the Rural Reconstruction Movement emphasized the integration of various aspects of reform—education, economy, health, and politics. From this general point of view, the improvement of rural health could not be separated from the social and economic conditions. On the basis of Charles W. Hayford's excellent research on Yan Yangchu and the Rural Reconstruction Movement, I would like to point out that many innovative features of the health-care system developed by Chen Zhiqian in Ding County were at once a substantiation and an innovative constituent of the general philosophy of the Rural Reconstruction Movement. For example, in both health care and education, the Rural Reconstruction Movement strove to develop a hierarchic, cost-effective organization that made extensive use of laypeople.[34] Proponents of the Rural Reconstruction Movement realized that the various aspects of village life constituted an interrelated whole that therefore had to be tackled with an integrated reform program—not one limited to a single dimension of reform, such as literacy, land tenure, political revolution, or village industry. Thus, a great deal of cross-disciplinary learning took place among people working on each of the fronts of the reconstruction. While Chen Zhiqian was undoubtedly a great medical innovator, his medical innovation was clearly inspired by the generalist approach of the Rural Reconstruction Movement.

The Ding County Model of Community Medicine

Chen Zhiqian himself and several authors of more recent studies have documented the process by which he created a prototype for State Medicine in Ding County.[35] For a detailed understanding of the development of State Medicine, Ka-che Yip's meticulous scholarship in *Health and National Reconstruction in Nationalist China* is particularly useful. Drawing on these scholarly achievements, in this section I highlight a relatively neglected dimension of his Ding County experiment. Instead of treating it as a preliminary stage in the development of State Medicine per se, I analyze how Chen strove to create a feasible model for community medicine in Ding County. From this perspective, I highlight the considerable philosophical differences between the Ding County experiment and State Medicine, even though the former did pave the way for the development of the latter.

The distinction between the Ding County experiment and State Medicine deserves to be drawn for two reasons. First, while it was widely praised as the forerunner of State Medicine, Chen's experiment strove to build upon and empower a self-governed local community by establishing local organizations. This belief in and commitment to the empowerment of rural communities was the central philosophy of the Rural Reconstruction Movement. As such, Chen's vision of State Medicine stood in stark contrast to the top-down vision of State Medicine embraced by the Nationalist government. Second, in terms of historical developments, because this experiment demonstrated that Chen had created a health-care system that even a poor region like Ding County could afford by itself, this achievement helped greatly to make the notion of State Medicine acceptable both to the biomedical profession and to the Nationalist government. In other words, Chen's bottom-up experiment at once paved the way for and constituted a rival to the top-down conception of State Medicine that later came to be endorsed by the Nationalist government. To highlight the often-neglected tension between Chen's experiment in Ding County and the top-down conception of State Medicine, I have used the subtitle that Chen used in his own memoir, "The Ding County Model of Community Medicine," as the heading for this section.[36]

While I have been emphasizing the influence of the Rural Reconstruction Movement on Chen's experiment in Ding County, I should note as well that Chen's contributions had a significant influence on the Rural Reconstruction Movement. According to Chen's recollections, it took just three years (1932–34) for his team to develop a health-care system within the economic reach of all village residents. Moreover, this system emerged as the consensus for the whole Rural Reconstruction Movement in 1934. In the report that he submitted to the Milbank Foundation in 1936, Chen proudly announced that the Ding County experiment had been declared as the leading example for any discussion on public health by a unanimous vote during the National Conference on Rural Reconstruction held on October 10, 1934.[37] By the end of this national meeting, its underlying principles and hierarchic organization were embraced by conference participants as the model for emulation.

In the published conference proceedings, Chen introduced his Ding County experiment to the public for the first time. Chen's article, "Health-Care Institutions in the Ding County Social Reform," which has not been used in previous historical studies, provides a revealing window into Chen's own conception of the experiment at that time and into the strategies he used to motivate his colleagues in the Rural Reconstruction Movement.

As Chen pointed out in the first paragraph of his article, the major difficulty in health-care construction continued to lie in the government's failure to accept the responsibility to protect the health of its citizens.[38] Having pointed out this fundamental problem, Chen was no longer interested in awakening, or blaming, either the government or the people by means of moral rhetoric. Instead, he wanted to promote a view of health care as a collective enterprise and to put it into practice by building a community-based health-care movement.

As a graduate of Peking Union Medical College, the most prestigious medical college in the Far East at that time, Chen made the following provocative announcement: "In terms of rural health work in today's China, we should by no means rely completely on experts (*zhuanjia*)."[39] At that time, it was a common sentiment that medicine—the first example of a modern profession, from which other professions-to-be drew inspiration[40]—was a field that belonged exclusively to professionals such as doctors and nurses. Moreover, these professionals should be rigorously trained and supervised to comply with internationally recognized standards. To expose the problems pertaining to the on-the-ground application of this commonsense principle in contemporary China, especially rural China, Chen posed three questions to his readers. First, since the things that could be done to improve China's rural health were so limited and simple, was it really necessary to involve doctors and nurses in every one of them? Second, since the salaries for doctors and nurses were rather high, could rural economies even afford their services? Third, since the training of doctors and nurses had been copied entirely from Europe, America, or Japan, did it really fit China's needs?

Once he had posed these three questions, Chen felt that the following conclusion was unavoidable: "In terms of constructing rural health, whenever it is possible to not use doctors and nurses, we should not use them."[41] A medical doctor by training, Chen could not help feeling that his own conclusion was extremely "paradoxical and ironic."[42]

Since these three questions—on feasible actions to improve rural health, affordability for the rural economy, and suitability to the needs of rural China—all took aspects of rural China as their departure point, this radical conclusion might sound like the logical result of the new emphasis on rural China. Nevertheless, it was by no means a result of pure logic. On the contrary, what made this conclusion empirically compelling was a series of groundbreaking social surveys and local experiments; viewed jointly, they provided the public with a nuanced picture of rural China that was at once strange, devastated, and full of hope.

According to Chen, his experiment had been built on a survey of health conditions in Ding County that had been done in 1930, before Chen had accepted his position there. Again, Chen benefited here from a distinctive feature of the Rural Reconstruction Movement, namely, its emphasis on the use of science in understanding the social reality. To realize this objective, the Rural Reconstruction Movement had created a Department of Social Survey, then headed by the pioneering Columbia-trained sociologist Li Jinghan (1895–1986).

As Tong Lam points out, the relocation to Ding County represented a drastic shift in the career of Li Jinghan.[43] The training that Li had received at Columbia University was intended for urban settings; he was thus forced to develop a new methodology in conducting social surveys in rural China. Moreover, according to Li, this was "a move from conducting surveys for the sake of pure knowledge to conducting them for social betterment."[44] Li had assumed the leadership of this department in 1928 and five years later published a detailed, eight-hundred-page report entitled *Social Survey of Ding County* (1933). This document subsequently "became a model study for the social survey movement in China."[45] Chen learned the following crucial facts from Li's 1930 survey:

> Nearly 30 percent of people who die in Ding County have received no medical attention whatsoever. Of the 472 villages in the district, 220 possess no medical facilities of any kind, and the other 252 can boast of little more than a self-made (and not infrequently illiterate) physician of the old type who prescribes drugs that he himself sells. Nevertheless, [the survey] revealed that the annual per capita expenditure for medications and medical attention of such an unreliable type is about 30 cents in this district.[46]

These quantitative facts provided crucial clues for Chen as he set up the objectives for his Ding County experiment. First of all, Li Jinghan's survey not only discovered the pitiful financial resources available to these villagers for medical expenses, it also made clear that, at least from Chen's point of view, these limited but valuable resources had so far been wasted on traditional medicine. Elsewhere, Chen explicitly stated, "In our endeavor to popularize the new style of medicine, including scientific hygiene, it is already a difficult objective to aim at sharing one-third of the amount of money that people spend on the old style of medicine."[47] As an experiment in community medicine, Chen's efforts did not aim at a dramatic increase of governmental investment in medicine, but at a transfer of limited community resources from traditional medicine to biomedicine. In fact, when Chen reported on his remarkable success in popularizing Jennerian vaccination in Ding County, he called attention to the fact that "several old-time professional vaccinators have recently been forced to leave this district."[48] Because his objective was a transfer of limited financial resources, Chen set up the following famous litmus test for rural health care: it had to function on a yearly budget of less than ten cents per person—less than one-third of what Ding villagers had spent on "old-style medicine."

By itself, Chen's second question, which concerned the affordability of health care in the rural economy, posed a seemingly unsolvable problem for China's rural health care: Villagers were too poor to afford the professional services of modern medicine. If one considers this problem along with the first question, however, a potential solution emerges. That is, although limited by their financial means, the medical services that these villagers could afford often did not need to involve modern-trained professionals. For example, one of the major causes of increased mortality was the prevalence of smallpox. As an effective remedy, vaccination against smallpox was so simple that it could easily be done by a layperson without the involvement of doctors and nurses.[49]

This practical attitude could sound like a self-serving tautology: people would not need services that they could not afford in the first place. Nevertheless, Chen proved that a remarkable decrease in the mortality rate could be achieved under such a highly limited budget, as long as resources were used strategically and collectively. In his 1936 report to the Milbank Foundation, Chen could claim that the crude death rate for the registration area had decreased from 31.6 percent in 1932 to 27.2 percent in 1933, and to 22.6 percent in 1934.[50] If we evaluate this achievement within Grant's framework of analyzing China's "medical problem," Chen's experiment cut China's excess death rate in half within just three years.

To provide a cost-effective health-care system that met the specific needs of the villagers, Chen and his colleagues scientifically analyzed the role of the medical professionals into its component parts to see which could be parceled out to people with short-term or mid-term training. The result of this reorganization was Chen's famous three-level health-care pyramid.[51]

At the top level of the county (*xian*), the District Health Center (*baojian yuan*) consisted of a hospital and administrative center, thereby serving as headquarters for both curative and preventive medicine. At the mid-level of the market town (*qu*), the Subdistrict Health Station (*baojian suo*) served as a point for distribution and supervision. The bottom level of this pyramid was the village (*cun*)—a crucial decision with great consequence. In China, subdistricts often consisted of a few dozen villages that were separated from each other by rather long distances with hardly any roads to connect them, which made it very difficult for either doctors or patients to travel between villages for consultations or treatments. Except for particularly critical situations, it was therefore impractical to expect people to seek health care outside their village.

The problem thus centered on the fact that it was economically impossible to rely on modern-trained physicians to serve the health-care needs of the villages. As Chen discovered from his survey, "The average village, with a population of about 700 and not more than $150 available for medical purposes, could not possibly support any known type of regular medical personnel; and yet under present conditions in the rural districts, the foundation of a community health system must be the village."[52] Forced by geographical and financial constraints, Chen developed the most daring innovation of his system at the base level, that is, the use of full-time farmers to serve as health-care personnel in the village. These key individuals were called "village health workers" (*baojianyuan*) and constituted a new kind of medical personnel who

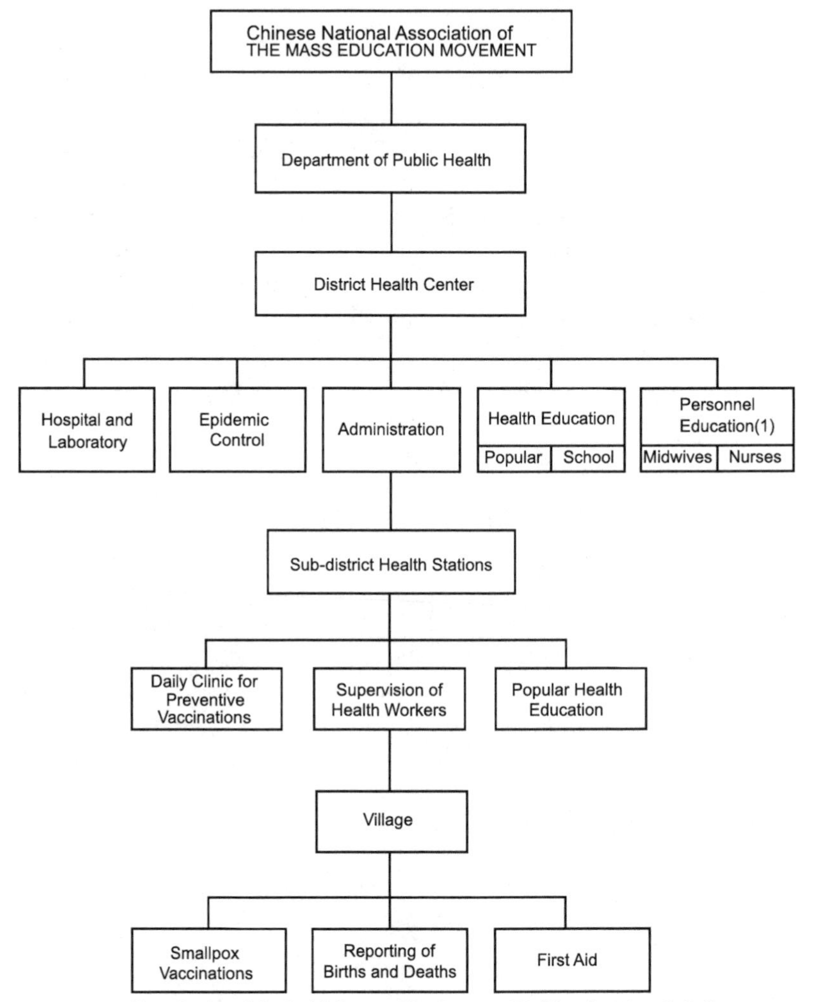

FIGURE 10.1 "Organization of the Health System, Ding County, 1933," flowchart from C. C. Chen, *Medicine in Rural China: A Personal Account* (Berkeley and Los Angeles: University of California Press, 1989), 82.

came from the village and stayed there to serve their fellow villagers. I return to this crucial innovation in a later section to trace the story of the village health worker from its beginning in the 1930s to the emergence of the barefoot doctor in the late 1960s.

Smallpox vaccination provided the most salient example for the success of this innovation. With the help of village health workers, one-seventh of the total population and three-fourths of newborn infants were vaccinated within seven years (by 1937).[53] As a result of this impressive effort, the population of Ding County was spared from massive smallpox epidemics that ravaged the neighboring counties.[54] In conclusion, Chen and his coauthor suggested, "Every health officer should start with smallpox in preventive immunization work and make it a good demonstration of organization thoroughness before proceeding to other types of preventive inoculation."[55]

Besides vindicating Grant's strategy of focusing on the most easily controllable infectious diseases, such as smallpox and gastrointestinal diseases, this achievement strongly supported Chen's strategy of using nonprofessional village health workers. Most importantly, in the year 1934, all these remarkable results were achieved with a per capita cost of 9.08 cents—even less than his litmus test of ten cents.[56] This experiment demonstrated that Chen's system was well within the economic reach of the villages, even in a place as poor and barren as Ding County.

Based on his experiments, Chen concluded as follows in 1934:

> According to our practical experiences in the villages, we think on the one hand that many rural health measures are extremely simple and that whenever it is possible to use laypeople to substitute for professionals, we should try our best to use them. On the other hand, we think that the modern urban-educated doctors and nurses do not fit the need of the rural villages. The latter is an issue related to the larger problem of medical education, which is beyond the scope of this article; the former is a fact to which we paid special attention when we conducted our work in rural health.[57]

As an answer to his third question regarding the suitability of urban-trained professionals for rural health-care needs, Chen pointed out that current medical personnel did not fit the needs of rural health care in two different senses. During the developmental phase of community medicine, Chen had focused on recruiting nonprofessional "village health workers" to take responsibility for relatively simple but important health measures. Nevertheless, these health workers provided just half of the solution. The key problem of expanding the health-care pyramid into a national network lay in the supervisory middle level, the Subdistrict Health Stations, which needed not minimally trained laypersons, but full-time professionals. After two more years of efforts, Chen felt even more strongly in 1936 that medical education was the crux of the problem and that it could not be solved at the local level. Recognizing that "no single medical school in this country is assuming the responsibility of producing physicians for subdistrict stations," Chen could not help concluding that "unless these provincial medical schools take an interest in the training of needed personnel, there will be no hope of developing rural health care in this country."[58] In short, China's rural health care still needed doctors and nurses, but they needed to be trained in a very different fashion from the so-called international norm. By this time, what Chen needed most from the national government was not a financial endorsement but a very special kind of trained personnel, who could not be produced without a fundamental reform of medical education and the medical profession. This was the one element that community medicine would never be able to produce by itself. For the overhaul of the medical education system, it had to look to the state.

State Medicine and the Chinese Medical Association

In comparison to the Rural Reconstruction Movement, the Chinese Medical Association was relatively late in offering its support to State Medicine. While most of its members must have been hearing about the idea of State Medicine for years, the association never discussed this crucial policy as a professional body until its fourth grand meeting, held in April of 1937. Within the framework of that meeting, a high-profile symposium was devoted to the topic of State Medicine. The papers presented there were later published as a special issue of the *Chinese Medical Journal.*

The panel of this symposium consisted of prominent leaders from the government, the medical profession, and the medical missionaries: Liu Ruiheng (director of the National Health Administration), Wu Liande (director of the National Quarantine Service), Frank Oldt (Canton Hospital), R. K. S. Lim (Peiping Union Medical College), and Chen Zhiqian himself. In the opening lecture, Liu Ruiheng did not hesitate to state clearly that, "the ultimate aim is to provide modern health care and medical services for all people," but "the goal will not be realized for many years to come."[59] Having said so, Liu enthusiastically reported that "In many parts of the country there have sprung up centers of activity along modern and uniform lines, which have definitely taken root and will become the foundations of a future system of State medicine [*sic*] for China. The most encouraging point is that almost all of these centers are being provided by the local, that is, provincial or county, governments."[60]

Liu's enthusiasm reflected a recent trend of events. Just a month prior to the meeting, the Nationalist government had passed a new regulation that authorized all county governments to spend 5 percent of their annual budget on the maintenance of health centers. This rule was based on the experimental achievements in places like Ding County. In addition, for the sake of training personnel for State Medicine, a new kind of medical school, the Chung-Cheng Medical School, was started in the same year, 1937.[61] The general impression was that the Nationalist government was willing to embrace State Medicine as an "ultimate objective" and supported experimental efforts on the local scale, but did not contemplate formulating a policy to realize this objective in the near future.

Explaining "the need for State Medicine," Wu Liande relied heavily on the experimental results from Ding County and duplicated many of the numerical data and arguments that had been developed by Chen Zhiqian.[62] Most importantly, Wu pointed out, if one accepted Chen's discovery that villagers had such limited resources to spare on medicine, "it would seem at first glance that it was almost impossible to realize such a plan [of creating a State Medical system] for fiscal reasons alone."[63] Turning this crucial constraint into a positive rationale, Wu continued, "However, profound studies of this problem have been made in several experimental rural areas, and it has been shown that when a medical organization capable to deal with both preventive and curative medicine has been extended over a district, the amount of money required is no more than ten cents per head annually, or just one-third of the sum actually spent by the people each year for their medical needs."[64]

From these public statements, it is clear that the Ding County experiment played a crucial role in persuading both the medical profession—especially the Chinese Medical Association—and the Nationalist government to support the policy of State Medicine. As Chen repeatedly emphasized, if people accepted as a litmus test the condition that "the per capita expenses of rural health care per year should be around ten cents,"[65] his system would become almost a necessary conclusion. While one conference participant indeed took issue with Chen's estimation,[66] many medical leaders and state officers started accepting the numerical data as the indisputable bottom line.

It was also this numerical data that established the feasibility of the once unrealistic objective of State Medicine. Since its beginnings, when John Grant had promoted the idea of State Medicine in the 1920s, the key problem that had paralyzed everyone had always been the issue of financing it. Almost a decade after its creation, the budget of the National Health Administration was still only 1.2 percent of the total national budget.[67] It was hence unrealistic to expect the Nationalist state to dramatically increase its investment in medicine while on the verge of a full-fledged war against Japan. To the anticipated question about budgetary constraints, Chen responded that "[the Ding County] experience shows that without opening up new items of expenditure, and without adding new burdens to the mass, it is possible to build a rational medical organization on the basis [of a yearly expense] of 10 cents per capita."[68] In light of Wu's statement, the Ding County experiment constituted a historic breakthrough for State Medicine because it demonstrated that even in rural China a well-organized community medicine was capable of sustaining itself under such severe financial constraints.

In contrast to Wu's supportive article, Frank Oldt's pointed paper "State Medicine Problems" offers a rare window into the discontent with which this project was received within the medical profession. To begin with, Oldt challenged the idea that State Medicine was the only form of medicine suited to China's needs, pointing out that "it was usually given in an axiomatic manner, as if it were self-evident and needed no discussion."[69] It is not surprising that Oldt felt this way; some advocates of State Medicine indeed based this crucial policy choice on logical necessity. The most salient and representative example is in the title of John Grant's article "State Medicine: A Logical Policy for China."[70] Concerning the radical transformations that were bound to result from this medical policy, Oldt could not help noticing that "reaction on the part of physicians has been surprisingly little. It has never been adequately discussed in the Chinese Medical Association meetings. No expression of attitude has been made by the body, and no action taken indicating [the] course to be pursued in solution of the problems involved."[71] It was for the sake of opening up a space for collective deliberation that Oldt purposely chose to highlight the problems with State Medicine that had been pushed to the periphery by the force of logical deduction.

Oldt's main concern was the unavoidable conflict between private practice and State Medicine: "What will be the place in state medicine [*sic*] for private practice, either by individuals, or groups, or institutions? Or if there is no place, what is to become of existing medical practice? This is a point which should have thorough discussion in the Medical Association, a plan worked out with the government, clearly stated and agreed upon and above all not done hastily."[72]

While these were very reasonable questions to ask, it is revealing that there was almost no response following this line of "problems" in the discussion section or in subsequent journal articles. At least in this historic meeting, the leading members of the Chinese Medical Association did not see themselves as a professional society with private interests that needed to be protected from the public interest and the state. Instead, as suggested by the title of Liu Ruiheng's opening speech, "Our Responsibility in Public Health," the director of the National Health Administration was not speaking as a government bureaucrat outside the profession, but as a member of the medical profession who represented the association in its service to the government. It is remarkable that these leading figures conceived of the relationship between the Chinese Medical Association and the state as such a close alliance. Instead of worrying about the problems caused by State Medicine, they committed themselves to pursuing this form of a national health-care system, which was bound to radically transform the nature of the medical profession, medical education, and the daily realities of practicing medicine. Given that the Nationalist state did not intend to colonize the medical profession but consistently delayed taking on "the responsibility for people's health," this event again supports my larger argument that in early twentieth-century China, it was the leaders of the medical profession who strove to link medicine to the state and consequently proposed the new vision of State Medicine, for the sake of both the state and their profession.

Among the published contributions to the special symposium, the third article, "State Medicine" formulated the most radical vision of State Medicine. It was presented by Chen Zhiqian and the much-admired scientist R. K. S. Lim, an overseas Chinese from Singapore who had received his medical degree from Edinburgh University and was later praised as "the Father of Chinese Physiology." The two authors gave the following definition:

> *State Medicine*, as the policy of the National Health Administration and of the Ministry of Education, may be defined as the rendering available for every member of the community, irrespective of any necessary relationship to the conditions of individual payment, of all the potentialities of preventive and curative medicine. Consequently, it is evident that the State must be responsible for all medical work, viz. curative (or clinical) as well as preventive (or social) medicine, for the provision of all cadres of personnel and for all types of establishments and supplies. Further, the service should be provided without charge to the people, and yet the total cost must be such as the people are able to bear indirectly through taxation.[73]

It is important to point out that this definition left many features of that system unspecified. For example, early on Lim and Chen mentioned the possibility of the government providing subsidies or higher salaries to keep well-trained personnel in the rural districts.[74] They were fully aware that, "The greatest difficulty which the Medical Service is likely to encounter, is the payment of sufficiently satisfactory salaries to hold its personnel."[75] This system sounded rather similar to the system adopted by the Japanese colonial government in Taiwan, which was also called State Medicine.[76] Nevertheless, Lim and Chen quickly refuted this possibility on the basis of budget constraints. Instead of offering higher salaries to keep medical personnel in the countryside, they strove to create a system that minimized the need for highly trained technical professionals and thereby reduced the costs of maintaining such a system. As this comparison indicates, the demands placed on the state differed radically between Nationalist China and Taiwan under Japanese colonial rule, even though both systems were called "state medicine."

Reading through Chen and Lim's article, one cannot but feel surprised that the authors did not ask the state to dramatically increase its funding for medicine. Instead, what they considered essential and demanded support from the state for were two aspects: first, the organization of a three-tiered health-care system and, second, the reform of the medical education system so that it could provide medical personnel suitably trained for serving in the various positions of State Medicine. Regarding these points, Lim and Chen did not hesitate to use strong words: "State (or provincial) medical service must be operated with the discipline expected of a 'military machine,' and with the economic management associated with an 'industrial enterprise.' Thus the traditional functions of the doctor, nurse, midwife, pharmacist, etc., and of traditional hospitals, clinics, and public health bureaus must be scrutinized, and if they do not fit adequately in the new scheme, they must be altered. If necessary, new types should be evolved to take the place of the old."[77] Therefore, what they needed from the state was a new kind of personnel, to be produced from an equally new kind of educational system.

The high-profile symposium in 1937, which served as the public forum for the positions on State Medicine discussed above, resulted in an official expression of support for State Medicine by the Chinese Medical Association. In the same year, however, the Japanese army conquered Beijing, marking the end of Chen's Ding County experiment. Even though the Japanese authorities expressed their appreciation for his public health project, Chen decided to secretly leave Beijing for southwestern China, without even the company of his family. As a result, one scholar concludes that the year 1937 also marked the end of the "Nationalist attempt to provide health services to the rural population."[78]

State Medicine and Local Self-Government

To understand why the Nationalist government increasingly committed itself to the policy of State Medicine, it is helpful to examine the crucial role played by Jin Baoshan (1893–1984)—a highly unusual figure in the field of health administration during the Republican period. Against the tension between the so-called Japanese-German and Anglo-American factions, Jin was a rare figure who could claim membership in both factions. After receiving his medical degree from Chiba Medical College in Japan, Jin spent a total of nine years in Japan (1911–19), working in the world-renowned Kitasato Institute of Infectious Disease. Nevertheless, Jin also formed a good rapport with the so-called Anglo-American faction: He worked with Wu Liande to control the second Manchurian plague (1920–21) and, with the support of a fellowship from the Rockefeller Foundation, received a master's degree in Public Health from Johns Hopkins University (1926–27). Through an invitation by John Grant, Jin then became the founder and director of the Hangzhou City Bureau of Public Health.[79] As Jin later recalled, when the Nationalists established the Ministry of Health in 1928, they were keen on balancing power between the two factions and therefore entrusted him with this important post.[80] By 1933, Jin had become vice director of the important Central Field Health Station under the National Economic Council. With Liu Ruiheng serving as nominal director, Jin was the person in charge of running this important institution.

When Jin published (under the name of P. Z. King) an article entitled "Thirty Years of Public Health Work in China" in 1944, he admitted that "no defined [national health] policy was adopted during the two preceding periods, though a proposal for the adoption of a program of state medicine was made during the National Health Conference in 1934. After the Nationalist Party adopted this policy at its Eighth Plenary Session of the Central Executive Council held in 1940, the National Health Administration plunged into implementing this policy."[81] Serving as director of the National Health Administration at that time, Jin recounted history in this way partly because he was one of the main figures involved in the governmental effort to realize the policy of State Medicine.[82]

The year 1934 is important not just because it was the year when the Ding County model was embraced by the Rural Reconstruction Movement; it was also the year in which the National Health Administration passed a regulation to build a public health infrastructure at the county level.[83] Again, this development in health care was actually part of a larger collaboration between the Nationalist government and the Mass Education Movement to build local self-government. In the field of health care, the National Health Administration and the movement pushed from opposite directions to establish public health at the county level. While Chen Zhiqian was building his model of community medicine from the bottom up, Jin and his colleagues were striving to extend medical administration down to the county and village levels. Chen Zhiqian agreed with Jin that the newly energized concern of the government for health at the county level constituted remarkable progress. Nevertheless, he noticed a very worrisome trend that the government "focused only on building organizations, instead of on developing practical ways for solving the real [health] problems."[84]

The government's motivation for this new concern for county-level health care was made explicit by Jin at the first national conference of the Rural Reconstruction Movement in 1933. As the deputy director of the National Health Administration at that time, Jin represented the government's attitude toward the Rural Reconstruction Movement, especially toward the aspects related to rural health. Instead of emphasizing the control of notifiable infectious diseases, Jin called attention to schistosomiasis and other local diseases. Because people were infected with schistosomiasis mostly when the water-borne larvae of the pathogenic parasite penetrated the skin, this disease was particularly prevalent among farmers who labored in flooded rice fields with bare feet, thus exposing themselves to infection. It is worth pointing out that because schistosomiasis was considered a rural disease, Chairman Mao, in order to show his concern for farmers and rural residents, launched a national campaign to combat schistosomiasis right after the Communist takeover of China in 1949.[85]

As Jin pointed out, the Nationalist government considered schistosomiasis as a top priority for two reasons. First, tens of millions of patients suffered from schistosomiasis along the shore of the Yangtze River; this disease thus had a direct impact on farmers' livelihoods. Second, efforts to control this disease could serve as a precursor for other governmental enterprises, since the success of the anti-schistosomiasis campaign could help to win farmers' approval and support. Jin elaborated as follows: "In the past, when Japan had just acquired Korea, the Japanese encountered great difficulties in getting along with the Korean people. Therefore, Japan devoted itself to building hospitals and improving public health; gradually, it succeeded in extending its influence into the fabric of [Korean] society. Japan has adopted a similar strategy in Taiwan. It is said that Japan has again adopted this strategy of using medical and public health enterprises to win over people's hearts after occupying the northeast region [of China]."[86] While this paragraph reads almost like an insertion into his official report, it was no accident that Jin had in mind the role of medicine in Japanese colonial rule.

It is quite likely that Jin became fully aware of the political function of medicine and hygiene because of his experiences at the prestigious Kitasato Institute of Infectious Disease, the institute that had provided a network of physicians for Japanese colonial medicine extending from Taiwan, Shanghai, and Korea to Manchuria.[87] As a result, he urged the Nationalist government to follow suit in its struggle against the Communist Party. In light of the fact that in February of 1934 the Nationalist Party had just launched its most important mass movement, the so-called New Life Movement (*xinshenghuo yundong*), which was also focused on transforming the hygienic habits of the Chinese people and thereby served as a tool for penetrating into the area once controlled by the Communists,[88] the timing was quite right for the National Health Administration to push for the creation of a system of public health at the county level.[89]

In 1936, as vice director of the Central Field Health Station, Jin wrote an article entitled "The Institution of State Medicine," which was published in several journals and broadcast on the radio.[90] Instead of promoting State Medicine as an unfortunate but practical policy suited to China's economic situation, Jin cited the famous American "Report of the Committee on the Cost of Medical Care" to highlight the pathological development of health-care enterprises in major capitalist societies, especially the United States. Against this pathology of modern medicine, Jin confidently asserted that China's innovative development of State Medicine "could provide a shortcut to assuming a leadership position in the global enterprise of health care, while it would also be our country's greatest contribution to the international community."[91] Just like Grant in the late 1920s, promoters of State Medicine were keenly aware that they were not following any specific foreign model of medical development but striving to create a novel system for China as well as for the world.

Having such high expectations for State Medicine, Jin nevertheless pointed out that contending views existed regarding its concrete meaning and content. While some considered it identical to public health, others saw it as part of the health-care enterprise controlled by the government. But Jin made it clear that what he meant by State Medicine was quite different; it was "a well-planned, systematically implemented health-care system completely sponsored by the state to serve all citizens."[92] In short, to address the pathology of modern health-care systems as they had developed within capitalist societies, Jin's answer was the state. Moreover, the five principles of State Medicine that Jin stated in this much-promoted lecture were all consistent with the Ding County model developed by Chen; in fact, Jin mentioned in this article that among all the experimental efforts concerning rural health, "the most comprehensive experiment on health-care institutions is no doubt being conducted in Ding County in Hebei Province."[93]

As Jin pointed out, after the adoption of a program of State Medicine during the National Health Conference in 1934, the second major step toward State Medicine was the historic resolution passed during the Eighth Plenary Session of the Fifth Central Committee of the Nationalist Party in April 1941. Nevertheless, instead of foregrounding the name of State Medicine, this proposal was entitled "Promoting Public Health Construction so as to Improve Citizens' Physical Strength and the Nation's Health."[94] The proposal was named with such abstract and noncontroversial terms that one cannot help but suspect that this was done by design. Nevertheless, the key phrase State Medicine did appear in the first of its recommendations for concrete action: "Our central government should adopt the objective of popularizing public health infrastructure and implementing a system of state medicine."[95]

Taking State Medicine as a distant "objective," this Nationalist proposal in fact focused on establishing public health machinery at the county and subcounty levels and on legalizing public health as part of local governments' annual budget. No mention was made of the Ding County model, nor of the importance of integrating curative and preventive medicine, nor of the innovative nature of the view of State Medicine as a "logical choice" for China's situation. Perhaps because this proposal was supported not for its value in promoting health per se, but as part of the effort to build the political machinery of local self-government, which was one of the central tasks of the Nationalist Party,[96] it reads like a plan to build up the public health branch of the administrative structure without involving any innovative experiments. Still, in response to the Nationalist government's official resolution to legally guarantee a public health budget within local governments, the National Health Administration happily emphasized that the suggested method of State Medicine in this proposal had been its policy for decades and that it would strive to build a hundred "public health hospitals" (*weishengyuan*) per year at the county level beginning in the following year (1942).[97]

Just like Chen Zhiqian's District Health Centers, which integrated curative and preventive medicine, these "public health hospitals" were designed to serve two roles at the same time: as government-funded hospitals and as branches of the medical administration subordinated to the county government.[98] By 1934, about 50 percent of the health budget in Ding County was devoted to the treatment of diseases; curative medicine was thus already an important part of its system.[99] Because the "public health hospitals" were meant to serve this dual function, their director had to be a well-trained physician as well as an experienced administrator.

When Chiang Kai-shek published *China's Destiny* (*Zhongguo zhi mingyun*) in 1943 to express his views about China's bright future after the war, he specified a future ten-year plan for economic construction, a grand plan that, he emphasized, was based upon Sun Yat-sen's *The International Development of China* (*Shiye jihua*). Unlike the situation in the 1920s, when the Nationalists had decided to downplay the role of health care because it was not listed in Sun's plan, this time Chiang planned to devote the largest number of college graduates to medicine (232,000), far exceeding the numbers devoted to the second largest discipline, civil engineering (90,000).[100] In addition, Chiang announced the seemingly far-fetched objective to build a hundred major public health hospitals, a thousand county-level public health hospitals, and eighty thousand public health hospitals at the subcounty level over the next ten years.[101] These numbers originally came from the National Health Administration's Three-Year Plan (for 1942–45). Without mentioning State Medicine, Chiang's Nationalist government thus virtually committed itself, at least on paper, to building the administrative infrastructure and preparing the medical personnel for an immense health-care system.

The Issue of Eliminating Village Health Workers

According to Jin Baoshan, the National Health Administration accelerated this process of building a public health infrastructure at the local level when the war broke out, and relentlessly continued with its efforts until the end of the war.[102] By 1945, the administration had succeeded in building at least one public health hospital in 938 counties, which amounted to around 70 percent of all the 1,361 counties still under Nationalist control at that time.[103] In terms of numbers, this was a remarkable achievement. The fact that the Nationalist government continued to increase its investments in the field of health care compellingly demonstrates that even during a devastating war against Japan,[104] it still strove to maintain its commitment to improving people's health care. Nevertheless, because its major objective was focused on extending the state's influence into the counties, it did not strive to reproduce the Ding County model, which had emphasized community empowerment and local initiative. As a revealing culmination of these conflicting objectives, the Nationalist government informed Yan Yangchu in 1945, through his old friend Song Ziwen (1894–1971), who was also known as T. V. Soong, that it would take care of the rural economy, health, and politics, reducing the Rural Reconstruction Movement to its old role of teaching people how to read.[105]

The internally circulated journal *State Medicine* provides a valuable window into the situation behind this impressive numerical growth and these conflicting objectives. Readers might be surprised to find that participants in a conference on county-level public health seriously discussed whether or not to abolish public health stations below the county level and the position of village health workers. The issue of eliminating village health workers is particularly surprising because in Chen Zhiqian's experience, the position of the village health worker had been at once a crucial innovation and the key to success in his Ding County model of community medicine. If people were seriously considering abandoning the position of the village health worker, the system under construction must have been very different from the Ding County model.

According to the record of this group meeting, "the situation of village health workers is no good almost everywhere; they sometimes even dare to open a business of traditional Chinese medicine. In the future, we should not continue establishing this position. In regard to the tasks that they are supposed to carry out, it is better to delegate these to public school teachers."[106] In appearance, the efforts by the National Health Administration carried on the name and the three-level health-care pyramid of the Ding County model; nevertheless, their negative evaluation of village health workers reveals a fundamentally different vision of both the exact nature of State Medicine and the appropriate way of constructing it.

To articulate this crucial difference, it is useful to go back in time a bit to consider an article published five years previously by Chen Wanli, who was the chairman of this group discussion. Entitled "How to Train the Village Health Worker," Chen's article provided a detailed comparison of the current situation of village health workers in counties in Zhejiang province with that proposed in Chen Zhiqian's original design. In Chen Zhiqian's vision, these village health workers were volunteers who had been recruited from among the local villagers. Fortified with minimal training and armed with a simple first-aid box, they were expected to perform basic but important duties, including (1) to record births and deaths in the village, (2) to vaccinate the inhabitants of the village against smallpox, (3) to construct their own well in accordance with an approved design for demonstration purposes, and (4) to render first aid, treat common skin diseases and simple surgical wounds, and so forth.[107] Knowing how crucial these people were for his system, Chen was very specific about how to choose these volunteers, especially since the easiest way of doing so would have spelled failure for his vision:

> In theory, the most appropriate way would be to allow the village leader to pick a villager. Under the current situation in our villages however, the village leader is very often the government's tool for collecting people's properties and appropriating them, paying no attention at all to local constructive enterprises. . . . As the result, persons who would have been chosen by the leader would tend to have a personal connection to him and be motivated by personal interests. Once they realized that there was little personal benefit to be had in this role, they would simply abandon their task. The responsibility of choosing village health workers should not be entrusted to the village leader.[108]

Having bluntly stated his distrust of governmental bureaucrats, Chen Zhiqian emphasized the importance of recruiting village health workers from the well-organized civil associations in the local villages, such as the Alumni Association of the Adult People's School of the Mass Education Movement. The local community, especially the organized associations, provided the mechanisms for effectively monitoring and rewarding these village health workers.[109] For Chen, such nongovernmental, local organizations constituted the essence and vitality of his vision of community medicine.

As Chen later recalled movingly in his memoirs,

> Persons examining the model of community medicine developed at Dingxian [i.e., Ding County] should consider it from the standpoint that it was, in fact, a community-based system, rather than that it used lay personnel (village health workers)

> as assistants within that system. The important points were not that the community-based system used voluntary village workers for certain tasks but that village health workers constituted the lowest tier of a health-maintenance system whose effective function began with them, and that their performance was monitored by a strong community organization responsible to ensure that quality was maintained.[110]

Chen concluded that "our solution, therefore, was to make the villagers themselves aware of the problems and arouse their sense of community responsibility and their motivation to work on the problem. That was the philosophy underlying the Dingxian model of community medicine."[111]

Without the support and involvement of "strong community organizations," it is no wonder that the National Health Administration found the idea of the village health worker unreliable, a problem that Chen Zhiqian had fully anticipated. Since the administration had defined its task narrowly as building the organizational infrastructure of State Medicine, it was not interested in cultivating "strong community organizations." Coming from this state-centered view of State Medicine, Chen Wanli reasoned that these kinds of civil organizations did not exist everywhere and that the best channel for recruiting village health workers was therefore from among the teachers of public schools. The reason for this choice was that this group of professionals was already entrusted with many responsibilities—civil administration, surveillance and police, economic and cultural affairs—within the *baojia* system,[112] a traditional, community-based system of law enforcement and civil control that the Nationalists had begun to introduce.[113]

Judging from the candid self-criticism that was published in this internally circulating journal, State Medicine without community support had encountered problems from many fronts. Because of the wartime economy, the National Health Administration had problems finding medical personnel to serve as directors for the county-level public health hospitals, since their salary could barely feed a family.[114] Even if a county was lucky enough to have a hospital director, this person had almost no resources for doing anything besides conducting physical examinations for newly recruited soldiers and opium smokers, and caring for wounded combat personnel.[115] These tasks had never been part of the Ding County model, but they nevertheless became routine challenges for the directors of public health hospitals, who also served as administrative officers of the governmental machinery. The author of one article observed with bitterness that in comparison to "the [Ding] County experiment," which had once been such an inspiration for many people, the current situation was no more than a lifeless and pale imitation. According to this author, some people in fact raised questions about the current strategy of focusing on increasing the number of public health hospitals, to which the central government just responded by stating the principle that "existence is better than nonexistence and more is better than less."[116] As limited resources were further spread thin in this pursuit of numerical growth, however, the staff at public health hospitals "were forced by reality to follow routines and do slipshod work. Consequently, 'public health hospital' became a derogatory term."[117]

In fact, many of these so-called "hospitals" that had been established in 978 counties by the Nationalist government did not even have any sickbeds.[118] In the second group discussion held in May of 1945, the same chairman, Chen Wanli, did not hesitate to admit that the pursuit of numerical growth had brought the public health enterprise to the verge of total collapse due to the lack of appropriate personnel to fill these positions. Chen therefore concluded, "In the future, we should stop building new public health hospitals for the counties that do not have any. With regard to the more than nine hundred public health hospitals already established, we should seriously investigate how to make them capable of carrying out their responsibilities. This is the most worrisome and urgent task for us."[119]

Three months later, Japan surrendered to the Allied nations after the bombing of Hiroshima (August 6) and Nagasaki (August 9). The Nationalists' ever-increasing commitment to health care culminated two years later on January 1, 1947, when the government promulgated a constitution for the Republic of China. Under the category of social security, item 157 stated, "For the sake of improving national health, the state should popularize public health, health-care enterprises, and State Medicine."

Chinese Medicine for Rural China

During the two decades that it took the government and the community of Western medicine to gradually embrace the idea of State Medicine, advocates of Chinese medicine responded with both strong resistance and remarkable enthusiasm. In response to the announcement by the Hunan provincial government of its ten-year project of establishing State Medicine in January of 1934, the Association of National Medicine in Changsha (the provincial capital of Hunan) strongly opposed the idea of turning health care from a private practice into a public service monopolized by the state. What disturbed these traditional practitioners most was the fact that the Hunan provincial government would fill all the positions of State Medicine with graduates from the Western-style Hunan-Yale medical college, and that practitioners of Chinese medicine would thus be excluded from such a system.[120] The association further argued that its members would have no problem at all carrying out the eight tasks of rural health and that their traditional skill of variolation was more reliable and effective than the Western method of vaccination against smallpox.[121] Dominated by hard-liners, the association saw neither an opportunity in the state's deeper involvement in health care, nor the need to reform their practices to take advantage of that opportunity.

Against this doubly negative response, though, others saw this new emphasis on rural health both as an opportunity for political advancement and a challenge related to their efforts in reforming Chinese medicine. For advocates of Chinese medicine, many of the rationales for rural health and State Medicine could easily be turned into powerful arguments for Chinese medicine. Practitioners of Chinese medicine were glad to hear from the mouths of biomedical practitioners that Western medicine was not compatible with China's national economic situation and that it was inappropriate "to sell a medicine that is actually based on European-American living standards to the poor and weak Chinese people."[122]

Among the responses to State Medicine along this line by traditional practitioners, the most systematic and comprehensive one was a book written by Zhu Dian in 1933 entitled *Constructing Three Thousand Village Hospitals*.[123] One of the most remarkable features of this book is the author's clear statement that his concern for rural China was very much inspired by the modern Social Survey Movement and the Rural Reconstruction Movement. In the very first page of this book, prominent names from the Social Survey Movement, such as Li Jinghan and Sidney D. Gamble, already appear with references. In sharp contrast to the stereotypical image of practitioners of Chinese medicine as conservative, backward-looking, and indifferent to national affairs, Zhu fashioned himself as a progressive intellectual, well-versed in cutting-edge research on social science, and seriously concerned with the national crisis. In an advertisement for this book inserted in a journal of Chinese medicine, Zhu's book is described as "a masterpiece of medical politics" (*weida zhi yixue zhengzhi zhuzhuo*).[124] While this advertisement was no doubt a self-congratulatory exaggeration, the book did in fact outline a political strategy focused on current concerns for both the public and the state—in other words, for rural China. As the political function of Western medicine expanded from safeguarding China's sovereignty to winning the hearts and minds of rural villagers, efforts to connect Chinese medicine with the state expanded accordingly from making its practice useful for coping with notifiable infectious diseases to serving the health needs of rural China.

After citing the relevant literature from social science to establish the seriousness of the rural health problem, Zhu went on to argue that practitioners of both Western and Chinese medicine had substantial problems that made their services unsatisfactory for the medical needs of the villages: Western medicine was too expensive, and Chinese medicine lacked a scientific foundation.[125] Zhu then posed some crucial questions: Between practitioners of Chinese and Western medicine as they were currently available, which were better qualified to go to the villages? Which of them was better equipped to function within the village environment? Which of them was more compatible with Chinese national characteristics? "Once we have answers to these questions, we can specify the basis for reforming medicine," concluded Zhu.[126]

In a way, this series of questions is similar to Chen Zhiqian's original three questions, from which he had derived the basic principles for his rural health experiment in Ding County. Both Chen and Zhu strove to develop a health-care service that would be compatible with the social and economic realities of rural China. Nevertheless, from this shared objective Zhu arrived at a radically different conclusion: In his eyes, Chinese medicine, which was currently available in the villages and trusted by villagers, would be better able to serve the role of providing health care for rural residents, as long as it underwent a number of necessary reforms. Instead of promoting State Medicine as the magic bullet for rural health, people like Zhu saw practitioners of Chinese medicine as the solution to this politically important problem, albeit after sufficient retraining and reform.

While Zhu did not specify the aspects in which practitioners of Chinese medicine should be retrained so as to be able to serve the villages, a few features do stand out in his plan. Zhu emphasized the need for a new way of delivering babies,[127] for smallpox vaccination (as against traditional variolation),[128] and for the hospitalization of patients suffering from acute infectious diseases, a measure that was unprecedented in premodern China. Another unique feature was his call for an ophthalmology department in each village hospital.[129] Just as Chen Zhiqian had pointed out, while trachoma was so prevalent in rural China that it constituted a medical crisis, graduates of modern medical schools received almost no training in treating it because people in developed countries were no longer suffering from this disease and therefore gave little attention to it in their medical curricula.[130] Both Zhu and Chen strove to alleviate the characteristic problem of eye diseases in the villages.

In spite of the fact that Zhu's proposed design of village hospitals resembled that of the Ding County experiment, he had no intention of turning practitioners of Chinese medicine into "village health workers," which is precisely what was to happen later during the Communist era. Zhu also considered it important to recruit laypeople, whom he referred to as "health instructors" (*weisheng zhidaoyuan*). Nevertheless, Zhu emphasized that this role would better be served by women.[131] It is true that the village health workers envisioned by Chen Zhiqian can be seen as the predecessors of the Mao-era barefoot doctors, and that many barefoot doctors were practitioners of Chinese medicine. As Chen Zhiqian pointed out in his memoir,[132] however, most practitioners of Chinese medicine were never interested in serving as simple lay village health workers. Instead, they pushed the government to promulgate a legal Regulation for Chinese Medicine (in 1936) and strove to equip themselves with the knowledge and skills necessary for diagnosing and managing notifiable infectious diseases. They endeavored to transform themselves—both in a legal and a technical sense—into competent and state-sanctioned professionals for the peasants who had always trusted and relied on them.

As State Medicine gradually emerged as the most feasible policy for China, practitioners of Chinese medicine also began to look for a place for Chinese medicine within such a system. For example, Shi Yiren, the founder of the influential Shanxi Research Society of Medicine, whom we have already discussed in chapter 8, issued a call for papers in 1936 on the question "whether adopting Chinese medicine and pharmaceutics within the system of State Medicine could contribute to the economic and social well-being of the nation."[133] Among the lengthy and detailed award-winning articles published in response to this question, one of which was written by Shi himself, quite a few agreed that for serving rural residents, "in terms of therapy, Chinese medicine is more appropriate, but in terms of disease prevention, Western medicine is more appropriate."[134] These authors were keenly aware that even if they could claim Chinese medicine to be more compatible with the socioeconomic environment of rural China, it had to undergo substantial reform to acquire the capability of preventing and controlling infectious disease, a practical capability that was crucial to the health of China's rural population.

Perhaps the most concrete effort to connect Chinese medicine with State Medicine took place at the Jiangsu Provincial School of Medical Administration (*Jiangsu yizheng xueyuan*), which was founded by Chen Guofu in the fall of 1934.[135] In creating this school, of which he subsequently served as the first president, Chen Guofu cooperated closely with Hu Ding'an, a German-trained public health expert and the author of the *Plan for Public Health Administration in China*.[136] Chen deliberately chose to call this school a School of Medical Administration, rather than a School of Medicine because he wanted to highlight the fact that it was a very special kind of medical school, which "by no means focused on curative medicine, like other regular medical schools [in China]."[137] Instead, this school had the dual objective of "conducting experiments on institutional innovations in medical administration" and of "creating a new medicine for China that integrates Chinese and Western medicine."[138]

With regard to the first objective, Chen explicitly spelled out that the experimental efforts of this school were aimed at realizing the State Medicine system.[139] For that purpose, he designed a three-level division of medical practitioners to be trained within this school. Between the best-trained physicians of Western medicine, who served the urban areas, and the least-trained practitioners, who served the poor rural areas, Chen designed a two-year Hygiene Training School (*weisheng xunlian ban*) for the middle level, to teach licensed practitioners of Chinese medicine basic science, physiology, and the germ theory. Unlike the other departments of this school, which encountered serious recruitment problems because of their unconventional agenda, this training school was very popular with practitioners of Chinese medicine. Chen therefore urged other schools of Western medicine to follow suit and offer similar kinds of training courses.[140] Nevertheless, some members of the Committee for Medical Education openly criticized this school as "neither donkey nor horse,"[141] and Chen never tried to register his brainchild with the Ministry of Education, even though he was at that time in the powerful position of being the chairman of the Jiangsu Provincial Government. Four years after its founding, Chen resigned from his post as president of the school. He recalled the brief history of this school with sadness because "all of the ideals were terminated without realization."[142]

Many scholars have taken note of the curious fact that the advocates of State Medicine during the Republican period never considered involving practitioners of Chinese medicine, even though they were painfully aware of the lack of medical personnel. In fact, when Frank Oldt presented his article "State Medicine Problems" in 1937, he warned the audience that "there is always the possibility that the practitioners of Chinese medicine will demand recognition by the government and a place in State Medicine."[143] As this warning reveals, an unstated consensus existed among the promoters of State Medicine to guard this system against Chinese medicine. In this sense, even though we can see clear continuities between the Ding County experiment of community medicine, the Republican government's efforts to promote State Medicine, and the State Medicine that was ultimately implemented during the Communist era, a sharp discontinuity existed as well between the first two forms and the Communist version in terms of their opposing attitudes toward the role of traditional Chinese medicine in State Medicine.

The issue of whether or not to incorporate Chinese medicine was explicitly considered at a conference held by the Committee for Medical Education at the Ministry of Education in 1940. When this committee deliberated on proposals for State Medicine, its chairman, Hu Ding'an, pointed out that the objective of this conference was to reform the national medical education system, turning it into a six-tier system that would provide the six different levels of personnel for the operation of State Medicine. As part of this objective, a proposal for training practitioners of Chinese medicine was raised and considered. The committee decided to put this proposal on hold and "establish a research institute to investigate Chinese medicine with scientific methods. As this research will bear fruit after some time, [we will] incorporate these results into the curriculum."[144] During the entire Republican period, this was perhaps the moment when Chinese medicine was closest to being incorporated into State Medicine. Still, it was science that blocked the way to this integration. From the hindsight of history, systematic, large-scale integration became possible only after the rise of a new and alternative vision of science during the Communist era, a vision that allowed a "Chinese science" to differ from the version of science that had been developed in the capitalist West.

In 1947, when the Nationalist government promulgated its new constitution, which included an official governmental policy of State Medicine, many prominent leaders of Chinese medicine—Zhang Jianzhai (1880–1950), Ding Jiwan (1903–63), Lu Yuanlei (1894–1955), Shi Jinmo (1881–1969), and Ren Yingqiu (1914–84)—attended the election of the first members of the National Assembly and the Legislative

Yuan. In a special issue devoted to this historic election, a newspaper of Chinese medicine published a manifesto, addressed to the national electorate, that included five suggestions, the third of which was to "establish State Medicine and provide publicly funded health care."[145]

Ironically, at the time when the Nationalist government wrote the policy of State Medicine into its constitution, a policy that even practitioners of Chinese medicine embraced, Chen Zhiqian did not see this as a time for celebration but as one for a fundamental reformulation of China's public health policy. Chen's provocative article brings us back to Grant's definition of "China's Medical Problem," with which I began this chapter. Ever since Grant had formulated this definition in the 1920s, the reduction of China's "excess mortality rate" had been taken both as the standard measure and central objective of public health.[146] Having pursued this objective for two decades, Chen confessed that he finally realized that the real threat to China's national health was actually not its mortality rate but its birth rate. If China actually succeeded in reducing the mortality rate, there would be a horrible national population crisis lurking around the corner. Therefore, Chen concluded, "Compared with reducing the death rate, it is far more important and urgent to reduce the birth rate."[147]

Less than two years later, the Communist Party defeated the Nationalist Party and took over China. In comparison to the Nationalist state, the Communist state would prove to be a much more competent government for China's public health, especially for its rural population. Moreover, it was also a more powerful and supportive patron of Chinese medicine. Ironically, as a direct consequence of the Communist Party's success in implementing its vision of State Medicine and thereby solving "China's medical problem" as it had been defined by Grant, it created the daunting problem of overpopulation that Chen Zhiqian had warned his compatriots about in 1947.

11 Conclusion: Thinking with Modern Chinese Medicine

Within a month of the March 17 campaign that launched the National Medicine Movement in 1929, the journal *Annals of the Medical Profession* (*Yijie chunqiu*) dedicated a special issue to this campaign. In the preface, Zhang Zancheng, the journal's editor-in-chief and the main organizer of the campaign, made the following remarks: "We have documented the detailed process [of this campaign] to inform our fellow citizens and to preserve it as a token of memory. If someday, as time progresses, Chinese medicine is adopted by people all over the world, they will know how deeply deluded practitioners of Western medicine were at this time. We have collected these documents here for the reference of future historians of medicine."[1]

Instead of seeing themselves as the passive target of a "medical revolution," traditional practitioners at this point in time were keenly aware that they were making history for Chinese medicine. While they felt elated about this prospect, they also knew too well that their opponents and other modernizers would respond with skepticism, anger, and ridicule, or by dismissing it as an act of obscurantism and self-interest. As most advocates of Chinese medicine had great difficulty in articulating the positive value of their endeavor, Zhang published this issue in the hope that its real significance would be appreciated in the remote future, when "Chinese medicine [would be] adopted by people all over the world."[2]

With the exception of the Chen brothers, almost all the progressive intellectuals, Nationalist politicians, and leaders of Western medicine appearing in this book—Sun Yat-sen, Lu Xun, Hu Shih, Yu Yan, Fu Sinian, Liu Ruiheng, Wu Liande, and Chen Zhiqian—would have considered Zhang's wish as nothing but a pipedream, a pitiful reflection of his total ignorance about the Enlightenment and modernity. Nevertheless, had they been able to transport themselves into our current age, they would have been appalled to discover that to a certain degree Zhang's hope for the global future of Chinese medicine has in fact become a reality, a phenomenon that scholars have characterized as the "globalization of Chinese medicine."[3] Since the present book documents the struggle to which Zhang and his fellow practitioners dedicated their lives and draws on the materials that they documented with such care, in this concluding chapter I take up the challenge that Zhang posted for "future historians" eight decades ago, that is, how to make sense of the modern history of Chinese medicine that they helped to create, beginning in the spring of 1929.

To explore the implications of this history, I invite the reader to go beyond the historical narrative and to "think with the modern history of Chinese medicine." After overcoming the main obstacle for appreciating the meaning of this history, namely, the popular ways of framing this history as a purely political history, I recast the historical findings of this book as a heuristic tool for reflecting on a series of ever-expanding issues: (1) the relationship between medicine and the state, (2) the (im)possibility of productive crossbreeding between Chinese medicine and biomedicine, (3) the notion of "China's modernity," and finally (4) the "Great Divide" between modern and premodern.

Medicine and the State

To articulate the meaning of this history for the world history of science and medicine, we have to go beyond the conventional interpretation, which treats it as just a political history about how the state influenced the development of Chinese medicine. While helpful in certain contexts, this understanding is at the same time limiting and incomplete, because it devalues the active and collective endeavors of practitioners of Chinese medicine such as Zhang in creating the modern history of Chinese medicine. By implication, this understanding has added fuel to the idea that Chinese medicine is nothing but a recent construct created by the various Chinese governments for political purposes alone; as such, it would be nothing but a Chinese version of the Lysenko affair. As a result, this political understanding—while partially correct—reinforces the stereotype that the modern history of Chinese medicine does not belong in the history of science proper but merely reflects the twists and turns of political turmoil in modern China. To prevent this reductive, political understanding of the modern history of Chinese medicine, I would like to recast it as a general reflection on the relationship between medicine/science and the state.

The concept of the state that I have drawn on in this book breaks with the intellectual tradition that treats the state as an autonomous actor. Instead of being a story about governmental intervention, in many aspects the history documented in this book is close to the Foucauldian story in which "disciplinary practices come to colonize, compose, and transform the state."[4] Most of the time, it was the practitioners of Chinese and Western medicine who actively forged alliances with the state, mobilized the state, and, in the case of Western medicine, constructed a medical administration for the state and proposed the health-care policy of State Medicine, rather than the other way around. To adopt sociologist Pierre Bourdieu's analytical framework, during the period under discussion the practitioners of both Chinese and Western medicine actually served as "the agents of the state who constituted themselves into a state nobility by instituting the state."[5]

The second sense in which I use the concept of the state in this book concerns the new sources of power and interests generated by the state, or, to put this in Bourdieu's language, the "field of the state." Although conventional wisdom suggests that Chinese medicine was severely suppressed under the Nationalist government, by the end of the Nationalist period, practitioners of Chinese medicine, at least on paper, had achieved the legal status of being equal to the practitioners of Western medicine.[6] It is beyond doubt that state intervention posed a serious challenge to Chinese medicine; nevertheless, the creation of the Nationalist state opened up a whole new horizon of possibilities that had never been accessible to traditional practitioners before the Republican period. As the Chinese counterpart of those in the privileged Western medical profession, practitioners of Chinese medicine consequently demanded the following professional concessions from the state: (1) an official state organ to govern the affairs of Chinese medicine, which they themselves administered; (2) a state-sanctioned licensing system; and (3) the incorporation of Chinese medicine into the national education system. Paradoxically, the alliance between the state and Western medicine, which had caused the most severe challenge ever to Chinese medicine, ended up enabling Chinese medicine to transform itself into a prestigious modern profession, as well as getting its doctrines adopted as official knowledge sanctioned by the state.

In order to realize this unprecedented opportunity for collective social mobility, traditional practitioners started the process of self-politicization, aimed at building an alliance between Chinese medicine and the Nationalist state. Nothing better symbolizes their vision than the official name that they chose for their own profession—National Medicine—although many people continue to understand this name as merely an appeal to cultural nationalism, when it was actually an appeal to the Nationalist state as well. Moreover, at the time when they endeavored to make their profession useful to the state, the Nationalist state itself was in the process of self-exploration and radical transformation. It is in this practical and hence important sense that I have worked to document the history of the coevolution of Chinese medicine, Western medicine, and the state, especially those developments that led all three of them to places that none of them had ever reached before the modern period, that is, to rural China.

Despite their ideological differences, both the Nationalist and Communist parties strove to create a united and sovereign nation of China through the expansion of the state. In their efforts to expand the reach of the state, they encountered strong resistance from the local gentry, who were unwilling to give up control over police, militia, taxation, and other matters of local government. To unseat these local power brokers, who had been the political foundation in imperial China, the Nationalist state after the mid-1920s sought to connect with millions of poor and illiterate peasants whose lives no Chinese state had ever influenced directly before this time.[7] Without this political development, the Nationalist state would not have accepted the innovative vision of State Medicine in the late 1930s, an ambitious policy that was eventually written into the new constitution of 1947 as the national health-care policy. In response to this political concern over rural China and the promotion of State Medicine by biomedical physicians, some advocates and practitioners of Chinese medicine conducted experimental training sessions to teach their fellow practitioners the basic skills of modern public health-care services, so that they could serve rural residents. In light of these innovative endeavors during the Republican period, the later development of the much-acclaimed primary health-care system by the Communist government—one that made extensive use of traditional medicine—should not be considered purely an aspect of the history of Western medicine in China. The advocates and practitioners of Chinese medicine in the Nationalist era also served as the forerunners of this innovative health-care system.

The developments in Chinese medicine that were aimed at improving public health in rural China also show why it is inappropriate to understand the modern history of Chinese medicine only as the "modernization of Chinese medicine." While that concept implies a universal process of reforming the contents of its knowledge and its institutional structures, the specific history involved local and social innovations that allowed Chinese medicine to become an integral part of the health-care system for rural China. Instead of viewing their efforts at innovation as part of a universal process of modernization, the advocates of Chinese medicine at that time strove to make Chinese medicine an integral part of the Nationalist government's project of expanding the state into rural society.

Although the practitioners of Chinese medicine were pursuing professional interests newly created by the state and were thereby willingly serving as agents of the state, they were not simply being forced to accept policies issued by the state. On the contrary, they actively participated in initiating, negotiating, and even designing the policies, regulations, and reform programs for Chinese medicine. In the numerous public debates and controversies that were published in dozens of their medical journals at this time, we can discern which changes they embraced whole-heartedly and which they were forced to accept as their own agenda for the time being. One crucial finding of this study is that their scholarly efforts of reforming Chinese medicine were inseparable from their political objective of assimilating Chinese medicine into the state. As shown in chapter 8, their controversial effort to unify the nosological nomenclature of Chinese medicine was a means for assimilating the schools of Chinese medicine into the national system of education. The general conclusion of this study expresses this kind of inseparable duality: It was precisely because Chinese medicine aspired to become the official National Medicine and to thrive in conjunction with the emerging Nationalist state that it embraced the task of Scientizing Chinese Medicine imposed by the state and thereby also accepted the daunting challenge of developing a new species of Chinese medicine that was "neither donkey nor horse."

Creation of Values

It is for the sake of highlighting the cognitive aspect of this "inseparable duality" between political strategy and scholarly innovation that I have chosen the title *Neither Donkey nor Horse* as the title of the present book. As mentioned in the introduction, this expression provides the key to the question posed in the beginning of this book, that is, How did Chinese medicine transform from being seen as an antithesis of modernity to being the most potent symbol and enterprise for China's exploration of its own modernity? Instead of viewing this transition as a derivative of the political history of modern China, I argue that China's medical history had a life of its own and at times had its own influence on the ideological struggle over the definition of China's modernity and the Chinese state. The key development of this medical history was the rise of a new medicine that was seen by its detractors as being "neither donkey nor horse." Nevertheless, some of its hard-won achievements demonstrated in concrete terms that the relationship between Chinese medicine and modernity was not destined to be antithetical.

The expression "neither donkey nor horse" highlights the fact that this new medicine was defined by its critics as being impossible, pathological, and self-contradictory, a "mongrel" that was infertile and valueless. This deliberately derogatory expression accurately conveys the humiliation and emotional violence that its advocates had to endure during this time. Against this hegemonic discourse of modernity that denied the possibility of productive crossbreeding between the modern and the traditional in principle,[8] the definitive feature in the historical emergence of modern Chinese medicine was the fact that this new medicine took the discourse of modernity (and the accompanying knowledge of biomedicine) seriously and survived the resulting epistemic violence. In this sense, the historic rise of this "neither donkey nor horse" medicine constitutes a defining local innovation for the general history of modernity in China, fundamentally challenging the dominant understanding of modernity in terms of the modern sciences.

The expression "neither donkey nor horse" can help us understand the reformers' struggle because it spells out the central challenge that they had to cope with, that is, the fact that this new medicine risked being seen as a double betrayal of both science/biomedicine and Chinese medicine. The challenge to the reformers lay in the dilemma that they were seen as being committed to the simultaneous pursuit of two seemingly contradictory objectives. On the one hand, they strove under the banner of Scientizing Chinese Medicine to assimilate biomedical knowledge into Chinese medicine and to make Chinese medicine more standardized, systematic, and objective. On the other hand, they strove to preserve and develop the specific features and strengths of Chinese medicine and to safeguard its authenticity. In short, they intended to reassemble Chinese medicine by adapting from science while still retaining its own distinctive identity. According to the critics of this medical syncretism, because one could not pursue these two contradictory objectives at the same time, any efforts of integrating Chinese medicine and Western medicine were bound to betray either one or both of the parental traditions.

To appreciate the efforts to develop a medicine that was self-consciously "neither donkey nor horse," we have to understand that the two conventional demands—"preserve the authenticity of Chinese medicine" and "dissolve Chinese medicine through scientific integration"—were just two polarized extremes. Between these two mutually exclusive positions, a whole spectrum of possibilities existed for combining selective aspects of Chinese medicine with those of Western medicine. To understand how the reformers of Chinese medicine at the time explored the possibilities of developing this new Chinese medicine, a medicine that was bound to be seen as a betrayal of one or the other of its parental traditions, I suggest that we study these historical efforts as a historical process aiming at the "creation of values."

Much of the reformers' struggle to negotiate the modernist criticism of Chinese medicine can be understood as a historical effort to reclaim the value of Chinese medicine. In the name of scientific truth, especially the representationist conception of reality, adversaries of Chinese medicine viewed (1) Chinese medical theories as erroneous, and thus valueless, representations of the world, (2) Chinese drugs as merely raw materials of nature, and (3) traditional practitioners' experience in terms of the quasi-Darwinian concept of instinct. Taking for granted this modernist evaluation of Chinese medicine, the project of Scientizing Chinese Medicine often did not commit itself to appreciate and to develop the value of Chinese medicine.

For those who valued so little about Chinese medicine, it was not that difficult to pursue the first objective of Scientizing Chinese Medicine in a simple-minded fashion. As I have explained in chapter 9, because the practitioners of Western medicine were preoccupied with guarding against crossbreeding between two styles of medicine, during the Republican period they endorsed only one research program in the contested project of Scientizing Chinese Medicine. This was the conventional research protocol of "1–2–3–4–5," officially adopted by the program of Scientific Research on Nationally Produced Drugs. As a political strategy in this medical struggle, this research protocol at once built upon and reinscribed a particular regime of assessing value versus risk—Chinese drugs were just raw materials from nature and should therefore be investigated by biomedical scientists as potentially life-threatening "new" drugs. By devaluating *changshan*, and Chinese drugs in general, as nothing but raw materials of nature, this research protocol manifested what Warwick Anderson described as "perhaps the most distinctively colonial feature of colonial science": that is, "its history can *seem* purely a matter of extraction and appropriation, an insertion of previously valueless objects into a scientific regime with the messy influences of local sociality and politics erased" (emphasis in original).[9] Thankfully, in the anomalous case of *changshan*, the history of validating its antimalarial efficacy not only made visible the politics of knowledge involved but, more importantly, exposed the fact that any research protocol that proceeded in the fashion of "extraction and appropriation" was far from the most cost-effective program to realize the potential value of Chinese medicine.

Even the radical reformer Lu Yuanlei became concerned with the destructive consequences of this kind of scientization of Chinese medicine. Having attended the heated debates surrounding his proposal for reorganizing traditional medicine, Lu pointed out with a heartfelt sense of regret, "Limited by its degree of development, science is currently still unable to thoroughly understand all natural phenomena. Among the traditional methods of our National Medicine, many are effective in practice but difficult to comprehend with scientific theories. As a result, the objective of our reorganizing project should be to help Chinese medicine take advantage of science, but not to turn Chinese medicine into a sacrifice for science."[10]

Lu was saddened to see so much unnecessary "sacrifice" of Chinese medicine in the name of scientization. Motivated by the same sentiment, the reform-minded practitioners of Chinese medicine strove to negotiate the project of Scientizing Chinese Medicine and to rescue what they cherished as valuable elements of Chinese medicine from this project, to which they officially subscribed. More importantly, to cope with the accusation of double betrayal, they had to negotiate both the conception of "scientization" as well as the authenticity of Chinese medicine.

When the practitioners of Chinese medicine were forced to embrace the project of Scientizing Chinese Medicine as their own agenda, they strove at the same time to appropriate the disunity and heterogeneity of science. Instead of essentializing modern science as a homogeneous entity, as implied by the Chinese term *kexuehua* (to scientize), traditional practitioners actively explored the heterogeneity of the biomedical sciences so as to create the conditions for a productive crossbreeding between Chinese medicine and biomedicine. They became particularly interested in disciplines of modern medical sciences that were concerned with aspects as invisible and immaterial as the concept of *qi*-transformation, such as the nervous system,[11] immunology and resistance to disease, the lymphatic system,[12] and hormones.[13] One crucial finding of this study concerns the fact that many practitioners of Chinese medicine did not see it as being in total opposition to science and modernity but recognized the weaknesses and limitations of their practice and strove to accommodate selective elements of science and biomedicine with the grand intention of creating a new Chinese medicine.

Not only did these practitioners refuse to essentialize modern science as a homogeneous entity, they also refused to preserve Chinese medicine as an unchanged, homogeneous tradition. Instead, they strove to identify the valuable subtraditions from many competing "currents" of Chinese medicine and thereby to reassemble a modern Chinese medicine. As I argued in chapters 4 and 8, for example, they used the concept of experience to valorize the relatively more "empirical" traditions of medicine that had developed before the Song dynasty vis-à-vis the style of medicine during the Jin-Yuan period, and to elevate *The Treatise on Cold Damage* over the *Inner Canon of the Yellow Emperor* as a more experience-oriented and thus more important canon of Chinese medicine. Many practitioners were keenly aware that they had to reinvent and reassemble the tradition of Chinese medicine in order to perpetuate it. They purged Chinese medicine of many historical aspects that they considered to be unfortunate, but also reincorporated into it some aspects that had been marginalized in its recent history, such as acupuncture. As discussed at several points in this book, Japanese scholarship had provided crucial resources for this reassembling of modern Chinese medicine.

The selective appropriation of science and the reinvention of Chinese medicine were two mutually informed historical processes. Nothing illustrates this point more clearly than the assimilation of the germ theory of disease into Chinese medicine. As detailed in chapter 8, although practitioners of Chinese medicine recognized the germ theory as the crucial challenge to etiology in Chinese medicine, they strove to incorporate it into Chinese medicine for the specific purpose of improving their medicine's ability to prevent and control epidemic diseases, which, in John Grant's analysis, were the root cause of China's six million deaths per year in excess of the average death rate in the United States. On the other hand, in the name of preserving the "therapeutic value" of Chinese medicine, they argued for the necessity of preserving the traditional nomenclature of pattern (*zheng*) on the grounds of modern immunology and the concept of resistance to disease. In order to achieve their dual objectives of adapting their medicine to the new knowledge of the germ theory while preserving its proven therapeutic and diagnostic techniques, they developed the incipient form of what has since become the defining feature of Traditional Chinese Medicine, namely, "pattern differentiation and treatment determination." As emphasized by this formula, the function of medical theory lay in "intervening in the world" rather than in "representing the world"—in developing effective treatments rather than in understanding the causes of disease. Inasmuch as this formula transcended the confinement of the representationist conception of reality, it dismantled the foundation for viewing Chinese medicine and modernity as irreconcilable opposites.

The critics of "mongrel medicine" (*zazhongyi*) were right to point out that this medicine would have difficulties in reproducing itself; there was absolutely no guarantee that the efforts to create value for this medicine would succeed. Besides, what was going to be perceived as valuable in Chinese medicine was bound to be influenced by the specific sociopolitical situation and would therefore keep changing as time moved on. Nevertheless, these efforts opened the door to thinking of the crossbreeding not in terms of two mutually exclusive objectives, namely, authenticity and scientization, frozen in a logical deadlock that had confounded reformers since the nineteenth century, but in terms of appreciation, translation, and sacrifice, and, most importantly, in terms of values (epistemic, clinical, and socioeconomic). Bolstered by their achievements both in negotiating these modernist discourses and in reassembling Chinese medicine, the reformers took on and, most would argue, succeeded in the historic challenge of developing a medicine that was "neither donkey nor horse." In sharp contrast to the image that their opponents painted of them as backward-looking conservatives, the reform-minded practitioners of Chinese medicine were active agents engaged in realizing the modernity of Chinese medicine, and, by extension, in exploring China's modernity as a whole.

Medicine and China's Modernity: Nationalist versus Communist

At this point, readers familiar with the history of Chinese medicine during the Communist era might wonder about the historical significance and implications of the history documented in this book, which ends before the Communist takeover of China in 1949 and does not seek to account for medical practice and policy in the areas of Communist control before 1949. Within the historiography of modern China, this question of historical significance is by no means peculiar to the topic of Chinese medicine but concerns the general issue of continuity and discontinuity between the Republican and Communist periods. As William C. Kirby has pointed out, scholars attempting to understand Communist China, "have usually bypassed the Republican period on the assumption that it is an interregnum, without heirs."[14] If this approach is problematic in the general history of modern China, it nevertheless appears justifiable in the field of Chinese medicine.[15] Despite the collective struggles of traditional practitioners to win support from the Nationalist government and to reform Chinese medicine, what they ultimately achieved simply pales in comparison to the full-scale, government-supported institutionalization of Chinese medicine that began in the mid-1950s. As scholars have discovered how drastically the implemented medical policies, institutional framework, and standardized textbooks transformed Chinese medicine during this time, they have rightly emphasized that the medicine thus created since the 1950s should be clearly distinguished in name from its traditional predecessor, to mark its difference from the Chinese medicine practiced during the Republican and Late Imperial periods. The English name under which this new form has come to be known is Traditional Chinese Medicine, or simply TCM, which is somewhat ironic and perhaps intentionally misleading. In contrast to such a sharp break in the history of Chinese medicine between the Republican and Communist periods, the history documented here invites scholars to explore the underlying continuity between these two periods and, more importantly, to articulate the crucial difference between the contrasting ways in which these two regimes related themselves to the concept of modernity as defined by science.

Echoing the general understanding of the Communist regime, historians of Chinese medicine have rightly found that the state was much stronger during the Communist period than in earlier times and was capable of implementing comprehensive and supportive policies with regard to Chinese medicine. As a result of this finding, while these scholars have also recognized the importance of the events and processes that took place "when Chinese medicine encountered the state," they have tended to downplay the earlier encounters of Chinese medicine with the weaker Republican state that I have documented in this book. Moreover, when documenting the fundamental policy changes implemented by the Communist Party, often posted as personal decisions by the powerful chairman, Mao Zedong, these scholars have tended to accord little agency to the practitioners and advocates of Chinese medicine. Medical historian Kim Taylor has gone so far as to conclude that the objective of her otherwise groundbreaking book is to "show how at no stage in the development of Chinese medicine in PRC China did the medicine advance 'on its own,' but rather was led at all times by state policies."[16]

Taylor made the key point of her book crystal clear: "The role that Chinese medicine played in the Communist Revolution was passive."[17] I suspect that Taylor came to this conclusion because she followed the intellectual tradition of viewing the state as an autonomous actor and therefore conceived of the relationship between the state and Chinese medicine as a zero-sum game of power balance. From this perspective, recognizing the unprecedented intervention of the Communist state, even if it was for the purpose of supporting Chinese medicine, might have led her to see a lack of agency coming from the community of Chinese medicine. For the sake of breaking with this view of the state, I have emphasized the concept of the "field of the state" in the first section of this chapter.

The history documented in this book can explain a counterintuitive phenomenon: How was the Nationalist state, which was weak in comparison to the Communist state, able to wield such great influence over the development of Chinese medicine? More than anything else, it was for the sake of being recognized by the Nationalist state that traditional practitioners started embracing the project of Scientizing Chinese Medicine and took on the challenge to develop a medicine that was stigmatized as "neither donkey nor horse." The Nationalist state did not exert this influence by means of its power to force its medical policies on the community of Chinese medicine. Rather, it influenced Chinese medicine as a "field of the state," that is, by promising a series of professional interests, or advantages, to be realized as the state implemented a modern system of medical education and health care. To exaggerate slightly, it was precisely the relative weakness of the Republican state that made the vision of National Medicine attractive to the practitioners of Chinese medicine. It was after the traditional practitioners had proven their political power and succeeded in blocking Yu Yan's proposal that they endeavored to extend their influence over the state and thereby in turn fell under the influence of the state. As recounted in Aesop's fable "The North Wind and the Sun," powerful influence can sometimes be realized in rather gentle ways.

More importantly, once the state had exerted its influence by creating these new interests, medical practitioners acted with remarkable agency in their pursuit of these novel interests and in their efforts to reform Chinese medicine. No matter how limited these interests were in their practical scope during that time, the practitioners of Chinese medicine saw themselves as devoted to a movement for collective social mobility. This newly created state capital and the associated possibility for collective social mobility are the keys to understanding the modern history of Chinese medicine. At least during the Republican period, traditional practitioners were not passively responding to the state, but actively and collectively pursuing their upward mobility by way of the state. Once the state became even more powerful, resourceful, and supportive toward Chinese medicine during the Communist era, we would expect to see not less agency from Chinese medicine, but more mutual cooperation between the state and the community of Chinese medicine.

It is very difficult to document this kind of mutual cooperation, especially with regard to the active role played by traditional practitioners, because, in contrast to their visible roles during the Republican period, the leading practitioners who helped organize the profession of Chinese medicine in the 1950s worked largely anonymously.[18] Some evidence of this kind of productive interaction can nevertheless be found in Volker Scheid's work on the history of the Menghe current of Chinese medicine. As Scheid points out, quite a few leading reformers of Chinese medicine, including the crucial figures in this book, Lu Yuanlei and Shi Jinmo, who had been unable to fully implement their vision for reforming Chinese medicine during the Republican period, became high-ranking leaders and advisors for the Communist government. As a rule, moreover, they expressed heartfelt gratitude toward the Communist government for the kind of support that they had desired for Chinese medicine all their lives.[19] Against the apparent discontinuity in the state's attitude toward Chinese medicine, we find significant continuity in terms of the key leaders of Chinese medicine and their agenda for its reform across the political divide between the Republican and Communist eras. In other words, in order to see the continuity in the modern history of Chinese medicine, we have to examine its development from the viewpoint of its practitioners.

It is crucial here to shed light on the agency of traditional practitioners in reforming Chinese medicine, because otherwise we risk reducing its modern history—one that has the potential to challenge the European-centered history of science—to a derivative of political history. This risk unfortunately becomes salient when Taylor concludes that the creation of modern Chinese medicine in the 1950s was based more on "a careful manipulation of its value as a 'cultural legacy' within the particular political, social, and economic circumstance of the early PRC, than any consideration of its actual therapeutic value."[20] It is true that practitioners of Chinese medicine endeavored to fashion their profession into China's "national essence" in the Republican period and into the "people's cultural legacy" in the Communist period. Nevertheless, as I have demonstrated in the preceding chapters, their political maneuvers were by no means aimed at "preserving" the authentic tradition of Chinese medicine but at creating a new Chinese medicine that could be sanctioned by the state and recognized as a valuable contribution to global medicine. By emphasizing the agency of traditional practitioners as we trace their continued efforts in realizing this vision for "modernizing" Chinese medicine, we are able to see more clearly the continuity in the development of Chinese medicine between the Nationalist and Communist regimes. More importantly, we are able to appreciate the crucial fact that these two regimes developed their policies not toward an unchanged "tradition" of Chinese medicine but toward this emerging new Chinese medicine.

In terms of developing such a new Chinese medicine, the difference between the Nationalist and the Communist stances toward Chinese medicine does not lie in the degree of support that each offered. Rather, the difference is qualitative: the two regimes positioned themselves differently vis-à-vis the concept of modernity by adopting different attitudes toward the possibility of creating a new Chinese medicine that was "neither donkey nor horse." The Nationalist government never entertained the idea that China's modernity could be radically different from the one defined by the universality of modern science. Unlike their contemporaries in India, who promoted the idea of "Hindu science" and adopted it as a symbol for the modern Indian nation,[21] the leaders in the Nationalist Party never tried to base their official nationalism on either a traditional form of science or Chinese medicine, not to mention elevating the latter to the status of being a viable alternative to modern science. Instead, the sympathetic leaders in the Nationalist Party, such as the Chen brothers, demanded that practitioners of Chinese medicine embrace the project of Scientizing Chinese Medicine when they decided to support it by setting up the Institute of National Medicine in 1931. This project signaled the crucial development, and the bottom line, for the alliance between the Nationalist government and Chinese medicine: while the Nationalist state did not entertain the idea of Chinese medicine as either a cultural particularism of China or a radical alternative to modern science, it did endorse the controversial possibility that the relationship between Chinese medicine and modern science was not necessarily an antithetical one. Although the Nationalist state recognized this possibility in theory, it was up to the advocates of Chinese medicine to realize this possibility in practice, showing in concrete terms how Chinese medicine could be scientized so as to be a part of modernizing China.

While the Nationalist government did not support a traditional medicine that was coded as antimodern, its support of State Medicine made it possible to imagine a "modern medicine" for China that was radically different from the health-care services in Western societies. Designed to solve China's medical problems under its own specific socioeconomic circumstances, the policy of State Medicine was self-consciously a local innovation and therefore not a duplication of so-called "modern medicine." In fact, one of the key principles that Chen Zhiqian had formulated in the context of his Ding County model was, "Whenever it is possible to not use doctors and nurses, we should *not* use them."[22] The Ding County model thus strove to "de-professionalize" health care whenever possible, as Chen had become keenly aware of how unpractical it was to rely on experts and professionals to serve the rural residents of China. Instead of viewing the Ding County model as a betrayal of the professionalism of modern medicine, Chen—following in the footsteps of his visionary mentor John B. Grant—took pride in developing a vision of State Medicine that anticipated the global trends of medical development.[23]

In order to implement this kind of State Medicine, the Communist government announced its policy of "unifying Chinese and Western medicine" (*tuanjie zhongxiyi*) right after it came to power in 1949.[24] Instead of valuing Chinese medicine for either its therapeutic efficacy or ideological implications, at this early stage the Communists' policy was intended to take advantage of the manpower of the practitioners of Chinese medicine to provide a tentative solution to the nation's urgent medical needs. This objective might sound rather limited in itself, especially when compared with the full-fledged support that the Communist government later provided to Chinese medicine in the mid-1950s. Nevertheless, it constitutes a breakthrough in the history of Chinese medicine in comparison with the Republican period. As indicated in chapter 10, when the Nationalist government had faced the same daunting problem of manpower for realizing its vision of State Medicine in 1940, it had refused to make use of the practitioners of Chinese medicine, even though some advocates and traditional practitioners were eager to integrate Chinese medicine into the health-care system for rural China.

The Communists' policy of "unifying Chinese and Western medicine" thus elevated Chinese medicine from the status of private practice to that of a legitimate part of the national health-care system. By doing so, it realized some traditional practitioners' political vision of connecting (biomedically equipped) Chinese medicine with the health-care needs of rural China. This development vindicated the methodology of going "beyond the dual history of surviving tradition and developing modernity," which I emphasized in the introduction of the present book. It was precisely because modern medicine in China was decidedly not just a local duplication of European medicine that Chinese medicine had the opportunity to become a constitutive member of China's modern health-care system.

Even though the Communist Party had gone beyond the Nationalists in making use of the manpower of the practitioners of Chinese medicine in 1949, at this early stage it had no intention of supporting the continued development of Chinese medicine, not to mention the creation of a new Chinese medicine as a part of global medicine. As Kim Taylor has convincingly demonstrated, the Communist Party did not develop this radical policy until some time between 1954 and 1956. Taylor's important finding goes against conventional wisdom, which assumes that the Communists had held a principled stance toward helping the development of Chinese medicine since the early Yan'an period. In other words, the crucial breakthrough in the Communist era occurred when the Communist Party actively embraced the stigmatized project of "mongrel medicine," legitimizing it in Mao's policy announcement of 1956 as the "integration of Chinese and Western medicine" (*zhongxiyi jiehe*). As Taylor's book demonstrates clearly, neither Mao nor any other leader of the Communist Party ever considered, let alone promoted, this vision for Chinese medicine before the mid-1950s, even though the idea of a new and integrative medicine, characterized as being "neither donkey nor horse," had been cherished by leading practitioners of Chinese medicine since the 1930s.

The crucial question hence becomes why the Communist Party came to support the stigmatized project of "mongrel medicine," a project pursued by the leaders of Chinese medicine but never endorsed by the Nationalist Party. The answer to this question brings us back to the key distinction that I emphasized in the introduction of this book, that is, the distinction between two independent but related struggles: the policy struggle over the role of Chinese medicine in the national health-care system, on the one hand, and the ideological struggle over science as modernity, on the other. The Nationalist Party and the Communist

Party pursued different policies with regard to Chinese medicine, especially the idea of a new Chinese medicine that was "neither donkey nor horse," not because they were committed to different health-care policies but because they had developed different stances toward the concept of science as modernity.

At least two historical developments enabled the Communist Party to develop a stance toward "science as modernity" that differed radically from its Nationalist predecessor. First of all, the Chinese medical community that the Communist Party encountered in the 1950s had already been committed to the "scientization" of Chinese medicine for more than two decades. When the Nationalist state had promoted the project of Scientizing Chinese Medicine, it had not been for the sake of resolving the issue of medical policy but to create for itself an ideological middle ground between cultural nationalism and science as modernity. Despite the fact that its meaning had been notoriously vague and contested, the project of Scientizing Chinese Medicine had nevertheless motivated traditional practitioners to explore *in practice* the possibility of merging the two styles of medicine, validating the efficacy of Chinese medicine with scientific protocols, and standardizing Chinese medicine in order to make it a part of the national health and educational system. In return, these endeavors and achievements in reforming Chinese medicine had given substance to the project of Scientizing Chinese Medicine as the ideological middle ground craved by some Nationalist elites, such as the Chen brothers.

Second, while inheriting the slogan of Scientizing Chinese Medicine from the Nationalists,[25] the Communists radically transformed its meaning, because Marxist ideology allowed them to hold a rather different relationship to the concept of science and to the capitalist West. Scholars have yet to recognize the ambivalent attitude that the Chinese Communist Party held toward the concept of science. Fashioning themselves as promoters of scientific socialism, the Communists on the one hand considered themselves to be the true successors of the May Fourth Movement and pledged their strong faith in "Mr. Science." On the other hand, however, because Marxism emphasizes that the "superstructure" of a society, including science, is always determined by its "base," that is, the relationships of production, the Chinese Communists had much less difficulty in imagining a kind of science that was substantially different from what had originated from the "base" of capitalist societies. If the Nationalist government had opened up the possibility for Chinese medicine and science to coexist and thereby motivated traditional practitioners to appropriate the heterogeneity of science, the Marxist vision of science, especially its dialectical materialism,[26] further offered Chinese medicine the opportunity to fashion itself as "naive dialectical materialism," and thus to become a kind of alternative science to the bourgeois science of the capitalist West.

Thanks to the important ideological transformation brought about by the Communist regime, the Chinese medicine that had been criticized as "neither donkey nor horse" transformed itself from being a monstrous "mongrel medicine" into a much celebrated "new medicine," representing the rise of a "new China." To give a concrete example, the approach of "pattern differentiation and treatment determination," the incipient form of which I have documented in chapter 8, became closely associated with dialectical materialism and thereby became "a defining characteristic of Chinese medicine and a fulcrum from which the modern development [could] proceed."[27] Instead of being perceived as a betrayal of both modern science and Chinese medicine, this new medicine gave the Communist government a justification for its anti-Western stance and a concrete example to showcase to the world as evidence that socialist China had achieved something beyond imagination; that is, it had successfully developed its own science and therefore its own modernity. With the help of this newly envisioned Chinese medicine, Communist China transcended the notion of "European science as modernity." Nevertheless, when the Communists elevated Chinese medicine to the status of being a symbol for "China's modernity," they often conveniently downplayed earlier efforts to "scientize" Chinese medicine and the influence played by Japan in this process during the Nationalist period.

Chinese Medicine and Science and Technology Studies

Since the present book draws on the scholarship of science and technology studies, or science, technology, and society (STS), I conclude by recasting the history documented in this book in terms of the theoretical issues that this scholarly tradition addresses. In particular, I explore the implications of the modern history of Chinese medicine for Bruno Latour's analysis of the "modern Constitution," on which this book is built. In his provocative book *We Have Never been Modern*, Latour tries to understand the normative concept of modernity in terms of what he calls a "modern Constitution," that is, a constitution of the modern world that is unwritten but can be documented on paper, just like the political constitutions of modern nations. From his point of view, the crucial feature of this modern constitution is that its two poles—nature vs. culture, for instance—are mutually defined to be each other's logical opposite. In more concrete terms, we, as modern individuals, define culture in contrast to what we think we have to attribute to nature, and vice versa. As a result, when the two opposites are paired, there is nothing thinkable and recognizable in the middle; all entities should fit neatly into one or the other of the paired categories, nature vs. culture, object vs. subject, science vs. politics.

So far, Latour's modern constitution might still sound similar to conventional understandings of modernity. Nevertheless, Latour argues that precisely because the modern constitution takes a series of pure categories that have been defined as mutually exclusive and separates them into binary opposites, it allows the translation from one into the other and thereby creates numerous hybrids between entities from the series of paired categories. The strength of modernity lies not just in clearly differentiating the nature of its two opposite poles, but in creating hybrids between them. In other words, the more we moderns recognize nothing but entities in these pure opposite categories, the more we become capable of mixing entities and creating hybrids out of these two polar categories. By way of understanding modernity as a constitution, Latour highlights the normative function of modernity in recognizing only some entities and actions as eligible subjects of its regime. While the intermixing of the two poles takes place every day before our eyes, the modern constitution "renders the work of mediation that assembles hybrids invisible, unthinkable, unpresentable."[28]

Moreover, according to Latour, this series of "great divides" provides the crucial justification for Westerners to set themselves apart from the rest of the human race. To use Latour's language, the "internal divide" between nature and culture justifies the "external divide" between moderns and premoderns.[29] To express this point in the terms used by my historical actors, the modern constitution forced the advocates of modern medicine to set themselves apart from those who continued to believe in premodern Chinese medicine. They were forced to recognize such an irreconcilable separation because, unlike all other medical traditions, modern medicine uniquely succeeded in differentiating nature from culture and building its medical doctrine exclusively on the basis of natural reality. Unless the actors were able to recognize the hybrid entities and mediating practices within the "internal divide," it would be impossible for them to recognize the possibility of a hybrid medicine within the "external divide" between modern medicine and premodern Chinese medicine.

Even if Latour is right to claim that this modern constitution worked for Westerners until the late twentieth century, many non-Westerners had long noticed that this constitution did not work for them. In the history documented in the present book, the practitioners of Chinese medicine rightly attributed the critical weakness of Chinese medicine to its lack of connection with the state. Ironically, instead of pursuing a science detached from politics as designated by Latour's "modern Constitution," what the advocates of Chinese medicine had learned from their encounter with Western medicine was the urgent need to create an alliance between their profession and the state so that Chinese medicine could become more like modern biomedicine.[30] Heeding the call from the president of the Institute of National Medicine to "develop a political perspective" for Chinese medicine, the advocates of the National Medicine Movement gave birth to the self-politicization of Chinese medicine, a powerful movement whose influence has continued until this day. Ironically, their modernist efforts to connect Chinese medicine to politics in turn have led to the suspicion that the reputation of Chinese medicine is nothing but politically motivated propaganda supported by the Chinese state.

For a completely different reason, the advocates of Chinese medicine must have realized that the so-called modern constitution did not work for their agenda. Since Chinese medicine was criticized as being "neither donkey nor horse," its reform-minded advocates and practitioners were painfully aware that they were supporting a hybrid medicine that was judged as an oxymoron by this modern constitution. In this sense, they were forced to take a stance rather close to that of the "non-moderns" (or "amoderns"), as characterized by Latour, namely, "anyone who takes simultaneously into account the modern's Constitution and the popularization of hybrids that the Constitution rejects and allows to proliferate."[31] In fact, these reformers of Chinese medicine went further than just "taking into account" the modern constitution and the popularization of hybrids; beginning in the 1930s, they struggled to accomplish the seemingly impossible task of recognizing the modern constitution while developing what they themselves saw as a "mongrel medicine."

According to Latour, it was when the modern constitution no longer "worked" for Westerners in the late twentieth century that they started to recognize the translation and hybridization that had been denied by the "internal divide" of the modern constitution. For the practical purpose of drafting a new constitution that can better recognize and regulate the hybrids abounding in our contemporary world, scholars of STS offer the crucial intellectual resources to reflect critically on the notion of science, which, when conflated with nature, has served as the theoretical foundation of this modern constitution. In light of this understanding, it is fair to say that some non-Westerners have long discovered that this constitution does not fit the purpose of regulating their rapidly modernizing social lives. As soon as practitioners of Chinese medicine embraced as their own the modern constitution in the name of Scientizing Chinese Medicine, they immediately found it necessary to negotiate with this constitution so as to create a revised version that was better adapted to their purpose of realizing and developing the potential values of Chinese medicine. It is therefore no wonder that in their endeavor to negotiate with the modern constitution, they struggled with many of the same issues that concern the modern scholars of STS—the representationist conception of reality, the practice theory of knowledge, the disunity and heterogeneity of science, and the historical epistemology of objectivity, to name just a few that have been touched upon in this book. Unlike modern STS scholars, though, their arguments were not taken seriously by their contemporaries but were belittled as hopelessly antimodern.

Chen Guofu bitterly accepted the accusation that his research of *changshan* was "5–4–3–2–1," that is, in the reverse order of the standard research protocol. In support of this research protocol that began with a clinical trial to evaluate the efficacy of Chinese herbs, Tan Cizhong could claim nothing better than "this approach is illogical in principle but fruitful in practice."[32] It is worth mentioning that at around the same time in the late 1920s, the young medical scientist Du Congming (1893–1986) proposed in what was then the Japanese colony of Taiwan to study traditional herbs with the same kind of approach. Being the first Taiwanese medical doctor to have been trained at the prestigious Kyoto Imperial University, Dr. Du refused to view this approach as either backward-looking or illogical. Rather, he promoted it as being in alliance with cutting-edge developments in global medical research, that is, with the new discipline of experimental therapeutics that had emerged since the 1910s in leading research institutions such as the Rockefeller Institute of Medicine.[33] Even with such impeccable credentials in biomedicine, Du—just like Chen and Tan in China—had to endure the accusation of being antimodern.

In spite of such denigration, the success of *changshan* research proved to be a crucial stepping-stone in the search for antimalarial drugs among Chinese herbs. Thus it paved the way for the discovery of artemisinin in the 1960s, which won the prestigious Lasker Clinical Medical Research Award in 2011. This is the most prestigious award that Chinese scientists have ever received for modern medical research, and it demonstrates the importance of viewing the modern history of traditional Chinese medicine as an essential part of the development of modern health care and medical sciences in China. Moreover, Chen Guofu's reverse-order protocol has recently inspired contemporary scientists to design a systematic "reverse pharmacology" approach for developing an antimalarial phytomedicine.[34] Just like the *changshan* project, such innovative research efforts have substantiated the possibility of productive crossbreeding between modern science and Chinese medicine.

In comparison to the public controversy that had taken place at Sun Yat-sen's deathbed in the spring of 1925, within the following decades it became increasingly—although slowly and circuitously—possible to view Chinese medicine not merely as antithetical to modern biomedicine but to imagine productive crossbreeding between them. While many sociopolitical forces had helped to create this limited space of imagination, medical developments and related epistemological struggles made crucial contributions to its formation in two ways. On the one hand, projects to crossbreed the two styles of medicine became imaginable once more because of reform efforts from within, as practitioners of Chinese medicine incorporated the germ theories of epidemics into Chinese medicine and thereby developed the incipient from of "pattern differentiation and treatment determination." In the case of "experience," moreover, they created a distinctively local version of the modernist epistemology for understanding and reforming Chinese medicine. Equally important, though, was the innovative research by scientists from without, which helped make it possible to imagine a spectrum of research designs and clinical practices across the great divide between modern and premodern.

Although Latour's critiques of the "internal divide" and "external divide" of the modern constitution are closely connected, it has become much easier since the publication of his book in 1993 to recognize the hybrids that had been prohibited by the internal divide. For example, the once obscure phenomenon of the ozone hole discussed in his book has become a crucial issue on the political agenda of the international community and is now being taught in grade schools around the world. In contrast, Latour's vision for revising the external divide so as to transform "the tortuous relationship that we [moderns] have maintained with the other nature-cultures"[35] remains largely unfulfilled. Rejecting the twin positions of universalism and cultural relativism, Latour proposes a nonmodernist perspective that compares moderns and premoderns not in abstract terms, but in terms of "the [concrete] process of establishing relations" between the two.[36] Latour's emphasis on historically specific relationships echoes the postcolonial understanding of modernity in non-European countries; both refuse to view the discrepancy between the historical reality in non-Western societies and the general ideal of European modernity as simply a delay, a failure, or a distortion on the part of non-Western societies.[37] Nevertheless, Latour provides no empirical study in his book to substantiate his idea of "relationalism." As far as I know, few studies exist that examine the long-term historical processes through which non-Westerners have endeavored at once to recognize the modern constitution and to establish a more equal and reciprocal relationship between the two sides of the external divide of modernity. By way of its unique status as "survivor" of modernity and its eight-decade-long efforts to relate traditional practices and theories to modern science and biomedicine, the modern history of Chinese medicine can provide STS scholars with concrete example for reflection on the relatively neglected issue of modernity, especially the relationship between modern science and non-Western knowledge traditions. It is my hope that this modern history of Chinese medicine can help to make visible other "mongrel" practices that so far have been invisible to us. If "we" finally realize that "we have never been modern," this insight should enable us to better appreciate the historic efforts of the "Others" in creating much more diversified modernities for all of us.

Acknowledgments

Over the years of writing this book, I have received encouragement and suggestions from Carol Benedict, Francesca Bray, Karin Chemla, Benjamin Elman, Judith Farquhar, Charlotte Furth, Yung-sik Kim, Arthur Kleinman, Bruno Latour, Yuqun Liao, Shizu Sakai, Nathan Sivin, Paul Unschuld, and most especially Shigehisa Kuriyama and my teachers in Chicago, Robert J. Richards and Prasenjit Duara.

Thanks to teachers and friends who cared deeply about this book—Yungfa Chen, Angela K. C. Leung, Daiwie Fu, Pingyi Chu, and Chia-ling Wu—I made the difficult decision to leave my beloved Tsing-hua University (in Hsinchu, Taiwan) to move to the Institute of Modern History at the Academia Sinica in Taipei in 2008. There I have enjoyed the most stimulating and supportive environment that one could imagine for writing this book, most notably in the form of daily exchanges with many learned colleagues. In particular, this book has benefited from exchanges with Kevin (Ku-ming) Chang, Ping-yi Chu, Jungwon Jin, Paul Katz, Jender Lee, Jian-min Li, Shang-jen Li, Fu-shi Lin, Shi-yung Liu, Miaw-fen Lu, Sung-Chiao Shen, Daw-hwang Wang, Tsui-hua Yung, and Peter Zarrow. In addition, I would like to thank Che-chia Chang for co-organizing with me the Workshop on Medicine, Techno-Science, and Society, as well as the participants

Ko-wu Huang, Ning Jennifer Chang, Cheng-hua Wang, Ling-ling Lien, and Yu-chih Lai, for creating such a lively and genuinely interdisciplinary workshop. Thanks to Director Wen-ji Wang's invitation, I have had the honor of becoming a member of the Institute of Science, Technology, and Society at Yangming University—the first STS institute in Taiwan—where I benefited greatly from the extraordinary intellectual energy of my students and colleagues: Jia-shin Chen, Mei-Fang Fan, Wen-hua Kuo, Yi-ping Lin, Hung-jen Young, and especially Daiwie Fu.

Chinese medicine has introduced me not just to a fascinating field of research but also to an unusual group of brilliant scholars and wonderful friends. As I have joined the field of Chinese medical history relatively late in my academic career, I feel deep appreciation for the warm welcome, stimulating exchanges, and exciting scholarship that I have received from many teachers and fellow scholars around the world. In addition to my colleagues at the Academia Sinica, these include Linda L. Barnes, Gerard Bodeker, Miranda Brown, Chia-feng Chang, Yuan-ling Chao, Hsiu-fen Chen, Ka-wai Fan, Asaf Goldschmidt, Marta Hanson, T. J. Hinrichs, Elisabeth Hsu, Eric Karchmer, Ruth Rogaski, Hugh Shapiro, Vivienne Lo, Guoli Pi, Kim Taylor, Michele Thompson, Yi-li Wu, Nian-qun Yang, Xinzhong Yu, Da-qing Zhang, and most especially Bridie Andrews and Volker Scheid. Ever since Bridie shared with me her unpublished PhD dissertation, she has kept inspiring me with both her scholarship and generosity. Moreover, it was her questioning that first made me realize that my study of the Manchurian plague had transformed the scope and nature of my book project. As Volker shared with me a common interest in creating a dialogue between studies of Chinese medicine and science studies, I have benefited from his friendship as well as his insights through many in-depth discussions, critical suggestions, and collaborative projects over the years.

In comparison to other historians of science, historians of Chinese medicine perhaps constitute an unusually lucky group because many practitioners and research scientists of Chinese medicine have a strong interest in the history of their profession and related theoretical issues. I have learned a great deal from them and feel particularly grateful to Ted J. Kaptchuk, Hen-hong Chang, Yung-hsien Chang, and Yi-tsau Huang for their long-term support and for inviting me to present my work in their institutes and hospitals. I also want to thank Professor Jingshen Zheng and his colleagues at the Chinese Academy of Traditional Chinese Medicine in Beijing; this project became feasible on the day that he showed me piles of Chinese medical journals published during the late Qing and early Republican period—at that time mostly bundled in piles and covered with a thick layer of dust.

It is my great fortune that this book has benefited from comments and suggestions from scholars in the broader fields of East Asian Studies, History of Science and Medicine, and Science and Technology Studies. These scholars' interest and suggestions gave crucial support to a book that by necessity involves technical details of Chinese medicine. I feel grateful to Janet Browne, Susan Burns, Janet Chen, Mei-shia Chen, Ruei-lin Chen, Wan-wen Chu, Juliette Chung, Robert Culp, Christopher Cullen, Steven Epstein, Fa-ti Fan, Ian Hacking, Christopher Hamlin, Larissa Heinrich, Chen Hu, Kuan-chung Huang, Wataru Iijima, Joan Judge, Chin-ming Ka, Tong Lam, Eugenia Lean, Jongtae Lim, Chung-hsi Lin, Rebecca Nedostup, Dagmar Schäfer, Ori Sela, George Steinmetz, Wen-ching Sung, John Warner, Fan-sen Wang, Wen-shin Yeh, Chen-Pang Yeang, and Ka-che Yip.

Over the years, I have presented parts of this book in various settings outside of East Asia; this book owes a great deal to the comments and suggestions from conference participants—many of their names already mentioned above. In addition, I have been extremely fortunate to be included in the formation of the East Asian STS and medical history communities since the beginning of the new century, especially since the founding of the Asian Society for the History of Medicine in 2002, the formation of the East Asian STS Network in 2000, and the launch of the Taiwan-based international journal *East Asian Science, Technology and Society* (now published by Duke University Press) in 2006. Trained mainly in the field of modern China, I feel particularly grateful for the opportunities to present my project in various parts of this region and in front of engaging audiences whose interests do not lie specifically in Chinese medical history. I would like to express my gratitude for invitations and comments from Mary Bullock and Bridie Andrews (China Medical Board, Beijing), Fan Chen (Dongbei University, Shenyang), Gregory Clancey (National University of Singapore), Xi Gao (Fudan University, Shanghai), Sungook Hong (Seoul National University), Angela Leung (Hong Kong University), Wen-yuan Lin (Tsing-hua University, Hsinchu), Hideto Nakajima (Tokyo Institute of Technology), Tsung-Hsueh Lu (Cheng Gong University, Tainan), Dongwon Shin and Buhm Soon Park (Korea Advanced Institute of Science and Technology), Sang-Yong Song (Hanyang University, Seoul), Akihito Suzuki (Keio University), Togo Tsukahara (Kobe University), Kiichiro Tsutani (Tokyo University), and Wen-ji Wang (Yangming University, Taipei).

It is hard to imagine a better English editor for this book than Dr. Sabine Wilms. Being a historian, a teacher, and a professional translator of Chinese medicine herself, she has been more a collaborator than an

English editor. In her editorial work, she has achieved the impossible task of combining two contradictory objectives: a relentless pursuit of scholarly and linguistic perfection in combination with a sensitivity to my voice that has allowed this book to remain completely an expression of my own thinking. By encouraging me to include practitioners of Chinese medicine as a core audience for this book, she led me to take on a much more challenging yet deeply meaningful project than I had envisioned myself.

Marta Hanson, Henrietta Harrison, and Sigrid Schmalzer read through the whole manuscript for the University of Chicago Press and offered extensive comments and helpful suggestions. Warwick Anderson, Christopher Hamlin, and Volker Scheid most kindly accepted my invitation and lent their expertise to comment on selected chapters. I am deeply indebted to each of them, especially for the concrete suggestions on strengthening the book's narrative arch and for Warwick Anderson's critical comments. Special thanks to Eugenia Lean for her friendship and for introducing this book to the East Asian series of the Weatherhead Institute, Columbia University. If Dr. Karen Darling, editor at the University of Chicago Press, had not discovered my book project at an annual meeting of the American Association for the History of Medicine in 2007, this book might never have been published. Karen's enthusiasm supported me to keep working on this project for five more years; the final product owes a great deal to her admirable patience, professional guidance, and numerous encouraging exchanges over all these years. I also want to express my gratitude toward the great team at the Press, including Nick Murray, Julie Shawvan, Sarah Mabus, Micah Fehrenbacher, and especially Mary Gehl, for their inspiring professionalism.

With the exception of two years, one at the Harvard-Yenching Institute and another at the Institute for Advanced Study at Princeton, Taiwan has been the home base for this book. Around the time that Martial Law was lifted in 1987, many talented and idealistic young Taiwanese chose the humanities and social sciences as their calling. I feel very privileged to have been surrounded by these teachers, colleagues, and students, particularly those involved in the four communities of medical history; science, technology, and society (STS); modern East Asian history; and gender studies, whose female scholars have influenced me in important ways both intellectually and personally. In ways that might be difficult to discern directly, this historical study has been inspired, and challenged as well, by their passion to make scholarship a valuable component in creating a better world.

Preparation of the book manuscript was supported by a grant from

Taiwan's National Science Council, a recent innovation to assist in the publication of books in humanities and social sciences. I feel grateful both to this grant (99-2410-H-001-037-MY2) and to Dung-sheng Chen, who initiated this important opportunity when he took charge of its Division of Humanities and Social Science. I also received travel grants from the Rockefeller Archive Center to use its wonderful collection; I owed a great deal to its archivist Tom Rosenbaum for both his helpful suggestions and daily conversations over lunch. I am grateful for a Harvard-Yenching Fellowship in 2004–2005, even though by the end of that year I felt farther away from finishing the book than ever—because my research during that year fundamentally changed the scope of my project. Within my home institute, I often feel the need to work harder to justify the administrative and professional support I have received from its competent staff, especially its secretary, Shuling Chiang. I would like to thank Jung-ju Tsai for preparing the illustrations and my research assistant Mu-yen Chan for his meticulous work in helping me prepare the footnotes.

I am thankful to the publishers of my previously published articles for granting me permission to reuse segments from them. They include "From *Changshan* to a New Anti-Malarial Drug: Re-Networking Chinese Drugs and Excluding Traditional Doctors," *Social Studies of Science* 29, no. 3 (1999); "How Did Chinese Medicine Become Experiential? The Political Epistemology of *Jingyan*," *Positions: East Asian Cultures Critique*, 10, no. 2 (2002); "Sovereignty and the Microscope: Constituting Notifiable Infectious Disease and Containing the Manchurian Plague," in Angela Ki Che Leung and Charlotte Furth, eds., *Health and Hygiene in Chinese East Asia* (Durham, NC: Duke University Press, 2010); and "*Qi*-Transformation and the Steam Engine: The Incorporation of Western Anatomy and the Re-Conceptualization of the Body in Nineteenth Century Chinese Medicine," *Asian Medicine: Tradition and Modernity* 7, no. 2 (2013).

In the final stage of revising this book, I had the privilege of working at the Institute for Advanced Study, Princeton, in Nicola Di Cosmo's remarkable group of East Asian Studies. I benefited greatly from my conversations with scholars there, especially Heinrich von Staden and Christopher Hamlin. The staff at IAS, especially Marian Zelazny and Terrie Bramley, have done much more than I could ever have expected in creating a most productive place for scholars. I also want to thank the Chiang Ching-kuo Foundation for International Scholarly Exchange for its support in producing this book, especially its competent and helpful staff member Maggie Lin.

A colleague jokingly told me that he found one thing admirable about me: "You once said that your scholarly work always assumes your wife as the imaged audience; if she appears to lose interest, you move on to something else." Given that Jean spends most of her waking hours thinking about Cloud Computing, IOT (Internet of Things), and other cutting-edge technologies, her continued interest in and influence on my work are truly admirable. Thanks to Jean for remaining my trusted *tongxue* (schoolmate/companion) while ever renewing the rich meaning of that term.

Notes

CHAPTER 1

1. Lu Xun 魯迅, *Ji wai ji* 集外集 [A collection outside of collection] (Beijing: Renming wenxue chubanshe, 1973), 152.

2. The real situation is not as straightforward. Lu Xun should have known that Sun was indeed taking Chinese medicine during the final days of his life, given that the whole process was widely reported in the major national newspapers. Even the medical prescription ordered by Sun's doctor of Chinese medicine, Lu Zhong'an, was published in newspapers and received intensive comments and criticisms. Instead of commenting on the fact that Sun took Chinese medicine in the end, Lu emphasized the fact that Sun had previously refused it, even though his friends and relatives had urged him to do so. By presenting the story in this way, Lu perhaps wanted to suggest that when Sun took Chinese medicine in the end, he might already have lost his consciousness or control over the treatments he was receiving. For the pulse diagnosis and medicinal formula that Lu prescribed, see *Chenbao* 晨報 [Morning news], February 21, 1925, 3.

3. Ibid.

4. William L. Prensky, "Reston Helped Open a Door to Acupuncture," *New York Times*, December 14, 1995.

5. Frank Ninkovich, "The Rockefeller Foundation, China, and Cultural Change," *Journal of American History* 70, no. 4 (1984): 799–820, esp. 803.

6. James Reston, "Now, About My Operation in Peking," *New York Times*, July 26, 1971.

7. Margaret Chan, "Keynote Speech at the International

Seminar on Primary Health Care in Rural China," in *International Seminar on Primary Health Care in Rural China* (Beijing: World Health Organization, 2007).

8. For the governmental regulation of medical practice in the Song dynasty, see Asaf Goldschmidt, *The Evolution of Chinese Medicine: Song Dynasty, 960–1200* (London and New York: Routledge, 2009).

9. C. C. Chen, *Medicine in Rural China: A Personal Account* (Berkeley and Los Angeles: University of California Press, 1989), 3–4.

10. Joseph S. Alter, ed., *Asian Medicine and Globalization* (Philadelphia: University of Pennsylvania Press, 2005); Elisabeth Hsu, "Introduction for the Special Issue on the Globalization of Chinese Medicine and Mediation Practices," special issue, *East Asian Science, Technology and Society: An International Journal* 2, no. 4 (2009): 461–64.

11. Charlotte Furth, "Becoming Alternative? Modern Transformations of Chinese Medicine in China and the United States," *Canadian Bulletin of Modern History* 28, no. 1 (2011): 5–41; Volker Scheid and Hugh MacPherson, eds., *Integrating East Asian Medicine Into Contemporary Healthcare: Authenticity, Best Practice and the Evidence Mosaic* (Oxford: Elsevier, 2011).

12. For more information on the emergence of Traditional Chinese Medicine, see Judith Farquhar, *Knowing Practice: The Clinical Encounter of Chinese Medicine* (San Franciso: Westview Press, 1994); Kim Taylor, *Chinese Medicine in Early Communist China, 1945–63* (London: Routledge Curzon, 2005); and Volker Scheid, *Chinese Medicine in Contemporary China: Plurality and Synthesis* (Durham, NC: Duke University Press, 2002).

13. Besides my own dissertation, Bridie Andrews's dissertation "The Making of Modern Chinese Medicine, 1895–1937" is one excellent exception to this convention. Since Andrews designed her work to study the dual processes of "the sinification of Western medicine and the creation of the modern professional discipline of Chinese medicine," she rightly emphasizes the association between public health and state building (in the first process) and the reformulation of Chinese medicine through selective assimilation of Western-style medical institutions, theory, and practice (in the second process). Nevertheless, she presents these as two distinct processes and treats them in two separate parts of her thesis, and this organization is not helpful for seeing how the two historical processes became intertwined as practitioners of Chinese medicine started struggling against practitioners of Western medicine for recognition and support from the state in the late 1920s. See Bridie Andrews, "The Making of Modern Chinese Medicine, 1895–1937" (PhD diss., Cambridge University, 1996).

Other important exceptions include Deng Tietao 鄧鐵濤 and Cheng Zhifan 程之范, *Zhongguo yixue tongshi: Jindai juan* 中國醫學通史：近代卷 [General history of medicine in China: The modern era] (Beijing: Renmin weisheng chubanshe, 2000); and Yang Nianqun 楊念群, *Zaizao bingren* 再造病人 [Remaking "patients": Politics of space in the conflict between traditional Chinese medicine and Western medicine, 1832–1985] (Beijing: Zhongguo renmin daxue chubanshe, 2006). Deng's comprehensive and foundational study also separates the history of Chinese medicine and the history of Western medicine into two independent volumes.

14. Histories of biomedicine in China include, for example, AnElissa Lucas, *Chinese Medical Modernization: Comparative Policy Continuities, 1930s–1980s* (New York: Praeger, 1982); Ka-che Yip, *Health and National Reconstruction in Nationalist China: Development of Modern Health Service, 1928–1937* (Ann Arbor, MI: Association for Asian Studies, 1996); Mary Brown Bullock, *An American Transplant: The Rockefeller Foundation and Peking Union Medical College* (Berkeley and Los Angeles: University of California Press, 1980); Iris Borowy, ed., *Uneasy Encounters: The Politics of Medicine and Health in China, 1900–1937* (Frankfurt am Main: Peter Lang, 2009); Zhang Daqing 張大慶, *Zhongguo jindai jibing shehuishi* 中國近代疾病社會史 [Social history of disease in modern China: 1912–1937] (Jinan: Shangdong jioayu chubanshe, 2006). Histories of traditional Chinese medicine include, for example, Ralph Croizier, *Traditional Medicine in Modern China: Science, Nationalism, and the Tensions of Cultural Change* (Cambridge, MA: Harvard University Press, 1968); Zhao Hongjun 趙洪鈞, *Jindai zhongxiyi lunzhengshi* 近代中西醫論爭史 [A history of the struggle between Chinese and Western medicine in the modern period] (Hefei: Anhui kexue jishu chubanshe, 1989); Deng Tietao 鄧鐵濤, ed., *Zhongyi jindaishi* 中醫近代史 [Modern history of Chinese medicine] (Guangdong: Guangdong gaodeng jiaoyu chubanshe, 1999).

15. Benjamin A. Elman, *On Their Own Terms: Science in China, 1550–1900* (Cambridge, MA: Harvard University Press, 2005), 420.

16. Legally speaking, Chinese medicine has been a recognized part of China's modern health-care services since the promulgation of the Regulation on Chinese Medicine (*zhongyi tiaoli*) by the Republican government in 1936.

17. Elman, *On Their Own Terms*, 420.

18. Warwick Anderson, "Postcolonial Histories of Medicine," in *Medical History: The Stories and Their Meanings*, eds. John Harley Warner and Frank Huisman (Baltimore, MD: Johns Hopkins University Press, 2004), 285–307, esp. 299.

19. Dipesh Chakrabarty, *Provincializing Europe: Postcolonial Thought and Historical Difference* (Princeton, NJ: Princeton University Press, 2000), 27.

20. With regard to how the framework of modernization has plagued the writing and reading of the history of science in modern China, see Grace Shen, "Murky Waters: Thoughts on Desire, Utility, and the 'Sea of Modern Science,'" *Isis* 98, no. 3 (2007): 584–96, esp. 586.

21. John Grant, "State Medicine: A Logical Policy for China," *National Medical Journal of China* 14, no. 2 (1928): 65–80, esp. 65.

22. Croizier, *Traditional Medicine in Modern China*.

23. D. W. Y. Kwok, *Scientism in Chinese Thought, 1900–1950* (New Haven, CT: Yale University Press, 1965), 135–60.

24. Hui Wang, "The Fate Of 'Mr. Science' In China: The Concept of Science and Its Application in Modern Chinese Thought," *Positions: East Asian Culture Critiques* 3, no. 1 (1995): 1–68, esp. 33 and 37.

25. It is remarkable that the term *belief* (*xinyang*; 信仰) was deployed in

many aspects of culture, including politics, love, and health care, which just a few years prior had had nothing to do with belief. To give just a few examples, it is well-known that Sun Yat-sen emphasized the importance of *xinyang* in his Three People's Principles. In a love letter to his girlfriend, the famous writer Shen Congwen 沈從文 also emphasized that he had *xinyang* in her. In the context of the doctor-patient relationship, biomedical doctors strove to transform Chinese patients into "qualified patients" who had *xinyang* in both Western medicine and its health-care providers. The use of *xinyang* in these fields reveals that they were in the process of being transformed into a Western monotheistic system. For the rise of *xinyang* in doctor-patient relationships in Republican China, see Sean Hsiang-lin Lei 雷祥麟, "Fu zeren de yisheng yu you xinyang de bingren: Zhongxiyi lunzheng yu yibing guanxi zai minguo shiqi de zhuanbian" 負責任的醫生與有信仰的病人：中西醫論爭與醫病關係在民國時期的轉變 [Accountable doctors and believing patients: The debate between Chinese and Western medicine and the transformation of the doctor-patient relationship in the Republican period], *Xinshixue* 新史學 [New history] 14, no. 1 (2003): 45–96, esp. 92–95.

26. David Arnold, *The New Cambridge History of India, Volume 3, Part 5: Science, Technology and Medicine in Colonial India* (Cambridge: Cambridge University Press, 2000), 15.

27. Fu Sinian 傅斯年, "Suowei guoyi" 所謂國醫 [So-called National Medicine], *Duli pinglun* 獨立評論 (Independent reviews) no. 115 (1934): 17–20, esp. 17.

28. Arnold, *New Cambridge History of India*, 16.

29. It is worth pointing out that the conflation of science with nature and reality is not an isolated phenomenon peculiar to the history under investigation. The "science war" in the United States, as recently as the 1990s, was also in essence a debate about the nature of the relationship between science and reality. As Hacking points out, the key concern of natural scientists involved in that debate was that "constructionists [of science and technology studies] want to unmask metaphysics as a bolster for the authority of science." As in Republican China, what was at stake in this public debate over metaphysics was the cultural authority of science. See Ian Hacking, *The Social Construction of What?* (Cambridge, MA: Harvard University Press, 1999), 95.

30. During the same period, the Nationalist Party campaigned to eradicate all quasi-religious practices, which it labeled as "superstition." See Rebecca Nedostup, *Superstitious Regimes: Religion and the Politics of Chinese Modernity* (Cambridge, MA: Harvard University Asia Center, 2009).

31. To make matters even worse in the eyes of its opponents, Chinese medicine had begun to ally itself with the rising tide of cultural nationalism. From the viewpoint of progressive intellectuals, the cultural nationalists were to blame for obstructing China's difficult modernizing process by attempting to preserve various traditional Chinese institutions in the name of "national essence" (*guocui*). Assuming the universal nature of science and biomedicine, they found it particularly unbearable that the advocates of the National Medicine Movement dared to differentiate medicine on the basis of nationality, claiming Chinese medicine to be the cultural essence of China.

32. This is the first point that Tani E. Barlow emphasizes in her influential elaboration of colonial modernity; see "Introduction: On 'Colonial Modernity,'" in *Formations of Colonial Modernity in East Asia*, ed. Tani E. Barlow (Durham, NC: Duke University Press, 1997), 1–20, esp. 1.

33. Sheila Jasanoff, "Ordering Knowledge, Ordering Society," in *States of Knowledge: The Co-Production of Science and Social Order*, ed. Shelia Jasanoff (London: Routledge, 2004), 13–45.

34. Ian Hacking, *Representing and Intervening* (Cambridge: Cambridge University Press, 1983), 130–46.

35. The closest equivalent in English is "neither fish nor fowl," but this English expression does not address the issue of crossbreeding which is essential to the Chinese expression "neither donkey nor horse."

36. Ban Gu 班固, *Xinjaioben hanshu* 新校本漢書 [Newly proof-read edition of Hanshu], ed. Yang Jialuo 楊家駱 (Taipei: Dingwen chubanshe, 1986), 3616–17.

37. I spell out the important differences between my use of "neither donkey nor horse" medicine and the postcolonial conception of hybrid medicine in the concluding section of chapter 7.

38. Liang Qichao, *Intellectual Trends in the Ch'ing Period*, trans. Immanuel C. Y. Hsu (Cambridge, MA: Harvard University Press, 1959), 113.

39. Lu Yuanlei 陸淵雷, "Ni guoyiyao xueshu zhengli dagang caoan" 擬國醫藥學術整理大綱草案 [Proposal for putting in order Chinese medicine and pharmaceutics], *Shenzhou guoyi xuebao* 神州國醫學報 [Shenzhou bulletin of national medicine] 1, no. 1 (1932): 1–9, esp. 3.

40. For example, in response to the accusation that he supported mongrel medicine, Tan Cizhong explicitly denied the possibility of a mongrel medicine because Chinese medicine and science belonged to two distinct species (*zhong* 種). See Tan Cizhong 譚次仲, "Zhiwen shanghai zhongxi yiyao zazhongyi zhi puoshi yi pian" 質問上海中西醫藥雜種醫之剖視一篇 [Inquiry about the article on the dissection of mongrel medicine published in the Shanghai journal of medical research in China], *Zhongxi yiyao* 中西醫藥 [Journal of the medical research society of China] 3, no. 2 (1937): 102–4, esp. 102.

41. Volker Scheid and Sean Hsiang-lin Lei, "Institutionalization of Chinese Medicine," forthcoming.

42. K. Chimin Wong and Lien-teh Wu, *History of Chinese Medicine* (Taipei: Southern Materials Center, 1985), 770; orig. pub. 1932.

43. Nathan Sivin, *Traditional Medicine in Contemporary China* (Ann Arbor: University of Michigan Press, 1987), 21.

CHAPTER 2

1. C. C. Chen, *Medicine in Rural China: A Personal Account* (Berkeley and Los Angeles: University of California Press, 1989), 20; Ralph Croizier, *Traditional Medicine in Modern China: Science, Nationalism, and the Tensions of Cultural Change* (Cambridge, MA: Harvard University Press, 1968), 45–46; Carl F. Nathan, *Plague Prevention and Politics in Manchuria, 1910–1931* (Cambridge, MA: East Asian Research Center, Harvard University, 1967), 6;

John Z. Bowers, "The History of Public Health in China to 1937," in *Public Health in the People's Republic of China*, ed. Myron E. Wegman, Tsung-yi Lin, and Elizabeth F. Purcell (New York: Josiah Macy Jr. Foundation, 1973), 26–46, esp. 32.

2. Editorial, *National Medical Journal of China* 2, no. 1 (1916): 2.

3. Xi Liang 錫良, "Xuyan" 緒言 [Preface], in *Dongsansheng yishi baogaoshu* 東三省疫事報告書 [Report on the epidemic in the three eastern provinces], 2 vols., ed. Zhang Yuanqi 張元奇 (Fengtian: Fengtian fangyi zhongju, 1911), 4.

4. Charles Rosenberg, "Introduction. Framing Disease: Illness, Society, and History," in *Framing Disease: Studies in Cultural History*, ed. Charles Rosenberg and Janet Golden (New Brunswick: Rutgers University Press, 1992), xiii–xxvi, esp. xviii.

5. L. Fabian Hirst, *The Conquest of Plague* (Oxford: Clarendon Press, 1953), 220.

6. See Andrew Cunningham and Perry Williams, eds., *The Laboratory Revolution in Medicine* (Cambridge: Cambridge University Press, 1992).

7. Andrew Cunningham, "Transforming Plague: The Laboratory and the Identity of Infectious Disease," in Cunningham and Williams, *Laboratory Revolution in Medicine*, 209–44, esp. 234. The microorganism was later renamed *Yersinia pestis*.

8. Liande Wu, *Plague Fighter: Autobiography of a Chinese Physician* (Cambridge: W. Heffer & Sons, 1959), 18.

9. Nathan, *Plague Prevention and Politics*. See also Carsten Flohr, "The Plague Fighter: Wu Lien-Teh and the Beginning of the Chinese Public Health System," *Annals of Science* 53, no. 4 (1996): 360–81.

10. Ramon H. Myers, "Japanese Imperialism in Manchuria: The South Manchuria Railway Company, 1906–1933," in *The Japanese Informal Empire in China, 1895–1937*, ed. Peter Duus, Ramon H. Myers, and Mark R. Peattie (Princeton, NJ: Princeton University Press, 1989), 101–32.

11. William C. Summers, *The Great Manchurian Plague of 1910–1911: The Geopolitics of an Epidemic Disease* (New Haven, CT: Yale University Press), 17.

12. Nathan, *Plague Prevention and Politics*, 50.

13. Ruth Rogaski, *Hygienic Modernity: Meanings of Health and Disease in Treaty-Port China* (Berkeley and Los Angeles: University of California Press, 2004), 187.

14. Wu, *Plague Fighter*, 1.

15. Although Chinese people had long assumed that rats played a role in spreading the plague, it was after the Hong Kong outbreak that transnational scientific elites agreed that rat fleas transmitted the bubonic form of plague.

16. Wu, *Plague Fighter*, 12.

17. Hirst, *Conquest of Plague*, 221.

18. Wu, *Plague Fighter*, 22.

19. Ibid., 12.

20. In response to outbreaks of plagues in India, an International Sanitary Convention was passed in 1903 that provided for the first time "for the

destruction of rats on board ship as a protective measure against plague." Norman Howard-Jones, *The Scientific Background of the International Sanitary Conferences 1851–1938* (Geneva: World Health Organization, 1975), 85.

21. Anon., "Lun fangyi xingzheng yiji zhuyi bushu" 論防疫行政宜極注意捕鼠 [Antiplague administration should absolutely pay attention to catching rats], *Shengjing shibao* 盛京時報 [Shengjing times], January 21, 1911.

22. Summers, *The Great Manchurian Plague of 1910–1911*, 74.

23. Anon., "Beili boshi yanshuoci" 北里博士演說詞 [A lecture given by Dr. Kitasato], *Shengjing shibao* 盛京時報 [Shengjing times], Feburary 24, 1911.

24. Summers, *The Great Manchurian Plague of 1910–1911*, 74.

25. Nathan, *Plague Prevention and Politics*, 32.

26. Wu, *Plague Fighter*, 19.

27. Benedict, *Bubonic Plague*, 63–64.

28. According to both Wu Liande's recollection and the government report, the sanitary police established before the outbreak were not very useful in containing the plague. Wu specifically pointed out the need "to ensure better control by replacing, as far as possible, untrained police in the routine inspection and report work with trained medical staff. The police, thus released, could return to their proper lay duties" (Wu, *Plague Fighter*, 23). After the plague, Wu often openly criticized those sanitary police who knew nothing about modern methods of hygiene except street cleaning. For the reasons why sanitary police were preoccupied with street cleaning, see Yu Xinzhong's fascinating study on night soil, "The Treatment of Night Soil and Waste in Modern China," in *Health and Hygiene in Chinese East Asia*, ed. Angela Ki Che Leung and Charlotte Furth (Durham, NC: Duke University Press, 2010), 51–72.

29. Wu, *Plague Fighter*, 12.

30. Zhang Yuanqi 張元奇, ed., *Dongsansheng yishi baogaoshu* 東三省疫事報告書 [Report on the epidemic in the three eastern provinces], 2 vols., ed. Zhang (Fengtian: Fengtian fangyi zhongju, 1911), vol. 1, chap. 2, 112. This two-volume report restarts page numbers at the beginning of every chapter. In citing it, therefore, I give the chapter number after the volume number to help identify the cited pages.

31. Even ten years after the initial outbreak, when another pneumonic plague broke out in Manchuria in 1921, the diagnostic procedure did not require microscopes in every suspected case. Clearly, instead of offering a universal test for every diagnosis of plague, microscopes actually provided the certainty needed "in case[s] of doubt." J. W. H. Chen, "Pneumonic Plague in Harbin (Manchurian Epidemic, 1921)," *China Medical Journal* 37, no. 1 (1923): 7–17, esp. 13.

32. Wu, *Plague Fighter*, 27.

33. Xi, "Xuyan," 8.

34. Benedict, *Bubonic Plague*, 130.

35. Elizabeth Sinn, *Power and Charity: The Early History of Tung Wah Hospital* (Hong Kong: Oxford University Press, 1989), 164.

36. Mary P. Sutphen, "Not What, but Where: Bubonic Plague and the Reception of Germ Theories in Hong Kong and Calcutta, 1894–1897," *Journal of the History of Medicine* 52, no. 1 (1997): 81–113, esp. 93.

37. Wu Liande, *Dongsansheng fangyi shiwu zongchu daqüanshu* 東三省防疫事務總處大全書 [The complete report on prevention of plagues in the northern three provinces], vol. 4 (n.p.: Dongsansheng fangyi shiwu zongchu, 1924), 116.

38. In sharp contrast, according to J. M. Atkinson, the principal civil medical officer in Hong Kong during the plague there, Chinese residents dumped the bodies of the sick and deceased onto the streets in that city. It was estimated that the proportion of bodies of plague victims that was dumped increased from 25.1 percent in 1898 to 32.7 percent in 1903 as a result of a 1903 law requiring the disinfection of houses on either side of those in which plague-infected rats had been found. See J. M. Atkinson, *A Historical Survey of Plague in Hong Kong since Its Outbreak in 1894* (Hong Kong: n.p., 1907), 23.

39. This might be the precursor of the circulation of rumors about Japanese medicine in Manchuria; see Ruth Rogaski, "Vampires in Plagueland: The Multiple Meanings of *Weisheng* in Manchuria," in Leung and Furth, *Health and Hygiene in Chinese East Asia*, 132–59.

40. Anon., "Sheiwei yibing guo buke zhi ye?" 誰謂疫病果不可治耶？ [Says who that the plague is not treatable?], *Shengjing shibao* 盛京時報 [Shengjing times], Feburary 19, 1911; and Anon., "Jingyou rushi zhi zhongyi hu?" 竟有如是之中醫乎？ [How come there existed this kind of Chinese medicine?], *Shengjing shibao* 盛京時報 [Shengjing times], Feburary 23, 1911.

41. Wu, *Plague Fighter*, 25.

42. Sinn, *Power and Charity*, 170.

43. Zhang, *Dongsansheng yishi baogaoshu*, vol. 1, 33.

44. Liang Peiji 梁培基, "Shang fangbian yiyuan lun zhiyi fangyi shu"上方便醫院論治疫防疫書 [Letter on treating and preventing the plague submitted to the hospital], *Zhongxi yixue bao* 中西醫學報 [Journal for Chinese and Western medicine], no. 16 (1911): 1–7. It is in fact unlikely that this story took place during the plague outbreak of 1894. The 1894 epidemic began in May and ended in August; Dr. Kitasato discovered the plague bacillus on June 14 of the same year. If this event had taken place during 1894, the British public health officers would have had to put into practice this new scientific knowledge within just one month of the discovery. Since in the ten years following the first outbreak in 1894, bubonic plague recurred more or less regularly in Hong Kong, it is highly likely that this event took place at a later time, if it took place at all.

45. Zhang, *Dongsansheng yishi baogaoshu*, vol. 2, chap. 2, 11.

46. Sutphen, "Not What, but Where," 81–113, esp. 100–101.

47. Sinn, *Power and Charity*, 180.

48. Henry Blake, *Bubonic Plague in Hong Kong. Memorandum: On the Result of the Treatment of Patients in Their Own Houses and in Local Hospitals, During the Epidemic of 1903* (Hong Kong: Noronha, 1903), 6.

49. Ibid., 8.

50. Dugald Christie, *Thirty Years in Moukden, 1883–1913* (London: Constable, 1914), 250.

51. John Bowers, *Western Medicine in a Chinese Palace: Peking Union*

Medical College, 1917–1951 (Philadelphia, PA: Josiah Macy Jr. Foundation, 1972), 25.

52. Christie, *Thirty Years in Moukden*, 250.

53. If the Chinese doctors had worn gauze masks and strictly obeyed the modern rules of infection and isolation, just as Chinese doctors did in treating SARS patients in Guangdong in 2003, they would not have received such a fatal lesson in Manchuria. It is interesting to compare the two epidemics: both involved highly virulent infectious diseases, but they led to two very different evaluations of traditional Chinese medicine. See Martha E. Hanson, "Conceptual Blind Spots, Media Blindfolds: The Case of Sars and Traditional Chinese Medicine," in Leung and Furth, *Health and Hygiene in Chinese East Asia*, 228–54.

54. Xie Yongguang 謝永光, *Xianggang zhongyiyao shihua* 香港中醫史話 [A history of Chinese medicine in Hong Kong] (Hong Kong: Sanlian shudian, 1998), 297.

55. Hirst, *Conquest of Plague*, 220. See also Wu, *Plague Fighter*, 48.

56. Wu Youxing 吳有性, *Wenyi lun* 溫疫論 [On febrile epidemics]. Reprinted as *Wenyi lun buzheng* 溫疫論補正 [On febrile epidemics, with appendices] (Taipei: Xinwenfeng, 1985), 11; orig. pub. 1642.

57. Marta E. Hanson, *Speaking of Epidemics in Chinese Medicine: Disease and Geographic Imagination in Late Imperial China* (London: Routledge, 2011), 92–103. esp. 101.

58. See, for example, Nathan, *Plague Prevention and Politics*, 6.

59. "We Chinese have believed in an ancient system of medical practice, which the experience of centuries had found to be serviceable for many ailments, but the lessons taught by this epidemic, which until practically three or four months ago had been unknown in China, have been great, and have compelled several of us to revise our former ideas of this valuable branch of knowledge" (Xi, quoted in Wu, *Plague Fighter*, 49).

60. For example, Li Yushang 李玉尚, "Jindai zhongguo shuyi duiying jizhi" 近代中國鼠疫對應機制 [Coping mechanisms regarding plague in modern China], *Lishi yanjiu* 歷史研究 [Historical research], no. 1 (2002): 114–27.

61. Hirst, *Conquest of Plague*, 220–53.

62. I would like to thank Angela Leung for sharing with me her thoughts about this ambivalent assertion. Professor Leung has told me that many Western scholars found it puzzling that the Chinese concept of *chuanran* often developed along with nonepidemic diseases such as leprosy and smallpox, but not with the plague, which was seminal in the European notion of contagion. Her insightful remark stimulated me to think through the issue here, but the responsibility for this analysis is completely mine. Both Barbara Volkmar and T. J. Hinrichs discovered a medical/moral controversy between contagionists and anticontagionists with regard to the appropriate response to epidemics. The contagionists in the twelfth century seemed to argue that an epidemic could be transmitted by way of direct human contact. While a more detailed discussion of the differences between my argument and their important discoveries is beyond the scope of this chapter, I think they are partially caused

by the lack of differentiation between the terms *contagion* and *infection* in Chinese language.

63. Zhang, *Dongsansheng yishi baogaoshu*, vol. 1, chap. 5, 1–7.

64. Margaret Pelling, "The Meaning of Contagion: Reproduction, Medicine and Metaphor," in *Contagion: Historical and Cultural Studies*, ed. Alison Bashford and Claire Hooker (London: Routledge, 2001), 15–38, esp. 15.

65. Xi, "Xuyan," 5.

66. Fan Xingzhun 范行准, *Zhongguo yufang yixue sixiang shi* 中國預防醫學思想史 [A history of Chinese preventive medical thought] (Shanghai: Huadong Yiwu Shenghuo Chubanshe, 1953), 81–84.

67. Yu Botao 余伯陶, *Shuyi juewei* 鼠疫抉微 [Nuanced points about bubonic plague] (Shanghai: Shanghai guji chubanshe, 1997), 422; orig. pub. 1910.

68. Ibid., 423.

69. Ibid.

70. Ibid., 418.

71. Quoted in Carney T. Fisher, "Zhongguo lishi shang de shuyi" 中國歷史上的鼠疫 [Plague in Chinese history], in *Ji jian suo zhi: Zhongguo huan jing shi lun wen ji* 積漸所至: 中國環境史論文集 [Sediments of time: Environment and society in Chinese history], ed. Cuirong Liu and Mark Elvin (Taipei: Zhongyang yanjiuyuan jingji yanjiusuo, 1995), 673–747, esp. 706.

72. Ibid., 724.

73. Telegram from the viceroy of Zhili, January 28, 1911, quoted in Peng Weihao 彭偉皓, "Qingdai xuantong nianjian dongsansheng shuyi fangzhi yanjiu" 清代宣統年間東三省鼠疫防治研究 [A study of antiplague measures in Manchuria during the late Qing period] (master's thesis, Donghai University, 2007), 69.

74. Zhang, *Dongsansheng yishi baogaoshu*, vol. 1, chap. 5, 4.

75. This revealing phenomenon was pointed out by Bridie Andrews in "Tuberculosis and the Assimilation of Germ Theory in China, 1895–1937," *Journal of the History of Medicine and Allied Sciences* 52, no. 1 (1997): 114–57, esp. 131.

76. W. J. Simpson, *Report on the Causes and Continuance of Plague in Hong Kong and Suggestions as to Remedial Measures* (London: Waterlow and Sons, 1903).

77. Zhang, *Dongsansheng yishi baogaoshu*, vol. 1, chap. 1, 12–17.

78. Summers, *The Great Manchurian Plague of 1910–1911*, 74.

79. Ibid., 89–90.

80. Ibid., 91.

81. *International Plague Conference, Report of the International Plague Conference Held at Mukden, April 1911* (Manila: Bureau of Printing, 1912), 363.

82. David P. Fidler, *International Law and Infectious Diseases* (Oxford: Oxford University Press, 1999), 12. I would like to thank an anonymous reviewer and Charlotte Furth for raising the question of the international origin of notifiable infectious diseases, and Ruth Rogaski for this reference.

83. Ibid., 30.

84. *International Plague Conference*, 362.

85. Ibid., 397.

86. Quoted in Erwin H. Ackerknecht, *A Short History of Medicine* (Baltimore, MD: John Hopkins University Press, 1982), 211.

87. Wataru Iljima 飯島渉, *Pesuto to gendai Chūgoku* ペストと現代中國 [Plague and modern China: Institutionalization of public health and social change] (Tokyo: Kembun shuppan, 2000), 188.

88. Quoted in Anon., "Beili boshi yanshuoci."

89. Wu, *Plague Fighter*, 51. I would like to thank Ian Hacking for posing this factual question to me. It motivated me to think through both the new scientific knowledge created (and contested) during this event and the role that this knowledge played in shaping the course of the event.

90. Summers, *The Great Manchurian Plague of 1910–1911*, 74.

91. Quoted in Wu Yu-lin, *Memories of Dr. Wu Lien-Teh: Plague Fighter* (Singapore: World Scientific Publishing, 1995), 96–97.

92. Xi, "Xuyan," *tian* 57. (This book includes Chinese characters—such as *tian*—in its page numbers.)

93. I would like to thank my colleague Chen Yongfa for sharing with me his admiration for Xi Liang after hearing my talk. His response inspired me to think through the historical implications of this medical event.

94. Because Confucian values were against isolating the sick, the Chinese state condemned families who abandoned their relatives during epidemics. Angela Ki Che Leung, "Organized Medicine in Ming-Qing China: State and Private Medical Institutions in the Lower Yangzi Region," *Late Imperial China* 8, no. 1 (1987): 134–66, esp. 144; Fan, *Zhongguo yufang yixue sixiang shi*, 91–100.

95. Xi, "Xuyan," *tian* 3.

96. Ibid.

97. Ibid.

98. Ibid.

99. Mark Gamsa, "The Epidemics of Pneumonic Plague in Manchuria 1900–1911," *Past and Present* 190, no. 1 (2006): 147–83, esp. 166.

CHAPTER 3

1. Yu Yan 余巖, "Ruhe nengshi zhongguo kexueyi zhi puji" 如何能使中國科學醫之普及 [How to popularize scientific medicine in China], *Shenbao yixue zhoukan* 申報醫學週刊 [Shenbao medical weekly], nos. 109–111 (1935).

2. Ibid.

3. Here I use the term *strategy* in the Foucauldian sense that "the logic [of strategies] is perfectly clear, the aims decipherable, and yet it is often the case that no one is there to have invented them, and few can be said to have formulated them." Michael Foucault, *The History of Sexuality* (New York: Vintage Books, 1990), 95.

4. K. Chimin Wong and Lien-teh Wu, *History of Chinese Medicine* (Taipei: Southern Materials Center, 1985), 589; orig. pub. 1932.

5. Ma Kanwen 馬堪溫, "Qing daoguangdi jin zhenjiu yu taiyiyuan kao"

清道光帝禁針灸於太醫院考 [An investigation into the reasons for the ban on acupuncture and moxibustion from the Imperial Medical Academy by Emperor Daoguang], *Shanghai zhongyiyao zazhi* 上海中醫藥雜誌 [Journal of Shanghai University for Traditional Chinese Medicine] 36, no. 4 (2002): 38–40.

6. George Macartney, *An Embassy to China: Being the Journal Kept by Lord Macartney During His Embassy to the Emperor Chien-Lung, 1793–1794* (London: Longmans; and St. Clair Shores, MI: Scholarly Press, 1972), 284.

7. Anon., *China Centenary Missionary Conference Records Held at Shanghai, April 25 to May 8, 1907* (New York: American Tract Society, 1907), 109.

8. Lien-teh Wu, "Past and Present Trends in the Medical History of China," *Chinese Medical Journal* 53, no. 4 (1938): 313–22, esp. 318.

9. W. G. Lennox, "A Self-Survey by Mission Hospital in China," *Chinese Medical Journal* 46 (1932): 484–534.

10. James L. Maxwell, "A Century of Medical Mission in China," *China Medical Journal* 34 (1925): 636–50.

11. Anon., *China Centenary Missionary Conference*, 110.

12. As Tong Lam points out, since the late nineteenth century, the lack of precise numbers for calculating the size of China's population was taken as the strongest evidence that the Chinese did not care about "facts." Tong Lam, *A Passion for Facts: Techno-Scientific Reasoning, Social Surveys, and the Chinese Nation in the Early Twentieth Century* (Berkeley and Los Angeles: University of California Press, 2011).

13. The China Medical Commission of the Rockefeller Foundation, *Medicine in China* (Chicago: University of Chiacgo Press, 1914), 1. The same concern about promoting public health in China was mentioned in Harold Balme, *China and Modern Medicine: A Study in Medical Missionary Development* (London: United Council for Missionary Education Publication, 1921), 170–71.

14. Paul A. Varg, *Missionaries, Chinese, and Diplomats: The American Protestant Missionary Movement in China, 1890–1952* (Princeton, NJ: Princeton University Press, 1958), 92.

15. G. H. Choa, *"Heal the Sick" Was Their Motto: The Protestant Medical Missionaries in China* (Hong Kong: Chinese University Press, 1990), 112.

16. Anon., *China Centenary Missionary Conference Addresses* (Shanghai: Methodist Publishing House, 1907), 28.

17. The China Medical Missionary Association had discussed whether full membership should be granted to Chinese graduates in China since 1905 but did not reach that conclusion until the end of the Qing dynasty. Since Yan Fuqing received his M.D. degree from Yale Medical School and then worked in *Xiangya* (Yale-in-China) Medical School, he became the first Chinese in 1910 to be admitted to active membership. In that same year, Wu Liande was nominated as an honorary member, because he was not affiliated with Christian institutions. Wong and Wu, *History of Chinese Medicine*, 562–63.

18. Li Jingwei 李經緯, *Xixue dongjian yu zhongguo jindai yixue sichao* 西學東漸與中國近代醫學思潮 [The diffusion of Western learning into the East

and modern medical thoughts in China] (Hubei: Kexue jishu chuabanshe, 1990), 59.

19. Chang Che-chia 張哲嘉, "Qingmo baike quanshu zhong de yixue lunshu" 清末百科全書中的醫學論述 [The ideas and contents in the section on medicine of the late Qing encyclopedia], *Taiwan wenxue yanjiu jikan* 台灣文學研究集刊 [NTU studies in Taiwan literature] 2 (2006): 59–78.

20. Shigehisa Kuriyama, "Between Mind and Eye: Japanese Anatomy in the Eighteenth Century," in *Paths to Asian Medical Knowledge*, ed. Charles Leslie and Allan Young (Berkeley and Los Angeles: University of California Press, 1992), 21–43.

21. Ruth Rogaski, *Hygienic Modernity: Meanings of Health and Disease in Treaty-Port China* (Berkeley and Los Angeles: University of California Press, 2004), 141.

22. For Japanese colonial medicine in Taiwan, see Mike Shiyong Liu, *Prescribing Colonization: The Role of Medical Practices and Policies in Japan-Ruled Taiwan, 1895–1945* (Ann Arbor, MI: Association for Asian Studies, 2009); and Fan Yanqiu 范燕秋, *Yibing, yixue yu zhimin xiandaixing-rzhi Taiwan yixueshi* 疫病、醫學與殖民現代性一日治台灣醫學史 [Epidemics, medicine and colonial modernity—A history of medicine during the Japanese colonial period] (Taipei: Daoxiang chubanshe, 2005). Also see articles in Li Shangjen 李尚仁, ed., *Diguo yu xiandai yixue* 帝國與現代醫學 [Empire and modern medicine] (Taipei: Lianjing chubanshe, 2008). For the case in Manchuria, see Robert John Perrins, "Doctors, Disease, and Development: Engineering Colonial Public Health in Southern Manchuria, 1905–1926," in *Building a Modern Nation: Science, Technology, and Medicine in the Meiji Era and Beyond*, ed. Morris Low (New York: Palgrave Macmillan, 2005), 103–32; Ruth Rogaski, "Vampires in Plagueland: The Multiple Meanings of *Weisheng* in Manchuria," in *Health and Hygiene in Chinese East Asia*, ed. Angela Ki Che Leung and Charlotte Furth (Durham, NC: Duke University Press, 2010), 132–59; Mariam Kingsberg, "Legitimating Empire, Legitimating Nation: The Scientific Study of Opium Addiction in Japanese Manchuria," *Journal of Japanese Studies* 38, no. 2 (2012): 325–51.

23. Wang Yi 王儀, "Yu guoren yan yishi shu" 與國人言醫事書 [A discussion with fellow citizens on medical matters], *Yiyao xuebao* 醫藥學報 [Journal of medicine and pharmacy] 2 (1907): 1–10, esp. 3.

24. Morris Low "Colonial Modernity and Networks in the Japanese Empire: The Role of Goto Shinpei." *Historia Scientiarum* 19, no. 3 (2010): 197–208.

25. Wang, "Yu guoren yan yishi shu," 6.

26. Yu Xinzhong, "Treatment of Night Soil and Waste in Modern China and Remarks on the Development of Modern Concepts of Public Health," in Leung and Furth, *Health and Hygiene in Chinese East Asia*, 51–72.

27. Lien-teh Wu, *Plague Fighter: Autobiography of a Chinese Physician* (Cambridge: W. Heffer & Sons, 1959), 49.

28. Benjamin A. Elman, *On Their Own Terms: Science in China, 1550–1900* (Cambridge, MA: Harvard University Press, 2005), esp. chapter 10.

29. Warwick Anderson and Hans Pols, "Scientific Patriotism: Medical Science and National Self-Fashioning in Southeast Asia," *Comparative Studies in Society and History* 54, no. 1 (2012): 93–113, esp. 96.

30. Wu, *Plague Fighter*, 279.

31. Wu Liande (Lien-teh Wu) 伍連德, *Dongsansheng fangyi shiwu zongchu daqüanshu* 東三省防疫事務總處大全書 [The complete report on prevention of plagues in the northern three provinces] vol. 4 ([No city]: Dongsansheng fangyi shiwu zongchu, 1924), 113.

32. Ibid.

33. Sanetô Keishû 實藤惠秀, *Zhongguoren liuxue riben shi* 中國人留學日本史 [History of Chinese overseas study in Japan], trans. Tan Ruqian 譚汝謙 and Lin Qiyan 林啟彥 (Hong Kong: Chinese University Press, 1982), 68–71.

34. Until the 1920s, well-to-do Chinese families still encouraged their youngsters to study law in preparation for governmental positions and therefore did not urge them to study the natural sciences. See Nathan Sivin, "Preface," in *Science and Medicine in Twentieth-Century China: Research and Education*, ed. John Z. Bowers, J. William Hess, and Nathan Sivin (Ann Arbor: Center for Chinese Studies, University of Michigan, 1988), xi–xxxvi.

35. Scheid, *Currents of Tradition in Chinese Medicine 1626–2006*, esp. chap. 2; and Yuanling Chao, *Medicine and Society in Late Imperial China: A Study of Physicians in Suzhou, 1600–1850* (New York: Peter Lang, 2009).

36. Balme, *China and Modern Medicine*, 62.

37. John Z. Bowers, *Western Medicine in a Chinese Palace: Peking Union Medical College, 1917–1951* (Philadelphia, PA: Josiah Macy Jr. Foundation, 1972), 17.

38. Balme, *China and Modern Medicine*, 109.

39. Anon., "Quan xiyi xiaoyin" 勸習醫小引 [An encouragement for the study of medicine], *Zhonghua yibao* 中華醫報 [Chinese medical news] 1, no. 1 (1912): 1.

40. Shu Xincheng 舒新城, Jindai Zhongguo Liuxue Shi 近代中國留學史 (A history of Chinese overseas study in recent times) (Shanghai: Zhonghua Shuju, 1989 (1933)), 28–33.

41. Ibid., 76.

42. Further study is needed to find out why both the Qing government and overseas Chinese students in Japan were not particularly attracted to medicine, which was so essential to Japan's experience of modernization. Nevertheless, the situation started to change in 1908, when the Qing government contracted with the Japanese Ministry of Education to send students to five designated Japanese colleges for postgraduate studies. One of them was Chiba Medical College, where ten Chinese students were sent every year. See Sanetô, *Zhongguoren liuxue riben shi*, 50.

43. In comparison, Japanese physicians and medical scientists founded two professional associations, the Meiji Medical Association and the Great Japan Medical Society, in the same year of 1893. See James R. Bartholomew, *The Formation of Science in Japan* (New Haven, CT: Yale University Press, 1989), 87–88.

44. Wong and Wu, *History of Chinese Medicine*, 604.

45. It is noteworthy that during the Republican period, quite a few prominent medical leaders and scientists were originally British citizens, such as Wu Liande from Malay, Lin Kesheng 林可勝 (R. K. S. Lim, 1897–1969) from Singapore, and Li Shufen 李樹芬 (1887–1966) from Hong Kong.

46. Larissa N. Heinrich, "Handmaids to the Gospel: Lam Qua's Medical Portraiture," in *Tokens of Exchange: The Problem of Translation in Global Circulation*, ed. Lydia H. Liu (Durham, NC: Duke University Press, 1999), 239–75.

47. Wong and Wu, *History of Chinese Medicine*, 605.

48. Mary Brown Bullock, *An American Transplant: The Rockefeller Foundation and Peking Union Medical College* (Berkeley and Los Angeles: University of California Press, 1980), 142; Ka-cheYip, *Health and National Reconstruction in Nationalist China: Development of Modern Health Service, 1928–1937* (Ann Arbor, MI: Association for Asian Studies), 19; and Arthur M. Kleinman, "The Background and Development of Public Health in China: An Exploratory Essay," in *Public Health in the People's Republic of China: A Report of Conference*, ed. Myron E. Wegman, Tsung-yi Lin, and Elizabeth F. Purcell (New York: Josiah Macy Jr. Foundation, 1973), 5–25, esp. 12.

49. Liping Bu, "Public Health and Modernization: The First Campaigns in China, 1915–16," *Social History of Medicine* 22, no. 2 (2009): 305–19.

50. E. S. Tyan, "A Plea for a Campaign of Public Health Education in China," *China Medical Journal* 29 (1915): 230–34.

51. Anon., "Proceedings of China Medical Missionary Association Conference," *China Medical Journal* 37 (1923): 301.

52. Anon., "Preventive Medicine," *China Medical Journal* 38, no. 1 (1924): 44–47, esp. 45.

53. Maxwell, "Medical Missions in China: A Time for Re-Statement of Principle," 584–85.

54. Yip, *Health and National Reconstruction*, 101.

55. Lien-teh Wu, "Some Problems before the Medical Profession of China," *National Medical Journal of China* 3 (1917): 5–9.

56. It is worth mentioning that the new role of public health officer substantially resolved the low-status problem of medicine that had troubled the medical missionaries for decades. To put it simply, the field of public health could enable its experts, such as Wu, to hold important positions within the government. When answering the question he posed to himself in his *Reminiscences* (see n. 114 below), "How it was that such a large percentage of PUMC graduates selected public health," John B. Grant, the widely respected father of Chinese Public Health, attributed this remarkable phenomenon to the "psychology" that, in terms of evaluating success, Chinese people preferred governmental positions to money or anything else. See Lien-teh Wu, "Some Problems before the Medical Profession of China," *National Medical Journal of China* 3 (1917): 5–9.

57. By now Wu was widely regarded as a pioneer of public health in China, but he was trained as a physician and medical scientist, not a public health expert. When John B. Grant was sent by the Rockefeller Foundation's China Medical Board to evaluate health services in Manchuria in 1921, he

was extremely critical of both the North Manchurian Plague Service and its leader. With regard to the service, Grant concluded that it had "failed utterly to make any attempt to carry out its stated object of introducing general sanitary measures in the regions in which it works." As to the reasons for this institutional failure, Grant attributed it to Wu's one-man show and suggested that "Dr. Wu never had any Public Health training. His knowledge of the field of Public Health administration is really superficial, although he has long been looked upon as one of the leaders of Public Health in China." If we recall the title of Wu's autobiography, *Plague Fighter*, we realize that the role Wu aspired to serve was indeed quite different from that of an architect or constructor of a public health system. See John Grant, "North Manchurian Plague Prevention Service," RF folder 347, box 55, series 2, RG 5, 1921, Rockefeller Foundation Archive Center, Sleepy Willow, NY.

58. Elected as its first president, Yan Fuqing was the son of an Episcopalian minister whose biomedically trained sons became prominent leaders in various fields of China's modernization, including medicine, engineering, and diplomacy. After receiving his M.D. degree from the Yale School of Medicine (1909), Yan became the first biomedically trained Chinese doctor to join Yale-in-China in 1910 and thereby the first active Chinese member of the China Medical Missionary Association. The recruitment of Yan was highly unusual and became a hallmark policy for the Yale-in-China mission. In cooperation with Edward H. Hume (1876–1957), Yan served as the first dean of Xiangya Medical College in Changsha at the beginning of his influential career as a leader and educator. See Nancy E. Chapman and Jessica C. Plumb, *The Yale-China Association: A Centennial History* (Hong Kong: Chinese University Press, 2002), 20; and Qian Yimin 錢益民 and Yan Zhiyuan 顏志淵, *Yan Fuqing zhuan* 顏福慶傳 [A biography of Yan Fuqing] (Shanghai: Fudan daxue chubanshe, 2007).

59. F. C. Yen, "Presidential Address," *National Medical Journal of China* 2 (1916): 4–9, esp. 8.

60. This parallel shift is illustrated by Wu's chapter title, "Substantial Progress in Public Health and Other Medical Activities Mainly under Direction of Chinese Physicians," in Wong and Wu, *History of Chinese Medicine*, 656.

61. James C. Thomson Jr., *While China Faced West: American Reformers in Nationalist China, 1928–1937* (Cambridge, MA: Harvard University Press, 1969), 39.

62. When Welch and Dr. Abraham Flexner were invited to join the Chinese Medical Mission in 1915, they were asked "to advise the Board in regard to the measures to be taken for the promotion of public health and medical education in the Republic of China." Bowers, *Western Medicine in a Chinese Palace*, 49. See also Mary E. Ferguson, *China Medical Board and Peking Union Medical College: A Chronicle of Fruitful Collaboration, 1914–51* (New York: China Medical Board of New York, 1970), 16.

63. Ferguson, *China Medical Board*, 20.

64. China Medical Commission of the Rockefeller Foundation, *Medicine in China*, 91.

65. Concerning Grant's background and his career during the Republican period, see Bullock, *An American Transplant*, 134–61.

66. John B. Grant, "A Proposal for a Department of Hygiene," RF folder 531, box 75, series 2, RG 2, 1923, Rockefeller Foundation Archive Center, Sleepy Willow, NY.

67. Ibid., 6.

68. Ibid.

69. Thomson, *While China Faced West*, 1–18 and 43–75.

70. Yip, *Health and National Reconstruction*, 28.

71. Y. F. Chang, "Medicine and Public Health Service under the Nationalist Government," *National Medical Journal* 15 (1929): 114–16.

72. Lien-teh Wu, "A Survey of Public Health Activities in China since the Republic," *National Medical Journal of China* 15, no. 1 (1917): 1–6, esp. 4.

73. Wu, *Plague Fighter*, 23.

74. To understand why street cleaning became the symbol of hygiene in late Qing and early Republican China, see Yu Xinzhong, "The Treatment of Night Soil and Waste in Modern China," in Leung and Furth, *Health and Hygiene in Chinese East Asia*, 51–72.

75. Wong and Wu, *History of Chinese Medicine*, 664.

76. Frank Ninkovich, "The Rockefeller Foundation, China, and Cultural Change," *Journal of American History* 70, no. 4 (1984): 799–820, esp. 803.

77. Ibid., 804.

78. Roger Greene, "Memorandum on Grant's Plan for a Hygiene Program for the P.U.M.C. Submitted by Him on October 8th," RF folder 531, box 75, RG2, 1924, Rockefeller Foundation Archive Center, Sleepy Willow, NY.

79. Instead of taking public health as his top priority, throughout his two decades of involvement with the Rockefeller Foundation in China, Greene focused on instilling the "spirit of science" into the Chinese minds.

80. Greene, "Memorandum on Grant's Plan," 1–2.

81. Ibid., 2.

82. Ibid., 2–3.

83. In fact, just three years later (December 1927) Greene withdrew his reservation about public health and supported Grant's proposal for building a modern public health service in Beijing. In contrast to the past, when "we were not in a position to suggest competent personnel," Greene felt that Grant had built up sufficient personal contacts and prepared competent personnel. Moreover, given the fact that "the only change that is conceivable at Peking in the near future is the coming in of a Nationalist government," Greene observed that most of Chinese public health men were "Southerners and in sympathy with the Nationalist movement" (Roger Greene, "Letter to George E. Vincent," RF series 601 J, 1927, Rockefeller Foundation Archive Center, Sleepy Willow, NY.

84. When John Grant and his colleagues started their innovative health demonstration station in Peking, they very soon discovered that "an absence of public health in China is due principally to the lack of knowledge on the part of Cabinet Ministers and higher administrative government officials of

the scope of modern public health." See John Grant, "Annual Report, 1924–25," 3.

85. According to Mary Brown Bullock, Liu Ruiheng participated in the drafting of Grant's memorandum, but she might have mistaken the year as 1928 instead of 1927; see Bullock, *An American Transplant*, 152.

86. Liu Ruiheng (also known as J. Heng Liu) received his medical training from Harvard University in 1913. He served as president of PUMC from 1926 to 1934 and as president of the National Medical Association of China from 1926 to 1928. Although Liu was trained in surgery, he served as the deputy minister for the Ministry of Health (1929–30) and then as its minister (1930–35). For his activities, publications, and people's recollections of him, see Irene Ssu-chin Liu 劉似錦, *Liu Ruiheng boshi yu zhongguo yiyao ji weisheng shiye* 劉瑞恆博士與中國醫藥及衛生事業 [Dr. J. Heng Liu and the development of medicine and health care in China] (Taipei: Taiwan shangwu yinshuguan, 1989).

87. Michael H. Hunt, "The American Remission of the Boxer Indemnity: A Reappraisal," *Journal of Asian Studies* 31, no. 3 (1972): 539–60, esp. 556.

88. Their discussion focused on the paper "The Use of Portion of the British Indemnity Fund for Public Health Work in China," presented by its first president, Yan Fuqing 顏福慶.

89. Wong and Wu, *History of Chinese Medicine*, 668–69.

90. The Association for the Advancement of Public Health in China, *On the Need of a Public Health Organization in China* (Beijing: Association for the Advancement of Public Health in China, 1926), 19.

91. Ibid., 20.

92. Ibid., 22.

93. Robert E. Bedeski, *State-Building in Modern China: The Kuomintang in the Prewar Period* (Berkeley and Los Angeles: Institute of East Asian Studies, University of California, 1981); Joshua A. Fogel, ed., *The Teleology of the Modern Nation-State: Japan and China* (Philadelphia: University of Pennsylvania Press, 2005).

94. John B. Grant, "Public Health and Medical Events During 1927 and 1928," in *The China Year Book, 1929–30*, ed. H. G. W. Woodhead (Shanghai: Christian Literature Society, 1928), 111–33, esp. 128.

95. Ibid., 111.

96. Marianne Bastid, "Servitude or Liberation? The Introduction of Foreign Educational Practices and Systems to China," in *China's Education and the Industrialized World*, ed. Ruth Hayhoe and Marianne Bastid (New York: M. E. Sharpe, 1987), 3–20, esp. 14.

97. Qian and Yan, *Yan Fuqing zhuan*, 68.

98. Duara's important study on Chinese state building concludes that since 1900 the Chinese state has gradually been caught up in a logic of "modernizing legitimatization." See Prasenjit Duara, *Culture, Power, and the State: Rural North China, 1900–1942* (Stanford, CA: Stanford University Press, 1988).

99. John B. Grant, "Provisional National Health Council." RF folder 529, box 75, China Medical Board, 1927, Rockefeller Foundation Archive Center, Sleepy Willow, NY.

100. Guo Tingyi 郭廷以, *Zhonghua minguo shishi rizhi* 中華民國史事日誌 [A chronology of the Republic of China], vol. 2 (Taipei: Institute of Modern History, Academia Sinica, 1984), 157.

101. Yip, *Health and National Reconstruction*, 45.

102. According to the exchange of letters between Grant and Yan, Grant first proposed the idea of a Ministry of Health to Yan. They cooperated in persuading the Nationalist leaders in Wuhan, and for a short while it looked possible that Yan would serve as the first Minister of Health. See Qian and Yan, *Yan Fuqing zhuan*, 92–93.

103. Grant, "Provisional National Health Council," 1.

104. Ibid.

105. Ibid., 2.

106. "Through government it would be possible to give the population a minimum degree of the benefits of curative medical science at a relative small cost, which will take decades to become available if left to the evolution of a private curative medical profession." Ibid., 6.

107. Ibid.

108. Ibid., 4.

109. Another remarkable feature shared by the two proposals was their acute historical consciousness; both proposals situated themselves in the context of historical developments. Liu's proposal viewed the contemporary Chinese situation as comparable to that of Britain on the eve of the Poor Low Reform in 1842; he foresaw that public health in China would go through the "normal course" of development from the cities, to provincial areas, and finally to "effective national health administration." In sharp contrast, Grant's memorandum urged China to skip these evolutionary stages and jump to adopt the most advanced health policy recently developed in Britain after the First World War, especially its establishment of an independent Ministry of Health in 1919. It was hoped that the most advanced state medical administration would provide the engine for transforming the premodern social and medical environment in China. More than persuading the Nationalist elites to adopt his medical policy and to create a new branch of the government, Grant endeavored to make them subscribe to his innovative vision of the state. See the Association for the Advancement of Public Health in China, *On the Need of a Public Health Organization in China*, 15.

110. George Rosen, *A History of Public Health* (Baltimore, MD: Johns Hopkins University, 1993), 439–53; orig. pub. 1958.

111. Grant, "Provisional National Health Council," 5.

112. John Grant, "State Medicine: A Logical Policy for China," *National Medical Journal of China* 14, no. 2 (1928): 65–80.

113. The Chinese translation of this memorandum was further extended and published in the *National Medical Journal of China* in July 1927; see Yan Fuqing 顏福慶, "Guomin zhengfu yingshe zhongyang weishengbu zhi jianyi" 國民政府應設中央衛生部之建議 [Suggestion for the Nationalist Government to establish a Ministry of Health], *Zhonghua yixue zazhi* 中華醫學雜誌 [National medical journal] 13, no. 4 (1927): 229–40.

114. Ibid., 233.

115. Xue served as the governor of Beijing in 1924 and thus became acquainted with John Grant. See John B. Grant, *The Reminiscences of Doctor John B. Grant*, Columbia University Oral History Collection (Glen Rock, NJ: Microfilming Corp. of America, 1977), 262.

116. Son of an American missionary, John Grant was born in China in 1890. While he was not a Chinese by blood, he whole-heartedly believed in and supported the idea that the development of public health should be led by Chinese nationals. Rather than relying on foreigners, Grant concluded that it was far more important to support Chinese efforts that are "60% efficient than Western ones that are 100%." Instead of being just a member of the Western medical establishment in China, Grant made great contributions to the growth of both the Chinese medical profession and the enterprise of public health led by the Chinese. Take for example the proposal for a Ministry of Health. Although Yan Fuqing's proposal was a translation of Grant's English proposal, the published Chinese article listed Yan as the sole author. To honor his contribution, I associate him with the group of what I call the first generation of Chinese practitioners of Western medicine. Personally, I felt very moved when reading the exchange of letters between him and his Chinese students during the devastating period of the Japanese invasion. It would be a worthy project to write a Chinese biography for John B. Grant.

117. Elizabeth Fee and Dorothy Porter, "Public Health, Preventive Medicine and Professionalization: England and America in the Nineteenth Century," in *Medicine in Society: Historical Essays*, ed. Andrew Wear (Cambridge: Cambridge University Press, 1992), 249–76, esp. 249.

118. Wu Liande 伍連德, "Haigang jianyi guanlichu lueshi" 海港檢疫管理處略史 [Brief history of the Office for Seaport Quarantine], *Yishi huikan* 醫事匯刊 [Collected papers on medical matters], no. 11 (1932): 3–7.

119. Pang Jingzhou 龐京周, *Shanghaishi jinshinianlai yiyao niaokan* 上海市近十年來醫藥鳥瞰 [Birds' eye view report of the recent ten years of the medical and pharamceutical situation in Shanghai] (Shanghai: Zhongguo kexue gongsi, 1933), 69.

120. Pierre Bourdieu, "Rethinking the State: Genesis and Structure of the Bureaucratic Field," *Sociological Theory* 12, no. 1 (1994): 1–18, esp. 16.

121. This shift of medical leadership was crucially aided by the larger wave of xenophobia. In the name of the Rights Recovery Movement, these Chinese practitioners of Western medicine joined forces with nationalism, demanding a return to China of its education rights. In January 1928, for the first time ever, members trained in various countries all attended the meeting of the National Medical Association. They passed the following resolutions: (1) The period is passing in which foreign medical institutions can aid materially in the development of medicine in China without the participation of Chinese in leadership roles. (2) The time has passed for the establishment in China of entirely foreign medical institutions. For the conservation of existing foreign medical institutions, there should be undertaken a statesmanlike policy which would make them integral parts of the community. See John Grant, "Public Health and Medical Events During 1932," *China Year Book* (1933): 172.

Meanwhile, the Nationalist government's Ministry of Education promulgated regulations concerning Western-owned educational institutions, requiring that all institutions must have a Chinese director and a local managerial body whose composition included a majority of Chinese members. It was in such an atmosphere that Liu Ruiheng became the first Chinese president of the Peking Union Medical College in 1926. See Bullock, *An American Transplant*, 59.

CHAPTER 4

1. Deng Tietao 鄧鐵濤, ed., *Zhongyi jindaishi* 中醫近代史 [Modern history of Chinese medicine] (Guangdong: Guangdong gaodeng jiaoyu chubanshe, 1999), 127.

2. Sean Hsiang-lin Lei, "Yu Yan," in *Dictionary of Medical Biography*, ed. W. F. Bynum and Helen Bynum (Westport, CT: Greenwood Press, 2006), 1341–42. For a detailed biography of Yu Yan, see Zu Shuxian 祖述憲, ed., *Yu Yunxiu zhongyi yanjiu yu pipan* 余云岫中醫研究與批判 [Yu Yunxiu's research and critique of Chinese medicine] (Hefei: Anhui daxue chubanshe, 2006), 1–5.

3. For Tang's biography and a detailed study of his works, see Pi Guoli 皮國立, *Yitong zhongxi: Tang Zonghai yu jiandai zhongyi weiji* 醫通中西：唐宗海與近代中醫危機 [Communication between Chinese and Western medicine: Tang Zonghai and the crisis of Chinese medicine in modern times] (Taipei: Dongda tushu gongsi, 2006), esp. 21–36.

4. Tang Zhonghai 唐宗海, *Zhongxi huitong yijing jingyi* 中西匯通醫經精義 [Essential meanings of the medical canons: (Approached) through the convergence and assimilation of Chinese and Western medicine] (Taipei: Lixing shuju, 1987), 1; orig. pub. 1892.

5. Sean H.-L. Lei, "Qi-Transformation and the Steam Engine: The Incorporation of Western Anatomy and the Re-Conceptualization of the Body in Nineteenth Century Chinese Medicine," *Asian Medicine: Tradition and Modernity* 7, no. 2 (2013): 1–39.

6. Catherine Despeux, "Visual Representations of the Body in Chinese Medical and Daoist Texts from the Song to the Qing Period," *Asian Medicine-Tradition and Modernity* 1, no. 1 (2005): 10–53, esp. 47.

7. Tang, *Zhongxi huitong yijing jingyi*, 1.

8. Benjamin A. Elman, *On Their Own Terms: Science in China, 1550–1900* (Cambridge, MA: Harvard University Press, 2005), 295.

9. Benjamin Hobson, *Quanti xinlun* 全體新論 [Treatise on physiology] (Guangzhou: Huiai yiguan, 1851), 57.

10. Tang, *Zhongxi huitong yijing jingyi*, 27.

11. For a detailed reconstruction of Tang's argument that the Triple Burner is roughly identical to the peritoneum, see Lei, "*Qi*-transformation and the Steam Engine."

12. Manfred Porkert and Christian Ullmann, *Chinese Medicine*, trans. Mark Howson (New York: Henry Holt, 1982), 123.

13. Bridie J. Andrews, "Wang Qingren and the History of Chinese Anatomy," *Journal of Chinese Medicine* 35 (January 1991), 30–36, esp. 33.

14. Nathan Sivin, *Traditional Medicine in Contemporary China* (Ann Arbor: University of Michigan Press, 1987), 137.

15. Ibid., 134.

16. Tang, *Zhongxi huitong yijing jingyi*, 111.

17. Ibid.

18. Ibid., 32.

19. Ning Jennifer Chang 張寧, "Nao wei yishen zhi zhu: Cong 'ailuo bunao zhi' kan jindai zhongguo shentiguan de bianhua" 腦為一身之主：從「艾羅補腦汁」看近代中國身體觀的變化 [From heart to brain: Ailuo brain tonic and the new concept of the body in late Qing China], *Zhongyang yanjiuyuan jindaishi yanjiusuo jikan* 中央研究院近代史研究所集刊 [Bulletin of the Institute of Modern History, Academia Sinica] 74 (December 2011): 1–40.

20. Tang, *Zhongxi huitong yijing jingyi*, 91.

21. Dauphin William Osgood, *Quanti chanwei* 全體闡微 [Anatomy, descriptive and surgical] (Fuzhou: Meihua shuguan, 1881), 1.

22. Tang, *Zhongxi huitong yijing jingyi*, 111.

23. Ibid.

24. Charlotte Furth, "The Sage as Rebel: The Inner World of Chang Ping-Lin," in *The Limits of Change: Essays on Conservative Alternatives in Republican China*, ed. Charlotte Furth (Cambridge, MA: Harvard University Press, 1976). For a concise discussion of Zhang's position on reforming Chinese medicine and his relationship with Yu Yan, see Volker Scheid, *Currents of Tradition in Chinese Medicine, 1626–2006* (Seattle: Eastland Press, 2007), 209–13.

25. Qian Xinzhong 錢信忠, *Zhongguo chuantong yiyaoxue fazhan yu xianzhuang* 中國傳統醫藥學發展與現狀 [The Development and present situation of traditional Chinese medicine and pharmacy in China] (Taipei: Qingchun chubanshe), 43.

26. Yu Yan 余巖, "Lingsu shangdui" 靈素商兌 [A critique of the *Divine Pivot* and *Basic Questions*] in his *Yixue geming lunwenxuan* 醫學革命論文選 [Collected essays on medical revolution] (Taipei: Yiwen yinshu guan, 1976), 89–130; orig. pub. 1917.

27. Ibid., 89.

28. In Chinese medicine, the "five viscera" (*wu zang* 五臟) are the pulmonary, hepatic, cardiac, splenetic, and renal systems. The "six bowels" (*liu fu* 六腑) refer to the gall bladder, stomach, large intestine, small intestine, Triple Burner, and urinary bladder systems.

29. Yu Yan 余巖, "Yixue geming guoqu gongzuo xianzai qingshi he weilai de celeu" 醫學革命過去工作現在情勢和未來的策略 [Medical revolution: Past efforts, present situation, and future strategy], *Zhonghua yixue zazhi* 中華醫學雜誌 [Chinese medical journal] 20, no. 1 (1933): 11–23.

30. Yu Yan 余巖, "Kexue de guochan yaowu yanjiu zhi diyibu" 科學的國產藥物研究之第一步 [First step in scientifically studying nationally produced drugs], *Xueyi* 學藝 [Studies and arts] 2, no. 4 (1920): 1–8, esp. 3.

31. Ding Fubao 丁福保 launched a similar critique of Chinese medicine a few years earlier than Yu Yan. See Liu Xuan 劉玄, "Tongsu zhishi yu xiandaixing: Ding Fubao yu jingdai Shangshai yixue zhishi de dazhongchuanbo"

通俗知識與現代性：丁福保與近代上海醫學知識 [Popular knowledge and modernity: Ding Fubao and the transmission of medical knowledge in modern Shanghai] (PhD diss., Chinese University of Hong Kong, 2013).

32. Ibid., 3–5. In addition to that article, Yu Yan addressed the same problem in his "Woguo yixue geming zhi pohuai yu jianshe xu" 我國醫學革命之破壞與建設（續）[Construction and destruction in our national medical revolution: Part two], *Yiyao pinglun* 醫藥評論 [Medical review], no. 9 (1929): 17–21.

33. Yu, "Kexue de guochan yaowu yanjiu zhi diyibu," 5.

34. Yu Yan 余巖, "Woguo yixue geming zhi pohuai yu jianshe" 我國醫學革命之破壞與建設 [Construction and destruction in our national medical revolution], *Yiyao pinglun* 醫藥評論 [Medical review], no. 8 (1929): 1–17, esp. 13.

35. Yu, "Kexue de guochan yaowu yanjiu zhi diyibu," 6.

36. Ibid., 6–7.

37. Asaf Goldschmidt, *The Evolution of Chinese Medicine: Song Dynasty, 960–1200* (London: Routledge, 2009).

38. Ibid., 56.

39. See Robert Hymes, "Not Quite Gentlemen? Doctors in Song and Yuan," *Chinese Science* 8 (1987): 9–76; Chen Yuan-Peng 陳元朋, "Songdai de ruyi jianping Robert P. Hymes youguan songyuan yizhe diwei de lundian宋代的儒醫—兼評 Robert P. Hymes有關宋元醫者地位的論點 [Ju-I of the Sung dynasty—with comment on Robert P. Hymes' 'Doctors in Sung and Yuan')," *Xin shixue*新史學 [New history] 6, no. 1 (1995): 179–201.

40. Yu Yan's letter to Tan Cizhong, in Tan Cizhong 譚次仲, *Yixue geming lunzhan* 醫學革命論戰 [Polemics of medical revolution] (Hong Kong: Qiushi chubanshe, 1952), 59; orig. pub. 1931.

41. Lorraine Daston, "Baconian Facts, Academic Civility, and the Prehistory of Objectivity," in *Rethinking Objectivity*, ed. A. Megill (Durham, NC: Duke University Press, 1994), 37–63.

42. Yu Yan 余巖, "Bo Yu Jianquan jingmai xieguan butong shuo" 駁俞鑑泉經脈血管不同說 [Rebuttal of Yu Jianquan's 'On the non-identity of jingmai and blood vessels'], *Tongde yoyaoxue* 同德醫藥學 (Tongde medicine and pharmacology) 7, no. 4 (1924): 15–20, esp. 15.

43. Yu, "Lingsu shangdui," 110.

44. Yu Jianquan 俞鑑泉, "Jingmai xueguan butong shuo er" 經脈血管不同說二 [On the non-identity between jingmai and blood vessels, part two] *Sansan Yibao* 三三醫報 [Three three medical news] 1, no. 29 (1924): 1.

45. Tang Zonghai 唐宗海, *Bencao wenda* 本草問答 [Questions and answers on materia medica] (Taipei: Lixing shuju, 1987), 2; orig. pub. 1880.

46. Yu Jianquan, "Jingmai xueguan butong shuo er," pt. 2, 1.

47. Yu Jianquan, "Jingmai xueguan butong shuo" 經脈血管不同說 [On the non-identity of jingmai and blood vessels], *Sansan Yibao* 三三醫報 [Three three medical news] 1, no. 12 (1923): 1.

48. Yu Jianquan, "Jingmai xueguan butong shuo," pt. 2, 2.

49. Yu Jianquan, "Jingmai xueguan butong shuo," 2.

50. Finally, to support my reading of Yu Jianquan's argument, I would like

to point out that in his view blood did not circulate within *jingmai* 經脈; only *jingqi* 經氣 did. Because of this unprecedented idea of separating *qi* and blood into two distinct systems, Yu had to explain how *qi* influenced blood, a phenomenon often mentioned in the *Inner Canon*. According to his interpretation, *jingmai* does not include *xuemai* 血脈, which is comparable, even identical, to the Western system of blood vessels. However, circulating in *jingmai*, *jingqi* can still influence blood, because blood is everywhere in the body, not just in the blood vessels. In support of this argument, Yu cited an everyday phenomenon. As one pinches the skin, blood pours out from that point even if that point is unconnected to any known blood vessel. For a critical discussion on whether *qi* and blood were ever considered as separately distributed, an idea that Joseph Needham had once supported, see Sivin, *Traditional Medicine in Contemporary China*, 119 and 437–38.

51. Carl F. Schmidt, "Pharmacology in a Changing World," *Annual Review of Physiology* 23 (March 1961): 3.

52. John Parascandola, *The Development of American Pharmacology: John J. Abel and the Shaping of a Discipline* (Baltimore, MD: Johns Hopkins University Press, 1992), 8.

53. 飯沼信子 Nobuko Linuma, *Nagai Nagayoshi to Terēze: Nihon yakugaku no kaiso* 長井長義とテレーゼ: 日本薬学の開祖 [Nagai Nagayoshi and Therese: The founding master of Japanese pharmacology] (Tōkyō: Nihon Yakugakkai, 2003).

54. John Black Grant, *The Reminiscences of Doctor John B. Grant*, Columbia University Oral History Collection (Glen Rock: Microfilming Corp. of America, 1977), 503.

55. In his article about the Department of Pharmacology, B. E. Read made it clear that the research of this department focused exclusively on Chinese materia medica. See Bernard E. Read, "Peking Union Medical College Department of Pharmacology," in *Problem of Medical Education* (New York: Rockefeller Foundation, 1925), 5–8.

56. Henry S. Houghton's letter to Edwin R. Embree, May 31, 1920, folder 868, box 120, China Medical Board, Rockefeller Foundation Archive Center, Sleepy Willow, NY.

57. Mary Brown Bullock, *An American Transplant: The Rockefeller Foundation and Peking Union Medical College* (Berkeley and Los Angeles: University of California Press, 1980), 8.

58. Carl F. Schmidt, "The Old and the New in Therapeutics," *Circulation Research: An Official Journal of the American Heart Association* 13, no. 4 (1960): 690.

59. Anon., *Chengbao* 晨報 (Morning news), February 19, 1925.

60. Schmidt, "Pharmacology in a Changing World," 7–8.

61. K. K. Chen and C. F. Schmidt, "The Action of Ephedrine, the Active Principle of the Chinese Drug, *Ma Huang*," *Journal of Pharmacology and Experimental Therapeutics* 24, no. 5 (1924): 339–57.

62. K. K. Chen, "Researches on Chinese Materia Medica," *Journal of the American Pharmaceutical Association* 20, no. 2 (1931): 110–13, esp. 112.

63. A letter from C. F. Schmidt to H. S. Houghton, October 22, 1924,

folder 873, box 120, IV 2 B9, China Medical Board, Rockefeller Foundation Archive Center, Sleepy Willow, NY.

64. M. R. Lee, "The History of *Ephedra* (*ma-huang*)," *Journal of Royal College of Physicians of Edinburgh* 41, no. 1 (2011): 78–84, esp. 81.

65. K. K. Chen and C. F. Schmidt, "Ephedrine and Related Substances," *Medicine: Analytical Reviews of General Medicine, Neurology and Pediatrics* 9, no. 1 (1930): 1–131.

66. Schmidt, "Pharmacology in a Changing World," 5.

67. K. K. Chen, "Two Pharmacological Traditions: Notes from Experience." *Annual Review of Pharmacology and Toxicology* 21, (1981): 1–6, esp. 3. I would like to thank David Chen for suggesting this reference.

68. This biographical information is based on the account written by K. K. Chen himself. K. K. Chen, ed., *The American Society for Pharmacology and Experimental Therapeutics, Incorporated: The First Sixty Years* (Washington: Judd & Detweiler, 1969), 67–68. For more information about K. K. Chen, see Ding Guangsheng 丁光生, "Chen Kehui: Guoji zhuming yaoli xuejia" 陳克恢—國際著名藥理學家 [K. K. Chen: An internationally renowned pharmacologist], *Shenglixue jinzhan* 生理科學進展 [Progress in physiological sciences] 40, no. 4 (2009): 289–91.

69. I would like to thank Elisabeth Hsu for sharing with me her thoughts and criticism on the issue of this "empirical tradition" of Chinese materia medica after she read through my 2002 article on *jingyan*. Her comments have helped me to think through this issue and to develop an understanding of "empirical tradition" as actors' category and relational concept.

70. Charlotte Furth, "The Sage as Rebel: The Inner World of Chang Ping-Lin," in *The Limits of Change: Essays on Conservative Alternatives in Republican China*, ed. Charlotte Furth (Cambridge, MA: Harvard University Press, 1976), 113–15.

71. Zhang Taiyan 章太炎, "Lunzhongyi bofuan yu Wu Jianzhai shu" 論中醫剝復案與吳檢齋書 [Letter to Wu Jianzhai on the Book of the Development and Decline of Chinese Medicine], *Huaguo yuekan* 華國月刊 (Huaguo monthly) 3, no. 3 (1926): 3.

72. Benjamin Elman, *From Philosophy to Philology: Social and Intellectual Aspects of Change in Late Imperial China* (Cambridge, MA: Council on East Asian Studies, Harvard University, 1984).

73. Jia Chunhua 賈春華, "Gufangpai dui zhongguo jindai shanghanlun yanjiu de yingxiang" 古方派對中國近代傷寒論研究的影響 [Influence of the Ancient Formula School on *Shanghanlun* studies], *Beijing Zhongyiyao daxue xuebao* 北京中醫藥大學學報 [Journal of Beijing University of Traditional Chinese Medicine] 17, no. 4 (1994): 5–9.

74. Benjamin Elman, "Sinophiles and Sinophobes in Tokugawa Japan: Politics, Classicism, and Medicine During the Eighteenth Century," *East Asian Science, Technology and Medicine: An International Journal* 2, no. 1 (2008): 93–121, esp. 116.

75. Zhang Taiyan 章太炎, "Xu" 序 [Preface], in *Shanghanlun jinshi* 傷寒論今釋 [Contemporary interpretation of the Shanghanlun], ed. Lu Yuanlei 陸淵雷 (Taipei: Wenguang chubanshe, 1961), 1; orig. pub. 1931.

76. Yumoto Kyushi 湯本求真, *Huanghan yixue*皇漢醫學 [Japanese-style Chinese medicine], trans. Zhou Zixu 周子敘 (Taipei: Dongfang shuju, 1958), 3; orig. pub. 1929.

77. Tang Shiyan 湯士彥, "Zhonguoren yu zhongguo yixue" 中國人與中國醫學 [Chinese people and Chinese medicine], *Yijie chonqiu* 醫界春秋 [Annals of the medical profession], no. 39 (1929): 3.

78. Yu Yan 余巖, *Huanghan yixue piping* 皇漢醫學批評 [Critique of Japanese-style Chinese medicine] (Shanghai: Shehui yibaoguan, 1931).

79. Sean Hsiang-lin Lei, "How Did Chinese Medicine Become Experiential? The Political Epistemology of *Jingyan*," *Positions: East Asia Cultures Critique* 10, no. 2 (2002): 333–64.

80. As a response to the global spread of biomedicine, many indigenous medical practices, such as Japanese-style Chinese medicine, Ayurvedic medicine, and Turkish medicine, to name just a few, began to characterize themselves similarly as experience-centered. In light of these historical developments, the experience-based characterization of Chinese medicine is part of the worldwide diffusion of the modernist divide between scientific Western medicine and the experiential medicine of non-Western societies. With regard to Ayurvedic and Turkish medicine, see Deepak Kumar, "Unequal Contenders, Uneven Ground: Medical Encounters in British India, 1820–1920," in Andrew Cunningham and Bridie Andrews, eds. *Western Medicine as Contested Knowledge* (New York: Manchester University Press, 1997), 172–90; and Feza Gunergun, "The Turkish Response to Western Medicine and the Turkish Medical Historiography," paper presented at the International Symposium on the Comparative History of Medicine—East and West, Division of Medical History, The Taniguchi Foundation, Seoul, Korea, July, 1998.

81. Concerning Japanese enthusiasm about scientific research on traditional drugs, John Grant commented in 1977, "That had a tremendous appeal, political appeal to people in all Asian countries." (*Reminiscences of Doctor John B. Grant*, 503). Also see E. Leong Way, "Pharmacology," in *Sciences in Communist China*, ed. Sidney H. Gould (Washington, DC: American Association for the Advancement of Science, 1961), 364.

CHAPTER 5

1. C. C. Chen, *Medicine in Rural China: A Personal Account* (Berkeley and Los Angeles: University of California Press, 1989), 3–4.

2. Fan Shouyuan 范守淵, *Fanshi yilunji* 范氏醫論集 [Mr. Fan's collected essays on medicine] (Shanghai: Jiujiu yixueshe, 1947), 597.

3. Yu Yan 余巖, "Woguo yixue geming zhi pohuai yu jianshe" 我國醫學革命之破壞與建設 [Construction and destruction in our national medical revolution], *Yiyao pinglun* 醫藥評論 [Medical review], no. 8 (1929): 1–17; and "Woguo yixue geming zhi pohuai yu jianshe xu" 我國醫學革命之破壞與建設(續) [Construction and destruction in our national medical revolution: Part two], *Yiyao pinglun* 醫藥評論 [Medical review], no. 9 (1929): 17–21.

4. In addition to "Construction and Destruction in Our National Medical Revolution," Yu Yan wrote at least three other articles that included "medical

revolution" in their titles: "Jinhou yixue geming zhi fangce" 今後醫學革命之方策 [The strategy for a future medical revolution], *Yishi huikan* 醫事匯刊 [Collected essays on medical matters], no. 10 (1931): 33–34; "Yixue geming guoqu gongzuo xianzai qingshi he weilai de celve" 醫學革命過去工作現在情勢和未來的策略 [Medical revolution: The past efforts, present situation, and future strategy], *Zhonghua yixue zazhi* 中華醫學雜誌 [Chinese medical journal] 20, no. 1 (1933): 11–23; and "Yixue geming zhi zhenwei" 醫學革命之真偽 [The genuine and fake medical revolution], *Zhongxi yiyao* 中西醫藥 [Journal of the Medical Research Society of China] 2, no. 3 (1936): 30–31.

5. Jiang Shaoyuan 江紹原, a student of Hu Shih, specifically pointed out that the medical revolution emerged independent of the New Culture Movement.

6. Yu Yunxiu 余雲岫, "Shuangshijie Zhi Xinyi Yu Shehui" 雙十節之新醫與社會 [The new medicine and society on the "National Day of Double Ten"], *Xinyi yu shehui huikan* 新醫與社會彙刊 [Collected essays on new-style medicine and society], no. 1 (1928): 30–31, esp. 31.

7. Ibid.

8. Warwick Anderson and Hans Pols, "Scientific Patriotism: Medical Science and National Self-fashioning in Southeast Asia," *Comparative Studies in Society and History* 54, no. 1 (2012): 93–113, esp. 98–99.

9. Chiang Kai-shek (Jiang Jieshi) launched the Northern Expedition in July 1926, and the Nationalist Revolutionary Army entered Shanghai in March 1927. Therefore, at the time Yu Yan wrote "Construction and Destruction in our National Medical Revolution," which contained his first use of the term *medical revolution*, he was already living in an area controlled by the Nationalist Revolutionary Army.

10. Yu Yan 余巖, "Xu yi" 序一 [Preface number one], *Xinyi yu shehui huikan* 新醫與社會彙刊 [The collected papers from the new medicine and society], no. 1 (1928): 1–2.

11. Ruth Rogaski, *Hygienic Modernity: Meanings of Health and Disease in Treaty-Port China* (Berkeley: University of California Press, 2004), 9.

12. Yu Yan 余巖, "Ruhe nengshi zhongguo kexueyi zhi puji" 如何能使中國科學醫之普及 [How to popularize scientific medicine in China], *Shenbao yixue zhoukan* 申報醫學週刊 [Shenbao medical weekly], nos. 109–11 (1935).

13. For more information on these schools, see Deng Tietao 鄧鐵濤, ed., *Zhongyi jindai shi* 中醫近代史 [Modern history of Chinese medicine] (Guangdong: Guangdong gaodeng jioayu chubanshe, 1999), 133–36. Concerning the history of the Shanghai Technical College of Chinese Medicine and its influential founder, Ding Ganren 丁甘仁, teachers, and alumni, see Qiu Peiran 裘沛然, *Mingyi yaolan—Shanghai zhongyi xueryuan (Shanghai zhongyi chuanmen xuexiao) xiaoshi* 名醫搖籃⊠上海中醫學院（上海中醫專門學校）校史 [The cradle of famous doctors: A history of Shanghai College of Chinese Medicine] (Shanghai: Shanghai zhongyiyao daxue chubanshe, 1998). See also Volker Scheid, *Currents of Tradition in Chinese Medicine, 1626–2006* (Seattle, WA: Eastland Press), 223–77, esp. 235–39.

14. K. Chimin Wong and Lien-teh Wu, *History of Chinese Medicine* (Taipei: Southern Materials Center, 1985), 160–61; orig. pub. 1932.

15. Marianne Bastid, *Educational Reform in Early Twentieth-Century China* (Ann Arbor: Center for Chinese Studies, University of Michigan, 1988), 85–86.

16. Asaf Goldschmidt, *The Evolution of Chinese Medicine: Song Dynasty, 960–1200* (London: Routledge, 2009).

17. Bao Boyin 包伯寅, "Gaijin zhongyi yijianshu" 改進中醫意見書 [A statement on reforming Chinese medicine], *Yixue zazhi* 醫學雜誌 [The medicine magazine] 3 (1919): 65–75.

18. Liu Nongbo 劉農伯, "Zhongxiyi pingyi" 中西醫平議 [A fair comment on Chinese and Western medicine], *Yijie chunqiu* 醫界春秋 [Annals of the medical profession], no. 4 (1926): 2–3.

19. For a detailed biography of Pang Jingzhou, see Pang Zehan 龐曾涵, Gao Yiling 高憶陵, and Chi Zihua 池子華, "Cishan rensheng—Pang Jingzhou yixshi de shengping yu shiye" 慈善人生—龐京周醫師的生平與事業 [A life of charity: Dr. Pang Jingzhou's life and career], *Hongshizi yundong zhongxin dianzi qikan* 紅十字運動中心電子期刊 [Electronic journal of the Center for Red Cross Movement], no. 6 (2007), available online at *http://www.hszyj.net/article.asp?articleid=731*.

20. Yu Yan 余巖, "Jiuyi xuexiao xitongan boyi" 舊醫學校系統案駁議 [Denouncement of the proposal for old-style medical schools], *Zhonghua yixue zazhi* 中華醫學雜誌 [Chinese medical journal] 12, no. 1 (1926): 5–12.

21. Wong and Wu, *History of Chinese Medicine*, 159–68.

22. Ibid., 163–64.

23. Yu Yan 余巖, "Feizhi jiuyi yi saochu yishi weisheng zhi zhangaian" 廢止舊醫以掃除醫事衛生之障礙案 [Abolishing old medicine to sweep away the obstacles to medicine and public health], *Yijie chunqiu* 醫界春秋 [Annals of the medical profession], no. 34 (1929): 9–10.

24. Wong and Wu have provided a complete English translation of Yu Yan's proposal. On the basis of their translation, I have retranslated this important proposal. For Wong and Wu's original translation, see Wong and Wu, *History of Chinese Medicine*, 161–65.

25. Yu, "Feizhi jiuyi yi saochu yishi weisheng zhi zhangaian," 9.

26. Ibid.

27. Yu Yan余巖, "Kexue de guochan yaowu yanjiu zhi diyibu" 科學的國產藥物研究之第一步[First step in scientifically studying nationally produced drugs], *Xueyi* 學藝 [Studies and arts] 2, no. 4 (1920): 1–8, esp. 3–5.

28. See Yu Yan 余巖, "Shuangshijie zhi xinyi yu shehui" 雙十節之新醫與社會 [The new medicine and society on National Day of Double Ten], in *Xinyi yu shehui huikan* 新醫與社會彙刊 [Collected essays on the new medicine and society] (1936): 5–6; Fan, *Fanshi yilunji*, 406–10.

29. It was important to characterize "individualized medicine" as outdated because, for critics of Chinese medicine, individual patients and the traditional doctor-patient relationship were among the central targets of the medical revolution. To establish the cultural authority that was deemed necessary for the practice of modern medicine, modern physicians felt the need to remake their fellow citizens into "qualified patients," that is, the kind of modern

patient who was "obedient, passive, silent, willing to endure hardship, and had a quasi-religious belief in modern medicine." Since Chinese patients were the very target of the medical revolution, they should have absolutely no role in deciding the course of this revolution. See Sean Hsiang-lin Lei 雷祥麟, "Fuzeren de yisheng yu youxinyang de bingren: Zhongxiyi lunzheng yu yibing guanxi zai minguo shiqi de zhuanbian" 負責任的醫生與有信仰的病人：中西醫論爭與醫病關係在民國時期的轉變 [Accountable doctor and loyal patient: The debate between Chinese and Western medicine and the transformation of the doctor-patient relationship in the Republican period], *Xinshixue* 新史學 [New history] 14, no. 1 (2003): 45–96.

30. Yu, "Feizhi jiuyi yi saochu yishi weisheng zhi zhangaian," 9.

31. Ibid.

32. Pierre Bourdieu, *The Logic of Practice*, trans. Richard Nice (Stanford, CA: Stanford University Press, 1990), 123–39, esp. 136.

33. See Zhang Zhaitong 張在同 and Xian Rjin 咸日金, eds., *Minguo yiyao weisheng fagui xuanbian* 民國醫藥衛生法規選編 [Selected collection of hygiene laws in the Republican period] (Taian, Shandong: Shandong daxue chubanshe,1990).

34. Yu, "Feizhi jiuyi yi saochu yishi weisheng zhi zhangaian," 9.

35. Jiao Yitang 焦易堂, "Guoyi dangyou zhengzhi yanguang" 國醫當有政治眼光 [National Medicine should develop a political perspective], *Guoyi gongbao* 國醫公報 [Bulletin of National Medicine] 1, no. 4 (1933): 1–3, esp. 1.

36. Chen Cunren 陳存仁, *Yinyuan shidai shenghuoshi* 銀元時代生活史 [Life during the time of the silver dollar] (Guilin: Guangxi shifan daxue chubanshe, 2007), 132.

37. Wang Qizhang 汪企張, *Ershi nian lai zhongguo yishi chuyi* 二十年來中國醫事芻議 [Preliminary comments on the past twenty years of medical affairs in China] (Shanghai: Zgengliao yibaoshe, 1935), 198–200.

38. James C. Thomson Jr., *While China Faced West: American Reformers in Nationalist China, 1928–1937* (Cambridge, MA: Harvard University Press, 1969), 1–41; Prasenjit Duara, "Knowledge and Power in the Discourse of Modernity: The Campaigns against Popular Religion in Early Twentieth-Century China" *Journal of Asian Studies* 50, no. 1 (1991): 67–83.

39. Mary Brown Bullock, *An American Transplant: The Rockefeller Foundation and Peking Union Medical College* (Berkeley and Los Angeles: University of California Press, 1980), 150–61.

40. Pan Guijuan 潘桂娟 and Fan Zhenglun 樊正倫, *Riben hanfang yixue* 日本漢方醫學 [Japanese kampo medicine] (Beijing: Zhongguo zhongyiyao chubanshe, 1994), 193–207; Zhao Hongjun 趙洪鈞, *Jindai zhongxiyi lunzhengshi* 近代中西醫論爭史 [A history of the struggle between Chinese and Western medicine in the modern period] (Hefei: Anhui kexue jishu chubanshe, 1989), 289–310; Bridie Andrews, "The Making of Modern Chinese Medicine, 1895–1937" (PhD diss., Cambridge University, 1996), 149–76.

41. Anon., "Jiaoweibu fenkeng guoyi guoyao zhi tongshi lu" 教衛部墳坑國醫國藥之痛史錄[Painful history of the destruction of Chinese Medicine and

pharmacy by the Ministries of Education and Health], *Xiandai guoyi* 現代國醫 [Modern national medicine] 1, no. 6 (1931): 5–16.

42. Ye Ruiyang 葉瑞陽, "Jinggao quanguo yixuehui (yufang fang riben xiaomie hanyi zhi zhizhao)" 敬告全國醫學會（預防仿日本消滅漢醫之執照） [A warning to medical associations nationwide (beware of emulating Japan in canceling Chinese doctors' licenses)], *Sansan yibao* 三三醫報 [Three three medical news] 2, no. 15 (1924): 1–2.

43. Ye Qishei 葉其誰, "Zhongyi zhi zibian" 中醫之自貶 [The self-humiliation of Chinese medicine], *Yijie chunqiu* 醫界春秋 [Annals of the medical profession], no. 5 (1926): 5–6.

44. Chen, *Yinyuan shidai shenghuoshi*, 110–11.

45. Pang Jingzhou 龐京周, *Shanghaishi jinshinian lai yiyao niaokan* 上海市近十年來醫藥鳥瞰 [A birds-eye view report of medicine and pharmacy in Shanghai in the last decade] (Shanghai: Zhongguo kexue gongsi, 1933), 79.

46. Quanguo yiyao zonghui 全國醫藥總會, *Quanguo yiyao tuanti zonglianhehui huiwu huibian* 全國醫藥團體總聯合會會務彙編 [The compiled documents of the affairs of the National Federation of Medical and Pharmaceutical Associations] (Shanghai: Quanguo yiyao zonghui, 1931), 37.

47. *Shenbao*, March 18, 1929.

48. Ibid.

49. Ralph Croizier, *Traditional Medicine in Modern China: Science, Nationalism, and the Tensions of Cultural Change* (Cambridge, MA: Harvard University Press, 1968).

50. Karl Gerth, *China Made: Consumer Culture and Creation of the Nation* (Cambridge, MA: Harvard University Asia Center, 2003).

51. Please see Pan Junxiang 潘均祥, ed., *Zhongguo jindai guohuo vendung* 中國近代國貨運動 [National Goods Movement in modern China] (Beijing: Zhongguo wenshi chubanshe, 1995).

52. *Shenbao*, March 17, 1929.

53. Anon., "Dahui qingxing" 大會情形 [Circumstances of the conference], *Yijie chunqiu* 醫界春秋 [Annals of the medical profession], no. 34 (1929): 18–45, esp. 43.

54. Ibid.

55. Quanguo yiyao zonghui 全國醫藥總會, *Quanguo yiyao tuanti daibiao dahui tian huilu* 全國醫藥團體代表大會提案會錄 [Collection of proposals from the National Convention of Medical and Pharmaceutical Associations] (Shanghai: Quanguo yiyao zonghui 全國醫藥總會 [National Federation of Medicine and Pharmacy], 1929).

56. Ibid., 6 and 21.

57. Ibid. See especially proposals 6, 8, 21, 91, 98, and 100.

58. Li Jian 李劍, "Quanguo yiyao tuanti lianhehui de chuangli, huodong ji lishi diwei" 全國醫藥團體聯合會的創立、活動及其9史地位 [The establishment, activities and historical position of the National Federation of Medical and Pharmaceutical Associations], *Zhongguo keji shiliao* 中國科技史料 [China historical materials of science and technology], no. 3 (1993): 67–75, esp. 68.

59. Ibid., 69–70.

60. Croizier, *Traditional Medicine in Modern China*, 4.

61. In fact, many associations and journals subsequently changed their names to replace "Chinese Medicine" with "National Medicine." The most salient example is the Shanghai Association of Chinese Medicine (*Shanghai zhongyi xiehui* 上海中醫協會), which was created out of a merger of three associations in December 1928; it changed its name to Shanghai Association of National Medicine (*Shanghai guoyi gonghui* 上海國醫公會) in March 1929, right after the campaign by traditional practitioners to protest Yu's proposal. See Qiu Peiran 裘沛然, *Minyi yaolan—Shanghai zhongyi xuryuan xiaoshi* 名醫搖籃—上海中醫學院（上海中醫專門學校）校史 [The cradle of famous doctors: A history of Shanghai College of Chinese Medicine] (Shanghai: Shanghai zhongyiyiao daxue chubanshe, 1998), 119–20.

62. I considered the possibility of translating *guoyi* as "state medicine," rather than "national medicine," but I decided not to do so for two reasons. First, the advocates of *guoyi* indeed tried to link *guoyi* with cultural nationalism, even though their major objective was to make Chinese medicine a part of the emerging state. More importantly, I analyze the rise of state medicine (*gongyi* 公醫), an important vision of the modern health-care system, in the second half of this book. Because *state medicine* was the English term that its advocates used in their official publications in English, I have decided to use *state medicine* in that context. To avoid the confusion of using the term *state medicine* to refer to two historical enterprises, I have decided to use *national medicine* as the English term for *guoyi*, but I try to remind the readers whenever possible that the National Medicine Movement should not be equated with or reduced to a movement of cultural nationalism.

63. Prasenjit Duara, *Culture, Power, and the State: Rural North China, 1900–1942* (Stanford, CA: Stanford University Press, 1988), 4.

64. Henrietta Harrison, *The Making of Republican Citizen: Political Ceremonies and Symbols in China, 1911–29* (Oxford: Oxford University Press, 2000), 1.

65. Anon., "Zhonghua yixuehui dahui jiyao" 中華醫學會大會紀要 [Minutes of the meetings of the Chinese Medical Association], *Zhonghua yixue zazhi* 中華醫學雜誌 [Chinese medical journal] 18, no. 6 (1932): 1140–47, esp. 1146.

66. Cheng Diren 程迪仁, "Zhide zhumu de yi fengxin" 值得注目的一封信 [A noteworthy letter], *Shenzhou guoyi xuebao* 神州國醫學報 [Shenzhou bulletin of national medicine] 1, no. 1 (1932): 13–17, esp. 13.

67. See Quanguo yiyao zonghui 全國醫藥總會, *Quanguo yiyao tuanti daibiao dahui ti'an huilu* 全國醫藥團體代表大會提案會錄 [Collection of proposals from the National Convention of Medical and Pharmaceutical Associations] (Shanghai: Quanguo yiyao zonghui, 1929), 65. In addition, proposals 54, 62, 69, 76, 81, and 91 all made the same suggestion.

68. *Shenbao*, March 25, 1929; and Li, "Quanguo yiyao tuanti lianhehui de chuangli, huodong ji lishi diwei," 70.

69. Chen, *Yinyuan shidai shenghuoshi*, 122.

70. Ibid., 127–37.

71. See *Shenbao*, March 26, 1929; and Chen, *Yinyuan shidai shenghuoshi*, 125.

72. See *Shenbao*, March 26, 1929.

73. Chen, *Yinyuan shidai shenghuoshi*, 124.

74. Tan Yankai 譚延闓, *Tan Yankai riji* 譚延闓日記 [Diary of Tan Yankai], March 22, 1929, from the Digital Databank of the Institute of Modern History, Academia Sinica, available online at *http://www.mh.sinica.edu.tw/PGDigitalDB_Detail.aspx?htmContentID=22*.

75. Tan, *Tan Yankai riji*, September 26, 1921.

76. Tan, *Tan Yankai riji*, March 23, 1925.

77. A similar attitude was shared by Chen Guofu 陳果夫, whose support for Chinese medicine was also based on his personal experience. Chen went even further than Tan by turning his personal experience as a patient into a public credential for his involvement in many aspects of medical affairs, including the struggle between Chinese medicine and Western medicine. See Lei, "Fuzeren de yisheng yu youxinyang de bingren," 84–92.

78. Anon., "Xue buzhang duiyu zhongyiyao cunfei wenti zhi tanhua" 薛部長對於中醫藥存廢問題之談話 [Minister Xue's statements concerning the issue of preserving or abolishing Chinese medicine], *Yijie chunqiu* 醫界春秋 [Annals of the medical profession], no. 34 (1929): 50–51, esp. 50.

79. Ibid., 51.

80. Chen, *Yinyuan shidai shenghuoshi*, 134.

81. Ibid.

82. Anon., "Zhongyang weisheng weiyuan huiyi yijue 'feizhi zhongyi an' yuanwen" 中央衛生委員會議議決「廢止中醫案」原文 [Original text of the "Proposal for Abolishing the Practice of Chinese Medicine" discussed and decided by the Central Board of Health], *Yijie chunqiu* 醫界春秋 [Annals of the medical profession], no. 34 (1929): 9–11, esp. 10–11.

83. Weisheng Bu 衛生部, "Dian" 電 [Telegraph], *Weisheng gongbao* 衛生公報 [Government bulletin on hygiene], no. 4 (1929): 1–4.

84. Anon., "Zhongyang weisheng weiyuan huiyi yijue 'feizhi zhongyi an' yuanwen," 10.

85. Anon., "Fu Chu Minyi dui xinjiu yiyao fenzheng zhi yijian" 附褚民誼對新舊醫藥紛爭之意見 [Appendix: Chu Minyi on the controversy between new and old medicine], *Yijie chunqiu* 醫界春秋 [Annals of the medical profession], no. 34 (1929): 32–34.

86. Anon., "Zhongyang weisheng weiyuan huiyi yijue 'feizhi zhongyi an' yuanwen," 10.

87. Deng, ed., *Zhongyi jindai shi*, 133–36 and 271–74.

88. Quanguo yiyao zonghui, *Quanguo yiyao tuanti daibiao dahui ti'an huilu*, 56–57.

89. Zhang Zanchen 張贊臣, "Xuyan" 緒言 [Preface], *Yijie chunqiu* 醫界春秋 [Annals of the medical profession], no. 34 (1929): Cover page.

90. Deng, ed., *Zhongyi jindai shi*, 287.

91. Shanghai weishengju, "Shanghai weishengju xunling" 上海衛生局訓令 [Official instruction from the Shanghai Bureau of Health], *Shenzhou guoyi xuebao* 神州國醫學報 (Shenzhou bulletin of national medicine) 5, no. 5 (1937): 40–41, esp. 41.

92. "Quanguo yiyao tuanti linshi daibiao dahui jiyao" 全國醫藥團體臨時

代表大會紀要 [Brief account of the provisional assembly of the National Medical and Pharmaceutical Associations], *Yijie chunqiu* 醫界春秋 [Annals of the medical profession], no. 42 (1931): 26.

93. Ibid., 28.

94. Anon., "Yishi xiaoxi" 醫事消息 [News about medical affairs], *Yijie chunqiu* 醫界春秋 [Annals of the medical profession], no. 43 (1930): 23.

95. Quanguo yiyao zonghui, "Guoyiguan wenti" 國醫館問題 [Issue concerning the Institute of National Medicine], in Quanguo yiyao zonghui, *Quanguo yiyao tuanti zonglianhehui huiwu huibian*, 86.

96. Ibid.

97. Quanguo yiyao zonghui, *Quanguo yiyao tuanti zonglianhehui huiwu huibian*, 88.

98. Ibid.

99. Ministry of Health 衛生部, Weisheng gongbao 衛生公報 [Goverment bulletin of hygiene] 2, no. 6 (1930): 69.

100. Quanguo yiyao zonghui, *Quanguo yiyao tuanti zonglianhehui huiwu huibian*, 102.

101. Jiao Yitang 焦亦堂 (1880–1950), a strong political supporter of Chinese medicine and the founding director of the Institute of National Medicine, asserted that "in the past, the very term 'National medicine' never existed in China." See Jiao Yitang 焦易堂, "Wei niding guoyi tiaoli jinggao guoren shu" 為擬訂國醫條例敬告國人書 [A public statement concerning the drafting of the regulation of national medicine], *Guoyi gongbao* 國醫公報 [Bulletin of national medicine] 1, no. 3 (1934): 1.

CHAPTER 6

1. Pang Jingzhou 龐京周, *Shanghaishi jinshinian lai yiyao niaokan* 上海市近十年來醫藥鳥瞰 [A bird's-eye view report of medicine and pharmacy in Shanghai in the last decade] (Shanghai: Zhongguo kexue gongsi, 1933), 11.

2. An informative biography of Pang Jingzhou appears in Chi Zihua 池子華, *Zhongguo hongshizi yundong shi sanlun* 中國紅十字運動史散論 [Essays on the history of the Red Cross Movement in China] (Hefei: Anhui renmin chubanshe, 2009).

3. Ibid., 3.

4. To enhance the comprehensiveness of this chart to English-speaking readers, I have added a number (in square brackets) for each English translation and its Chinese counterpart in the chart.

5. Pang, *Shanghaishi jinshinian lai yiyao niaokan*, 9.

6. Ibid., 50–51.

7. Lien-teh Wu, "The Problem of Veneral Diseases in China," *China Medical Journal* 15, no. 1 (1926): 28–36, esp. 29 and 34.

8. Pang, *Shanghaishi jinshinian lai yiyao niaokan*, 64.

9. Shanghai weishengju 上海衛生局, "Shanghai weishengju xunling" 上海衛生局訓令 [Official Instruction from the Shanghai Bureau of Health], *Shenzhou guoyi xuebao* 神州國醫學報 [Shenzhou bulletin of national medicine] 5, no. 5 (1937): 40–41, esp. 41.

10. Harold Balme, *China and Modern Medicine: A Study in Medical Missionary Development* (London: United Council for Missionary Education, 1921), 109.

11. Cheng Hanzhang 程瀚章, *Xiyi qianshuo* 西醫淺說 [Elementary introduction to Western medicine] (Shanghai: Shangwu yinshuguan, 1933), 70–71.

12. Anon., "Zhonghua xiyi gonghui xuanyan" 中華西醫公會宣言 [Manifesto of the Chinese Association of Western Medicine], *Shenbao* 申報, April 10, 1929.

13. K. Chimin Wong and Lien-teh Wu, *History of Chinese Medicine* (Taipei: Southern Materials Center, 1985), 781; orig. pub. 1932.

14. Pang, *Shanghaishi jinshinian lai yiyao niaokan*, 12.

15. Anon., "Zhonghu yixuehui gaikuo baogao" 中華醫學會概括報告 [General report about the National Medical Association of China], *Zhonghua yixue zazhi* 中華醫學雜誌 [Chinese medical journal] 18, no. 1 (1931): 181–83.

16. China Medical Commission of the Rockefeller Foundation, *Medicine in China* (New York: China Medical Commission of the Rockefeller Foundation, 1914), 8.

17. Zhu Jiqing 朱季清, "Wo guo linian lai gonggong weisheng xingzheng de shice" 我國歷年來公共衛生行政的失策 [Lapses in the administration of public hygiene in China over the years], *Zhongguo weisheng zazhi* 中國衛生雜誌 [Chinese journal of hygiene], 1933: 31–34; Zhang Daqing 張大慶, *Zhongguo jindai jibing shehuishi* 中國近代疾病社會史 [Social history of disease in modern China, 1912–1937] (Jinan: Shangdong jioayu chubanshe, 2006), 85–88.

18. Marianne Bastid, "Servitude or Liberation? The Introduction of Foreign Educational Practices and Systems to China," in *China's Education and the Industrialized World*, ed. Ruth Hayhoe and Marianne Bastid (New York: M. E. Sharpe, 1987), 3–20, esp. 11.

19. Jin Baoshan 金寶善, "Jiu zhongguo de xiyi paibie yu weisheng shiye de yanbian" 舊中國的西醫派別與衛生事業的演變 [On the cliques of Western medicine in old China and the evolution of the public health enterprise], in *Zhonghua wenshi zhiliao wenku* 中華文史資料文庫: 第十六卷 [Chinese collections of literary and historical documents, vol. 16] (Beijing: Zhongguo wenshi chubanshe, 1996), 844–50, esp. 848.

20. Pang, *Shanghaishi jinshinian lai yiyao niaokan*, 12.

21. Hu Dingan 胡定安, "Zhongguo yishi qiantu jidai jiejue zhi jige genben wenti" 中國醫事前途亟待解決之幾個根本問題 [Several fundamental problems crucial to medical progress in China], *Yishe huikan* 醫事匯刊 [Collected essays on medical matters], no. 18 (1934): 18–24.

22. Zhu Xiru 朱席儒 and Lasi Douyan 賴斗岩, "Wuguo xinyi rencai fenbu gaikuang" 吾國新醫人才分佈概況 [Distribution of modern-trained physicians in China], *Zhonghua yixue zazhi* 中華醫學雜誌 [National medical journal of China] 21, no. 2 (1935): 145–53, esp. 153.

23. Pang, *Shanghaishi jinshinian lai yiyao niaokan*, 94.

24. A. Stewart Allen, "Modern Medicine in China: Its Development and Its Difficulties," *Canadian Medical Association Journal* 56 (1947): 11–13.

25. Yuanling Chao, *Medicine and Society in Late Imperial China: A Study of Physicians in Suzhou, 1600–1850* (New York: Peter Lang, 2009). For details on how the scholar-physicians crafted this collective identity by writing about medical history, see Pingyi Chu 祝平一, "Song ming zhi ji de yishi yu ruyi" 宋明之際的醫史與「儒醫」 [Narrations of histories of medicine from the Song to the Ming and the rise of the Confucian physician], *Zhongyang yanjiuyuan lishi yuyan yanjiusuo jikan* 中央研究院歷史語言研究所集刊 [The bulletin of the Institute of History and Philology, Academia Sinica] 77, no. 3 (2006): 401–49.

26. Nathan Sivin, *Traditional Medicine in Contemporary China* (Ann Arbor: University of Michigan Press, 1987), 21.

27. Balme, *China and Modern Medicine*, 20.

28. Lo Vivienne, "But Is It [History of] Medicine? Twenty Years in the History of the Healing Arts of China," *Social History of Medicine* 22, no. 2 (2009): 283–303.

29. Yu Xinzhong 余新忠, "Qindai jiangna minsu yiliao xingwei tanxi" 清代江南民俗醫療的行為探析 [Exploration of folk medicine in Jiangnan area in the Qing dynasty], in *Qing yilai de jibing yiliao yu weisheng* 清以來的疾病醫療與衛生 [Disease, medicine, and hygiene since the Qing dynasty], ed. Yu Xinzhong (Beijing: Sanlian shudian, 2009): 91–108, esp. 100.

30. Yu Yan once made fun of the supporters of Chinese medicine by saying that since *wu* 巫 (shamanism) and *yi* 醫 (medicine) had traditionally been closely related, if Chinese medicine could be raised to the status of national essence in the name of tradition and history, then the government should also list fortune-telling and astrology as national essence as well. See Yu Yan 余巖, "Jiuyi xuexiao xitongan boyi" 舊醫學校系統案駁議 [Denouncement of the proposal for assimilating the schools of old medicine into the education system], *Zhonghua yixue zazhi* 中華醫學雜誌 [Chinese medical journal] 12, no. 1 (1926): 5–12.

31. Prasenjit Duara, "Knowledge and Power in the Discourse of Modernity: The Campaigns against Popular Religion in Early Twentieth-Century China," *Journal of Asian Studies* 50, no. 1 (1991): 67–83, esp. 78; Rebecca Nedostup, *Superstitious Regimes: Religion and the Politics of Chinese Modernity* (Cambridge, MA: Harvard University Asia Center, 2009).

32. Philip S. Cho, "Ritual and the Occult in Chinese Medicine and Religious Healing: The Development of Zhuyou Exorcism" (PhD diss., University of Pennsylvania, 2006).

33. Pang, *Shanghaishi jinshinian lai yiyao niaokan*, 17.

34. Yang Zemin 楊則民, "Zhongyi bianqian zhi shi de niaokan xu" 中醫變遷之史的鳥瞰(續) [A bird's-eye-view history of the evolution of Chinese medicine (a continuation)] *Guoyi gongbao* 國醫公報 [Bulletin of national medicine] 1, no. 9 (1933): 31–40, esp. 32.

35. Pang, *Shanghaishi jinshinian lai yiyao niaokan*, 20.

36. Ibid., 73.

37. Xiong Yuezhi 熊月之, "Lun Li Pingshu" 論李平書 [On Li Pingshu], *Shilin* 史林 [Historical review] 85, no. 3 (2005): 1–7.

38. Ibid., 3.

39. Li Pingshu 李平書, *Li Pingshu qishi zishu* 李平書七十自述 [The autobiography of Li Pingshu at the age of seventy] (Shanghai: Shanghai guji chubanshe, 1989), 51.

40. Ibid., 52.

41. Ibid., 54.

42. Deng Tietao 鄧鐵濤, ed., *Zhongyi jindaishi* 中醫近代史 [Modern history of Chinese medicine] (Guangdong: Guangdong gaodeng jiaoyu chubanshe, 1999), 133–36 and 271–74.

43. Volker Scheid, *Currents of Tradition in Chinese Medicine, 1626–2006* (Seattle: Eastland Press, 2007), 223–42. After the 1929 demonstration, both the journal and the association changed their names to include the term *national medicine*.

44. Ibid., chaps. 9 and 10.

45. Ibid., 281.

46. Ibid., 208.

47. Ibid.

48. Pang, *Shanghaishi jinshinian lai yiyao niaokan*, 63.

49. Ibid., 13.

50. Zheng Jinsheng 鄭金生 and Li Jianmin 李建, "Xiandai zhongguo yixueshi yanjiu de yuanliu" 現代中國醫學史研究的源流 [The origin of modern historical research on Chinese medicine], *Dalu zazhi* 大陸雜誌 [Mainland] 95, no. 6 (1997): 26–35, esp. 27.

51. Sean Hsiang-lin Lei, "Writing Medical History for a Living Tradition; or, Rescuing Medical History from Both the Nation and Nature," paper presented at the workshop Restrospect of a Century of Republican Scholarship, Institute of Modern History, Academia Sinica, Taiwan, January 11–13, 2012.

52. Longxibuyi 隴西布衣, "Shanghai qige yixuexiao de jiaocheng yu xingwang" 上海七個醫學校的教程與興亡 [Curricula and careers of the seven medical schools in Shanghai], *Yijie chunqiu* 醫界春秋 [Annals of the medical profession], no. 20 (1928): 1–3, esp. 2.

53. Longxibuyi 隴西布衣, "Shanghai qige yixuexiao de jiaocheng yu xingwang xu" 上海七個醫學校的教程與興亡續 [Curricula and careers of the seven medical schools in Shanghai: Part two], *Yijie chunqiu* 醫界春秋 [Annals of the medical profession], no. 21 (1928): 1–4, esp. 2.

54. Pang, *Shanghaishi jinshinian lai yiyao niaokan*, 13.

55. Longxibuyi, "Shanghai qige yixuexiao de jiaocheng yu xingwang xu," 4.

56. Another contrast between the two was that acupuncture was included in the curriculum of the institute but not in that of Ding's College.

57. Zhang Ziheng 張子恆, "Shanghai liang zhongyixiao zhi bijiao guan" 上海兩中醫校之比較觀 [A comparison of two schools of Chinese Medicine in Shanghai], *Yijie chunqiu* 醫界春秋 [Annals of the medical profession], no. 27 (1928): 4.

58. Deng, *Zhongyi jindaishi*, 153–61.

59. Sanetô Keishû 實藤惠秀, *Zhongguoren liuxue riben shi* 中國人留學日本史 [History of Chinese overseas study in Japan], trans. Tan Ruqian 譚汝謙 and Lin Qiyan 林啟彥 (Hong Kong: Chinese University Press, 1982), 236.

60. Pang, *Shanghaishi jinshinian lai yiyao niaokan*, 100.

61. Bridie Andrews, "The Making of Modern Chinese Medicine, 1895–1937" (PhD diss., University of Cambridge, UK, 1996), 197.

62. Ye Jinqiu 葉勁秋, "Guanyu zhongyi jiaoyu" 關於中醫教育 [On Chinese medical education], *Zhongxi yiyao* 中西醫藥 [Journal of the Medical Research Society of China] 3, no. 7 (1939): 461–64, esp. 462.

63. Pang, *Shanghaishi jinshinian lai yiyao niaokan*, 13.

64. Shanghai weishengju, "Shanghai weishengju xunling," 40–41.

65. Pang, *Shanghaishi jinshinian lai yiyao niaokan*, 6.

66. Ibid., 125–26.

CHAPTER 7

1. Qian Xingzhong 錢信忠, *Zhongguo chuantong yiyaoxue fazhan yu xianzhuang* 中國傳統醫藥學發展與現狀 [Development and present situation of traditional Chinese medicine and pharmacy in China] (Taipei: Qingchun chubanshe, 1995), 44.

2. Judith Farquhar, *Knowing Practice: The Clinical Encounter of Chinese Medicine* (San Francisco: Westview Press, 1994), 17–19.

3. *Shenbao*, March 21, 1929.

4. Weishengbu 衛生部, "Dian" 電 [Telegram], *Weisheng gongbao* 衛生公報 [Government bulletin on hygiene], no. 4 (1929): 1–4.

5. Gu Tisheng 顧惕生, "Zhongyi kexuehua zhi shangdui" 中醫科學化之商兌 [Critique of the scientization of Chinese medicine], *Yijie chunqiu* 醫界春秋 [Annals of the medical profession], no. 44 (1930): 1–3, esp. 1.

6. This proposal was issued by Tan Yankai, Hu Hanmin, Chen Zhaoyin, Zhu Peide, Shao Yunchong, Chen Lifu, and Jiao Yitang. See Chen Yu 陳郁, "Zhongyiyao wenxian zhi linzhua" 中醫藥文獻之鱗爪 [Fragmentary documents on Chinese medicine and pharmacy], *Zhongyiyao*中醫藥 [Chinese medicine and pharmacy] 1, no. 5 (1960): 9–22, esp. 9.

7. Quanguo yiyao zonghui, 全國醫藥團體總聯合會, *Quanguo yiyao tuanti zonglianhehui huiwu huibian* 全國醫藥團體總聯合會會務彙編 [Compiled documents of the affairs of the National Federation of Medical and Pharmaceutical Associations] (Shanghai: Quanguo yiyao zonghui, 1931), 178.

8. Ibid., 179–80.

9. Ibid., 181.

10. Anon., "Zhongyang guoyiguan choubei dahui kaihui jilu" 中央國醫館籌備大會開會紀錄 [Record of the preparatory meeting for the Central Institute of National Medicine], *Guoyi gongbao* 國醫公報 [Bulletin of national medicine] 2, no. 2 (1934): 6–14, esp. 8.

11. Quanguo yiyao zonghui, *Quanguo yiyao tuanti zonglianhehui huiwu huibian*, 102.

12. Anon., "Zhongyang guoyiguan xuanyan" 中央國醫館宣言 [Manifesto of the Central Institute of National Medicine], *Guoyi gongbao* 國醫公報 [Bulletin of national medicine] 1 (1932): 1–7, esp. 7.

13. Anon., "Guoyiguan wenti" 國醫館問題 [Issue concerning the Institute

of National Medicine], in Quanguo yiyao zonghui,, *Quanguo yiyao tuanti zonglianhehui huiwu huibian*, 86.

14. John Fitzgerald, *Awakening China: Politics, Culture and Class in the Nationalist Revolution* (Stanford, CA: Stanford University Press, 1996), 55.

15. The Chinese Medical History Society (*zhonghua yishi xuehui* 中華醫史學會) was founded in 1936, and the Museum for the History of Chinese Medicine (*zhongguo yishi bowuguan* 中國醫史博物館) was opened in 1938.

16. See D. W. Y. Kwok, *Scientism in Chinese Thought, 1900–1950* (New Haven, CT: Yale University Press, 1965).

17. Yu Yan 余巖, "Jinhou yixue geming zhi fangce" 今後醫學革命之方策 [Strategy for a future medical revolution], Yishi huikan 醫事匯刊 [Collected essays on medical matters], no. 10 (1931): 33–34.

18. Yu Yan 余巖, "Du guoyiguan zhengli xueshu caoan zhi wojian" 讀國醫館整理學術草案之我見 [My opinion on the proposal for sorting out Chinese medicine as issued by the Institute of National Medicine], *Zhongxi yiyao* 中西醫藥 [Journal of the Medical Research Society of China] 2, no. 2 (1936): 178–92, esp. 178.

19. Yu Yan 余巖, "Dierban zixu" 第二版自序 [Preface to the second edition], in *Yushi yizhu* 余氏醫述 [Mr. Yu's essays on medicine], 2nd ed. (Shanghai: Shehui yibao she, 1932), 2; orig. pub. 1928. This work is also known as *Yixue geming lunwunxuan* 醫學革命論文選 [Collected essays on medical revolution].

20. For example, Yuan Fuchu had advocated scientizing Chinese medicine in 1924. However, in sharp contrast with the post-1929 project of scientization, Yuan Fuchu wanted to put "the theories of Yin-Yang and the Five Phases in the track of science." See Yuan Fuchu 袁復初, "Gaijin zhongyi zhi wojian" 改進中醫之我見 [My opinion on improving Chinese medicine], *Sansan yibao* 三三醫報 [Three three medical news] 2, nos. 12 and 13 (1924): 1–4 and 1–2.

21. Chen Peizhi 陳培之, "Zhongguo yixue kexue zhengli zhi wojian" 中國醫學科學整理之我見 [My opinion on the scientific reorganization of Chinese medicine], *Zhongxi yiyao* 中西醫藥 [Journal of the Medical Research Society of China] 2, no. 2 (1936): 146–48, esp. 147.

22. Hong Guanzhi 洪冠之, "Guanyu 'zhongyi kexu hua wenti' de shangque" 關於「中醫科學化問題」的商榷 [The discussion on the problem of scientizing Chinese medicine), *Zhong xi yiyao* 中西醫藥 [Journal of the Medical Research Society of China] 2, no. 2 (1936): 148–54, esp. 148.

23. Scholars have tended to marginalize those who have questioned the ultimate value of the Enlightenment project. In this respect our views are consistent with those of modernizing Chinese intellectuals and the modern nation state. See Prasenjit Duara, "Knowledge and Power in the Discourse of Modernity: The Campaigns against Popular Religion in Early Twentieth-Century China," *Journal of Asian Studies* 50, no. 1 (1991): 67–83, esp. 74.

24. Ibid.

25. Peng Guanghua 彭光華, "Zhongguo kexuehua yundong xiehui de chuangjian, huodong ji qi lishi diwei" 中國科學化運動協會的創建、活動及其歷史地位 [Founding, activities, and historical role of the China Scientization

Movement], *Zhongguo keji shiliao* 中國科技史料 [China historical materials of science and technology] 13, no. 1 (1992): 61–72.

26. Ralph Croizier, *Traditional Medicine in Modern China: Science, Nationalism, and the Tensions of Cultural Change* (Cambridge, MA: Harvard University Press, 1968), 92–99.

27. Chen Shou 陳首, *Zhongguo kexuehua yundong yanjiu* 中國科學化運動研究 [A study on the China Scientization Movement], PhD diss., Department of Philosophy, Beijing University, 2007. I would like to thank Chen Peiying 陳珮瑩 for sharing with me her copy of this dissertation.

28. In addition to their work in establishing the China Scientization Movement, Chen and Gu worked together later on as minister and vice minister of education for the Nationalist government in 1938, until the movement was terminated because of the Japanese invasion. See Gu Yuxiu 顧毓琇, "Zhongguo kexuehua de yiyi" 「中國科學化」的意義 [The meaning of [the] scientization of China], *Zhongshan wenhua jiaoyuguan jikan* 中山文化教育館季刊 [Quarterly of the Zhongshan Bureau of Culture and Education] 2, no. 2 (1935): 415–22; Chen Shou 陳首, "Kexue yu kexuehua: Gu Yuxiu de linian fenxi" 科學與科學化：顧毓琇的理念分析 [Science and scientization: An analysis of Gu Yuxiu's ideas], *Kexue jishu yu bianzhengfa* 科學技術與辯證法 [Science, technology and dialectics] 24, no. 4 (2007): 84–88.

29. Chen Shou, *Zhongguo kexuehua yundong yanjiu*, 99.

30. Chen Guofu 陳果夫, "Yixue de youzhi ji Zhongyi kexuehua de biyao" 醫學的幼稚及中醫科學化的必要 [The immature nature of medicine and the necessity of scientizing Chinese medicine], *Guoyi gongbao* 國醫公報 [Bulletin of national medicine], no. 10 (1933): 12–15.

31. Anon., "Zhongguo kexuehua yundong xiehui dierqi gongzhuo jihua dagang" 中國科學化運動協會第二期工作計畫大綱 [A working plan for the second phase of the China Scientization Movement], *Kexue de Zhongguo* 科學的中國 [Scientific China] 5, no. 5 (1935): 181–84. See also Gu, "Zhongguo kexuehua de yiyi," 418 and 422.

32. Anon., "Zhongguo kexuehua yundong xiehui dierqi gongzhuo jihua dagang," 182.

33. Although the term *kexuehua* existed previously, it was rarely used before the launch of the China Scientization Movement in 1932.

34. *Zhongyi kexuehua lunzheng* 中醫科學化論戰 [Polemics of scientizing Chinese medicine], *Zhong xi yiyao* 中西醫藥 [Journal of the Medical Research Society of China] 2, nos. 2–3 (1936).

35. Yu Yan 余巖, "Woguo yixue geming zhi pohuai yu jianshe" 我國醫學革命之破壞與建設 [Construction and destruction in our national medical revolution], *Yiyao Pinglun* 醫藥評論 [Medical review], no. 8 (1929): 1–17; and "Woguo yixue geming zhi pohuai yu jianshe (xu)" 我國醫學革命之破壞與建設（續） [Construction and destruction in our national medical revolution: Part two], *Yiyao pinglun* 醫藥評論 [Medical review], no. 9 (1929): 17–21.

36. Among these twenty-three papers, only ten were published here for the first time. The editor-in-chief collected the other thirteen papers from various medical journals. Hypothetically speaking, the editor-in-chief could have se-

lected articles published before 1929, if he had seen them as fitting the purpose of the special issues.

37. Anon., "Jiemu" 揭幕 [Inauguration], *Zhong xi yiyao* 中西醫藥 [Journal of the Medical Research Society of China] 2, no. 2 (1936): 91–92, esp. 91.

38. Anon., "Zhengwen: zhongyi kexuehua lunzhan" 徵文：中醫科學化論戰 [Call for papers: The polemics of scientizing Chinese medicine], *Zhong xi yiyao* 中西醫藥 [Journal of the Medical Research Society of China] 1, no. 4 (1935): 300.

39. For a brief account of Lu Yuanlei's views on medicine, see Chen Jianmin 陳健民, "Lu yuanlei xiansheng de xueshu sixiang" 陸淵雷先生的學術思想 [On Lu Yuanlei's scholarly thinking], *Zhonghua yishi zazhi* 中華醫史雜誌 [Chinese journal of medical history] 20, no. 2 (1990): 91–95.

40. Lu Yuanlei 陸淵雷, "Ni guoyiyao xueshu zhengli dagang caoan" 擬國醫藥學術整理大綱草案 [Proposal for putting in order Chinese medicine and pharmaceutics], *Shenzhou guoyi xuebao* 神州國醫學報 [Shenzhou Bulletin of National Medicine] 1, no. 1 (1932): 1–9.

41. Ibid., 2.

42. Lu Yuanlei 陸淵雷, "Cong genben shang tuifan qihua" 從根本上推翻氣化 [To overthrow qihua at its very foundation], *Zhongyi xinshengming* 中醫新生命 [The new life of Chinese medicine], no. 3 (1934): 38–50. Summarizing Lu's program for reforming Chinese medicine, a loyal follower urged doctors of Chinese medicine to make a choice between two mutually exclusive options: science vs. *qihua*. See Xie Songmu 謝頌穆, "Zhongyi wang hechu qu?" 中醫往何處去 [Where does Chinese medicine go from here?], *Zhongyi xin shengming* 中醫新生命 [The new life of Chinese medicine], no. 3 (1934): 1–7, esp. 1.

43. Xie, "Zhongyi wang hechu qu?," 1.

44. Lu, "Ni guoyiyao xueshu zhengli dagang caoan," 2.

45. Charles E. Rosenberg, "The Tyranny of Diagnosis: Specific Entities and Individual Experience," in *Our Present Complaint: American Medicine, Then and Now* (Baltimore, MD: Johns Hopkins University Press, 2007), 13.

46. Tan Cizhong claimed to be the individual "most committed to scientifically reforming Chinese medicine." See Zhao Hongjun 趙洪鈞, *Jindai zhongxiyi lunzhengshi* 近代中西醫論爭史 [A history of the struggle between Chinese and Western medicine in the modern period] (Hefei: Anhui kexue jishu chubanshe, 1989), 249.

47. These two books were first published in 1931. Later on, they were integrated into one book, entitled *The Polemic of Medical Revolution*. See Tan Cizhong 譚次仲, *Yixue geming lunzhan* 醫學革命論戰 [The polemic of medical revolution] (Hong Kong: Qiushi chubanshe, 1952 [1931]).

48. See Tan Cizhong 譚次仲, "Wei kexuehua zhonggao quanti zhongyijie" 為科學化忠告全體中醫界 [Some advice about scientization to the entire community of Chinese medicine], in Tan, *Yixue geming lunzhan*, 94–111; and Tan Cizhong 譚次仲, "Yu quanti minzhong lun zhongyi kexuehua zhi biyao" 與全體民眾論中醫科學化之必要 [Discussing the necessity of scientizing Chinese medicine with the entire populace], in Tan, *Yixue geming lunzhan*, 115–18.

49. Ibid., 526.

50. Tan, *Yixue geming lunzhan*, 63.

51. Fan Xingzhun 范行准, "Zazhongyi zhi puoshi" 雜種醫之剖視 [An anatomy of mongrel medicine], *Zhongxi yiyao* 中西醫藥 [Journal of the Medical Research Society of China] 2, no. 11 (1936): 703–12, esp. 708.

52. Tan, *Yixue geming lunzhan*, 28.

53. Zeng Juesou 曾覺叟, "Wen Yuyan dui Jiaoguanzhang wei niding guoyi tiaoli gao guoren shangqueshu zhi jiaozheng" 聞余巖對焦館長為擬定國醫條例告國人商榷書之矯正 [Rectification of Yu Yan's comments on director Jiao's announcement to his fellow citizens about the ordinance on national medicine], *Guanhua zazhi* 光華雜誌 [Splendor] 2, no. 2 (1935): 1–10, esp. 1–2.

54. Zeng Jueshou曾覺叟, "Zhi Lu Yuanlei shu" 致陸淵雷書 [A letter to Lu Yuanlei], *Yixue zazhi* 醫學雜誌 [Journal of medicine] 72 (1933): 74–79, esp. 75.

55. See Zhao, *Jindai zhongxiyi lunzhengshi*, 232–33.

56. Fan Xingzhun 范行准, "Guoyi de paihuai shidai" 國醫的徘徊世代 [The period of lingering for national medicine], *Guoyi pinglun* 國醫評論 [Review of national medicine] 1, no. 3 (1933): 1–9, esp. 4.

57. Fan Xingzhun 范行准, "Gei Dong Zhiren xiansheng" 給董志仁先生 [To Mr. Dong Zhiren], *Zhongxi yiyao* 中西醫藥 [Journal of the Medical Research Society of China] 2, no. 5 (1936): 308–12, esp. 311.

58. For more information on Xie Guan, see Volker Scheid, *Currents of Tradition in Chinese Medicine, 1626–2006* (Seattle, WA: Eastland Press, 2007), 377–83.

59. Anon., "Guoyiyao xueshu zhengli dagang caoan" 國醫藥學術整理大綱草案 [Proposal for putting in order Chinese medicine and pharmaceutics], *Guoyi gongbao* 國醫公報 [Government bulletin of national medicine] 2, no. 2 (1934): 1–6.

60. Yu, "Du guoyiguan zhengli xueshu caoan zhi wojian," 189.

61. Bridie J. Andrews, "Acupuncture and the Reinvention of Chinese Medicine," *American Pain Society Bulletin* 9, no. 3 (1999), n.p.

62. Hiromichi Yasui, "History of Japanese Acupuncture and Moxibustion," *Journal of Kampo, Acupuncture and Integrative Medicine* 1 (February 2010): 2–9, esp. 7.

63. Sun Yanru 孫晏如, "Zhenxing guoyi youxu tichang zhenjiu shuo" 振興國醫尤需提倡針灸說 [Argument to promote acupuncture in order to develop national medicine], *Yijie chunqiu* 醫界春秋 [Annals of the medical profession], no. 54 (1930): 2.

64. Anon., "Wei sheli guoyiguan an zhongyzng zhengzhi huiyi zhi guoming zhengfu yuanhan" 為設立國醫館案中央政治會議致國民政府原函 [Letter from the political committee to the Nationalist government concerning the creation of the Institute of National Medicine], *Guoyi gongbao*國醫公報 [Bulletin of national medicine] 1, no. 10 (1933): 3.

65. Anon, "Zhongyang guoyiguan zhengli guoyiyao xueshu biaozhun dagang caoan" 中央國醫館整理國醫藥學術標準大綱草案 [Proposal for standards used for sorting out Chinese medical and pharmaceutical learning issued by the Central Institute of National Medicine], *Guoyi gongbao* 國醫公報 [Bulletin of national medicine] 2, no. 2 (1934): 1–6, esp. 5.

66. Chen Bichuan 陳碧川, "Zhenjiu liaobingfa zai yiyao shang de jiazhi yu

bihai" 針灸療病法在醫藥上的價之與其弊害 [Medical value and drawbacks of acupuncture treatment], *Guoyi gongbao* 國醫公報 [Bulletin of national medicine] 2, no. 3 (1934): 53–54.

67. Liang Chunxu 梁春煦, "Zhongyang guoyiguan zhengli guoyiyao xueshu biaozhun dagang caoan jianping" 中央國醫館整理國醫藥學術標準大綱草案僭評 [Comment on the proposal for standards used for putting in order Chinese medicine and pharmaceutics issued by the Institute of National Medicine], *Guoyi gongbao* 國醫公報 [Bulletin of national medicine] 2, no. 2 (1934): 87–96, esp. 95.

68. Anon., "Zhongyang guoyiguan zhengli guoyiyao xueshu biaozhun dagang caoan," 5.

69. Huang Zhuzhai 黃竹齋, *Zhenjiu jingxue tukao*針灸經穴圖考 [Verified charts of acupuncture channels and points] (Taipei: Xinwenfeng chuabn youxian gongsi, 1970), 56; orig. pub. 1934.

70. Concerning Chen Dan'an's 承淡安career and his contributions to acupuncture, see Andrews, "Acupuncture and the Reinvention of Chinese Medicine," 2–3. For an advertisement of Chen's posters, see the advertisement "Xinshi shier jingxue guatu chuban" 新式十二經穴掛圖出版 [Publication of the modern-style charts of the twelve channels and points], *Yijie chunqiu* 醫界春秋 [Annals of the medical profession], no. 69 (1932): back cover.

71. Bridie Andrews, "The Making of Modern Chinese Medicine, 1895–1937," PhD diss., Department of History and Philosophy of Science, University of Cambridge, Cambridge, 1996, 279.

72. Ibid., 280.

73. Lin Fushi 林富士, "Zhouyou shi yi: Yi huangdi neijing suwen wei hexin wenben de taolun" 「祝由」釋義：以《黃帝內經素問》為核心文本的討論 [A philological explanation of zhuyou], Zhongyan yanjiuyuan lishu yuyan yanjiusuo jikan 中央研究院歷史語言研究所集刊 [Bulletin of the Institute of History and Philology, Academia Sinica] 83, no. 4 (2012): 671–738.

74. Philip Cho, "Ritual and the Occult in Chinese Medicine and Religious Healing: The Development of Zhuyou Exorcism," PhD diss., University of Pennsylvania, 2006; Chen Hsiu-fen 陳秀芬, "Dang bingren jiandao gui: Shilun mingqing yizhe duiyu xiesui de taidu" 當病人見到鬼: 試論明清醫者對於「邪祟」的態度 [When patients met ghosts: A preliminary survey of scholarly doctors' attitudes toward 'demonic affliction' in Ming-Qing China], *Guoli zhengzhi daxue lishi xuebao* 國立政治大學歷史學報 [The journal of history, National Chengchi University], no. 30 (2008): 43–86, esp. 71–76.

75. Zhang Rongming 張榮明, "Luelun zhongyi zhuyoushu de lishi fazhan" 略論中醫祝由術的歷史發展 [On the historical developemnt of zhuyou in Chinese medicine], *Yiguowen zhishi* 醫古文知識 [Knowledge of ancient medical literature], no. 3 (1995): 11–13, esp. 13.

76. The only article I found is Feng Weixin 馮薇馨, "Wushu duiyu xinli liaobing zhi biyao" 巫術對於心理療病之必要 [On the necessity of magic in treating disease], in *Zhongyi xunlun huibian* 中醫新論彙編 [Collected essays on new discussions of Chinese medicine], ed. Wang Shenxuan 王慎軒 (Suzhou: Suzhou guoyi shushe, 1932), 45.

77. Chen Guofu 陳果夫, "Jiangsu yizheng xueyuan de guoqu yu weilai"

江蘇醫政學院的過去與未來 [The past and future of the Jiangsu Provincial School of Medical Administration], in *Kukou tan yiyao* 苦口談醫藥 [Bitter chats on medicine], ed. Chen Guofu 陳果夫 (Taipei: Zhengzhong shuju, 1949), 47–66, esp. 62.

78. Chen Guofu 陳果夫, "Duiyu yixueyuan de qiwang" 對於醫學院的期望 [Expectations for medical schools], in *Kukou tan yiyao*, 66–75, esp. 73.

79. It might be useful to mention a comparable situation. When Westerners started using Chinese medicine to treat psychological disorders, they discovered that certain areas of acupuncture that they found useful in that regard were considered as "superstitious" and therefore had been purged from the standardized textbooks of Traditional Chinese medicine produced in the 1950s. Linda L. Barnes, "The Psychologizing of Chinese Healing Practices in the United States," *Culture, Medicine and Psychiatry* 22, no. 4 (1998): 413–43, esp. 421–25.

80. Anon., "Jiemu" 揭幕 [Inauguration], *Zhong xi yiyao* 中西醫藥 [Journal of the Medical Research Society of China] 2, no. 2 (1936): 91–92, esp. 91.

81. Yu, *Yushi yizhu*, 1.

82. Huang Kewu 黃克武, "Cong shenbao yiyao guanggao kan minchu shanghai de yiliao wenhua yu shehui shenghuo, 1912–1926" 從申報醫藥廣告看民初上海的醫療文化與社會生活 [Medical culture and social life in the early Republican era as seen through the medical advertisements of *Shenbao*, 1912–1926], *Zhongyang yanjiuyuan jindaishi yanjiusuo jikan* 中央研究院近代史研究所集刊 [Bulletin of the Institute of Modern History, Academia Sinica] 7, no. 2 (1988): 141–94.

83. For an attack from the first group, see Qin Bowei 秦伯未, "Zhongyi yu kexue" 中醫與科學 [Chinese medicine and science], *Zhongyi shijie* 中醫世界 [The world of Chinese medicine] 5, no. 2 (1933): 1–5, esp. 2. With regard to attacks from the third group, Fan Xingzhun, the famous historian of Chinese medicine, simply trashed Tan Cizhong's ten interpretations of the key concepts of Chinese medicine. See Fan, "Zazhongyi zhi puoshi."

84. Bill Ashcroft, Gareth Griffiths, and Helen Tiffin, *The Post-Colonial Reader*, 2nd ed. (London: Routledge, 2006), 137.

85. Lu, "Ni guoyiyao xueshu zhengli dagang caoan," 3.

86. Ibid.

87. While I see it as a general attitude among traditional practitioners to go beyond the defensive strategy of preserving Chinese medicine, it might be useful to provide an example that addressed this issue explicitly and denounced the idea of preserving Chinese medicine as the cultural essence of China. In an article from 1930 entitled "Chinese Medicine and Its Future," Yao Zhaopei asserts, "As the struggle between Chinese and Western medicine is intensifying each day, the idea of preserving Chinese medicine as national essence is becoming more and more popular accordingly. I feel that we need to take a careful look at this idea for two reasons. First, Chinese medicine can be developed further, but it cannot be preserved [as it is]. The idea of preserving national essence can only keep Chinese medicine from being abolished [by the government]. In light of the world trend of medical development, the effort of preservation is far from enough to keep Chinese medicine alive. In the best

scenario, the efforts of preservation might be able to extend the life of Chinese medicine by a few years. Sooner or later, it will go extinct. [To avoid this fate], Chinese medicine has no choice but to move forward. . . . If Chinese medicine is not effective in treating disease, or if it is effective but inferior to Western medicine, why don't we just abolish it rather than preserve it? If Chinese medicine is effective in treating disease and superior to Western medicine in terms of therapeutic efficacy, why should we confine ourselves to preserving Chinese medicine? Why don't we collect all the effective treatments [of Chinese medicine] into a book and devote it to the world! In that case, Chinese medicine can enjoy a position vis-à-vis Western medicine within the world medical market"; Yao Zhaopei 姚兆培, "Zhongyi yu qiantu" 中醫與前途 [Chinese medicine and its future], *Yijie chunqiu* 醫界春秋 [Annals of the medical profession], no. 45 (1930): 3.

88. Fan, "Zazhongyi zhi puoshi," 703.

89. An explicit formulation appears in Fan Xingzhun's criticism of mongrel medicine. "If we analyze the idea of 'scientizing Chinese medicine,' as it is being advocated by the promoters of mongrel medicine, from a logical point of view, we must presuppose the idea that Chinese medicine is a nonscientific entity, because only in this way can we reach the conclusion that it needs scientization" (ibid., 706).

90. For example, see Fu Sinian 傅斯年, "Zailun suowei 'guoyi'" 再論所謂「國醫」 [Another discussion on so-called "national medicine"], in *Fu Sinian quanji* 傅斯年全集 [Collected works of Fu Sinian] (Taipei: Lianjing chubanshe, 1980), 309–14, esp. 311; orig. pub. 1934.

91. Anon., "Jiemu," 92.

92. Fan, "Zazhongyi zhi puoshi," 706.

93. When doctors of Western medicine freely spoke their minds, as the editor-in-chief of these special issues did, they asserted that "the fact that Chinese medicine can't be scientized is a conclusion that has long been established." Therefore, "the problem of today is not whether Chinese medicine can be scientized but whether the population of practitioners of Chinese medicine can be reduced." In this sense, doctors of Western medicine really did not consider the problem of Scientizing Chinese Medicine as one worthy of serious consideration. What concerned them was whether the government would outlaw, or at least substantially contain, the practice of Chinese medicine in the name of science. That was why the editor explicitly connected the epistemological issue of scientizing Chinese medicine with the controversy over governmental policy on whether Chinese medicine should be abolished. See Song Daren 宋大仁, "Fu 'zailun zhongyi kexuehua wenti'" 覆「再論中醫科學化問題」 [Response to "On the problem of scientizing Chinese medicine"], *Zhongxi yiyao* 中西醫藥 [Journal of the Medical Research Society of China] 2, no. 8 (1936): 544–49, esp. 544–45.

CHAPTER 8

1. Ye Guhong 葉古紅, "Chuanranbing zhi guoyi liaofa" 傳染病之國醫療法 [Treatment methods of our national medicine for infectious diseases], *Guoyi*

gongbao 國醫公報 [Bulletin of national medicine] 2, no. 10 (1935): 67–71, esp. 67.

2. See, for example, the following two articles by Ye Guhong 葉古紅, "Zhonghua yiyao geming lun" 中華醫藥革命論 [On the medical revolution of Chinese medicine and pharmacy], *Yijie chunqiu* 醫界春秋 [Annals of the medical profession], no. 49 (1930): 1–3; and "Zhonghua yiyao geming lun" 中華醫藥革命論 [On the medical revolution of Chinese medicine and pharmacy], *Yiyao pinglun* 醫藥評論 [Medical review] 6, no. 1 (1934): 28–31.

3. Sanetō Keishū 實藤惠秀, *Zhongguoren liuxue riben shi* 中國人留學日本史 [A history of Chinese overseas study in Japan], trans. Tan Ruqian 譚汝謙 and Lin Qiyan 林啟彥 (Hong Kong: Chinese University Press, 1982), 235.

4. *huleila* 虎列剌 (cholera), *chili* 赤痢 (dysentery), *changzhi fusi* 腸窒扶斯 (typhoid fever), *tianran dou* 天然痘 (smallpox), *fazhen zhifusi* 發疹窒扶斯 (typhus exanthemata), *xinghongre* 猩紅熱 (scarlet fever), *shifu dili* 實扶的里 (diphtheria), and *baisituo* 百斯脫 (plague). Zhang Zaitong 張在同, ed., *Minguo yiyao weisheng fagui xuanbian* 民國醫藥衛生法規選編 [Selected collection of hygiene laws from the Republican period] (Taian, Shandong: Shandong daxue chubanshe, 1990 [1912–1948]), 10.

5. Liu Shiyong 劉士永, "'Qingjie,' 'weisheng,' yu 'baojian'—rizhi shiqi Taiwan shehui gonggong weisheng guannian zhi zhuanbian「清潔」,「衛生」與「保健」—日治時期臺灣社會公共衛生觀念之轉變 ["Cleanliness," "hygiene," and "health care"—Changing notions of public hygiene in Taiwanese Society during the Japanese occupation], *Taiwanshi yanjiu* 臺灣史研究 [Studies in Taiwanese history] 8, no. 1 (2001): 41–88, esp. 57.

6. Yu Yan 余巖, "Feizhi jiuyi yi saochu yishi weisheng zhi zhangaian" 廢止舊醫以掃除醫事衛生之障礙案 [Abolishing the old medicine to sweep away the obstacles to medicine and public health], *Yijie chunqiu* 醫界春秋 [Annals of the medical profession], no. 34 (1929): 9–10, esp. 9.

7. Anon., "Zhongyang weisheng weiyuan huiyi yijue 'feizhi zhongyi an' yuanwen" 中央衛生委員會議議決「廢止中醫案」原文 [Original text of the "Proposal for Abolishing the Practice of Chinese Medicine" discussed and decided by the Central Board of Health], *Yijie chunqiu* 醫界春秋 [Annals of the medical profession], no. 34 (1929): 9–11, esp. 10.

8. The information in parentheses is part of the original. The Chinese original identifies what it calls "pestilential diseases" (*yibing* 疫病) with the more commonly used term *chuanranbing*.

9. Quanguo yiyao zonghui 全國醫藥總會 [National Federation of Medicine and Pharmacy], *Quanguo yiyao tuanti daibiao dahui ti'an huilu* 全國醫藥團體代表大會提案會錄 [Collection of proposals raised in the joint national convention of medical and pharmaceutical associations] (Shanghai: Quanguo yiyao zonghui, 1929), 17–8.

10. Anon., "Jiangsu sheng zhongyi jianding guize" 江蘇省中醫檢定規則 [Tentative regulation concerning the examination and registration of Chinese medicine in Jiangsu Province], *Yijie chunqiu* 醫界春秋 [Annals of the medical profession], no. 89 (1934): 29–30, esp. 30.

11. Ibid., 29.

12. Literally translated, this term means "cold damage." It is a broad traditional disease category, but was equated by the government here with typhoid.

13. Literally "wind damage"; the term was used by the government here to refer to the biomedical disease influenza.

14. Anon., "Wuxian zhongyi gonghui yijue fandui jiangsu sheng zhongyi jianding cuize" 吳縣中醫公會議決反對江蘇省中醫檢定規則 [Resolution by the Union of Chinese Medicine of Wu County against the tentative regulation concerning the examination and registration of Chinese medicine in Jiangsu Province], *Yijie chunqiu* 醫界春秋 [Annals of the medical profession], no. 91 (1934): 39–40.

15. Volker Scheid, "Foreword," in *Warm Diseases: A Clinical Guide*, ed. Guohui Liu (Seattle, WA: Eastland Press, 2001), vii–x.

16. Shi Jinmo was respected as one of Beijing's four famous doctors of Chinese medicine. Shi devoted his youth to overthrowing the Qing dynasty and therefore was acquainted with the leaders of the Nationalist Party. Partially because of these political connections, he was invited by Jiao Yitang, president of the Institute of National Medicine and a high-ranking politician of the Nationalist Party, to serve as vice president of the institute. In this role, he became one of the main architects for reforming Chinese medicine. In 1932, he founded the Northern China Institute of National Medicine 華北國醫學院 (*huabei guoyi xueyuan*) in Beijing, in which modern medical disciplines and instruments were eagerly incorporated into the curriculum.

17. Shi Jinmo 施今墨, "Zhongyang guoyiguan xueshu zhenglihui tongyi bingming jianyishu" 中央國醫館學術整理會統一病名建議書 [Proposal for unifying disease nomenclature as suggested by the Academic Committee of the Institute of National Medicine], *Yijie chunqiu* 醫界春秋 [Annals of the medical profession], no. 81 (1933): 7–12, esp. 8.

18. Ibid., esp. 7.

19. Ibid.

20. Ibid.

21. Ibid, 8.

22. Marta E. Hanson, *Speaking of Epidemics in Chinese Medicine: Disease and the Geographic Imagination in Late Imperial China* (London: Routledge, 2011), 2.

23. Ibid., 102.

24. Ibid., 10.

25. Ibid.

26. Sean Hsiang-lin Lei, "Sovereignty and the Microscope: Constituting Notifiable Infectious Disease and Containing the Manchurian Plague (1910–1911)," in *Health and Hygiene in Chinese East Asia*, ed. Angela Leung and Charlotte Furth (Durham, NC: Duke University Press, 2010), 73–108, esp. 93.

27. For example, in one otherwise excellent book on the modern history of Chinese medicine, the author concludes that the most important developments to take place before 1949 were in the field of infectious disease. However, the author does not mention that this category itself was a revolutionary development in Chinese medicine. See Deng Tietao 鄧鐵濤, ed., *Zhongyi jindaishi*

中醫近代史 [Modern history of Chinese medicine] (Guangdong: Guangdong gaodeng jiaoyu chubanshe, 1999), 407–12.

28. Shi, "Zhongyang guoyiguan xueshu zhenglihui tongyi bingming jianyishu," 11.

29. Ibid.

30. For a detailed, comprehensive study about Yun's position with regard to germ theory, see Pi Guoli 皮國立, *"Qi" yu "xijun" de jindai zhongguo yiliaoshi* 「氣」與「細菌」的近代中國醫療史 [The modern Chinese medical history of pathogenic *qi* and bacteria] (Taipei: Guoli zhongguo yiyao yanjiusuo, 2012), esp. chap. 6.

31. Yun Tieqiao 惲鐵樵, "Duiyu tongyi bingming jianyishu zhi shangque" 對於統一病名建議書之商榷 [A discussion concerning the proposal for unifying disease names], *Yijie chunqiu* 醫界春秋 [Annals of the medical profession], no. 81 (1933): 20–24, esp. 22.

32. Ibid.

33. Concerning the relationship between germ theory and Chinese medicine, see Bridie Andrews, "Tuberculosis and the Assimilation of Germ Theory in China, 1895–1937," *Journal of the History of Medicine and Allied Sciences* 52, no. 1 (1997): 114–57.

34. Deng, *Zhongyi jindaishi*, 88–89.

35. Shi Yiren 時逸人, *Zhongguo shilingbing xue* 中國時令病學 [Chinese study of seasonal diseases] (Hong Kong: Qianqingtang shuwu, 1951), 7; orig. pub. 1930.

36. Tian Erkang 田爾康, "Zhongyi duiyu zhengzhi jixing chuanranbingzheng shi lieju qi chang bing jiuzheng qi duan" 中醫對於急性傳染病症試列舉其長並糾正其短 [Listing the strengths and rectifying the weaknesses in the diagnosis and treatment of acute infectious diseases with Chinese medicine], *Yixue zazhi* 醫學雜誌 [Journal of medicine], no. 69 (1933): 5–6.

37. Ibid.

38. Shi Yiren 時逸人, *Zhongyi chuanranbing xue* 中醫傳染病學 [Epidemiology of Chinese medicine] (Taipei: Lixing shuju, 1959), 97.

39. Ibid.

40. Cover of *Yixue zazhi* 醫學雜誌 [Journal of medicine], no. 69 (1933).

41. As his contemporaries mistook Pasteur's argument for germs as the *necessary* cause of disease as an argument for germs as the *sufficient* cause, they found his ideas absurd, since people often had germs but never manifested symptoms of illness. Understandably, when advocates of Chinese medicine passionately argued that the six *qi* were the root cause of disease and that germs were just a by-product of this process, they rehearsed many debates that had taken place half a century earlier in Europe. Carter has rightly emphasized that in both non-Western medical traditions and Western medicine prior to the middle of the nineteenth century, "one inevitably finds an almost exclusive use of causes (of diseases) that are sufficient but not necessary." See K. Codell Carter, "The Development of Pasteur's Concept of Disease Causation and the Emergence of Specific Causes in Nineteenth-Century Medicine," *Bulletin for the History of Medicine* 65, no. 4 (1991): 528–48, esp. 528.

42. Deng, *Zhongyi jindaishi*, 301.

43. Zhang Zanchen 張贊臣, "Guanyu tongyi bingming zhi suojian (My opinion on unifying the nomenclature of diseases) 關於統一病名之所見," *Yijie chunqiu* 醫界春秋 (Annals of the medical profession), no. 81 (1933): 2. The information presented in this paragraph and the following discussion on Zhang Taiyan and Watanabe supports but also complicates Scheid's statement that "contemporary distinctions between symptoms, patterns, and diseases were never clearly defined before the 1950s." See Volker Scheid, *Chinese Medicine in Contemporary China: Plurality and Synthesis* (Durham, NC: Duke University Press, 2002), 205.

44. The two words are *symptom* (*zheng* 症) and *pattern* (*zheng* 證).

45. Scheid, *Chinese Medicine in Contemporary China*, 205–6.

46. Ibid., 206.

47. Watanabe Hiroshi 渡邊熙, *Hehan yixue zhi zhensui* 和漢醫學之真髓 [True essence of Japanese-Chinese medicine], trans. Shen Songnian 沈松年 (Shanghai: Changming yixue shuju, 1931), 7; orig. pub. 1928.

48. Watanabe Hiroshi 渡邊熙, "Riben yixue boshi Dubianxi tiyi gedaxue tianshe hanyi jiangzuoshu" 日本醫學博士渡邊熙提議各大學添設漢醫講座書 [A proposal by the Japanese doctor Dubianxi for setting up a university chair for kampo medicine], *Yixue zazhi* 醫學雜誌 [Journal of medicine], no. 72 (1933): 70–72, esp. 71.

49. While Watanabe made this statement famous, and it was subsequently used by many advocates of Chinese medicine to create a contrast between the two styles of medicine, he made quite a few mistakes in this quotation. It is highly possible that Watanabe confused and combined Joseph Skoda (1805–81) with René-Théophile-Hyacinthe Laennec (1781–1826), the widely acclaimed inventor of the stethoscope. Watanabe's article entitled "Skoda's Erroneous Opinions" opens with the statement that "around 1805, Skoda, the inventor of today's stethoscope, said . . ." Skoda was born in 1805, so there is no way that he could have made that statement in 1805. Besides, Skoda is not credited with the invention of the stethoscope. Precisely because the link between Skoda and the quotation is incorrect, this link can serve as a useful clue in tracing the transmission of this opinion from Chinese thinkers back to Watanabe. See Watanabe, *Hehan yixue zhi zhensui*, 26.

50. Barry G. Firkin and J. A. Whitworth, *Dictionary of Medical Eponyms* (New York: Parthenon, 2002), 374.

51. Watanabe, *Hehan yixue zhi zhensui*, 2.

52. Zhang Taiyan 章太炎, "Xu" 序 [Preface], in *Shanghanlun jinshi* 傷寒論今釋 [Contemporary interpretation of the shanghanlun], ed. Lu Yuanlei 陸淵雷 (Taipei: Wenguang chubanshe, 1961 [1931]); orig. pub. 1931.

53. Shi Yiren 時逸人, "Renshen zhi xia yixue zhuanxiao disiban biyelu xu" 壬申之夏醫學專校第四班畢業錄序 [Preface for the graduation book of the fourth class of the Special College of Medicine in the summer of 1932], *Yixue zazhi* 醫學雜誌 [Journal of medicine], no. 65 (1932): 1–3, esp. 2.

54. Wu Hanqian 吳漢遷, "An" 按 [Note], *Yixue zazhi* 醫學雜誌 [Journal of medicine], no. 72 (1933): 72.

55. Lu Yuanlei, ed., *Shanghanlun jinshi* [Contemporary interpretation of the shanghanlun], 2. While his statement looks like an exact mirror image

of Skoda's, Zhang Taiyan mentioned neither Watanabe nor Skoda by name. Nevertheless, even other people recognized that in this preface Zhang had paraphrased Watanabe's quote of Skoda. See Zeng Juesou 曾覺叟, "Zhi Zhang Taiyan shu" 致章太炎書 [Letter to Zhang Taiyan], *Yixue zazhi* 醫學雜誌 [Journal of medicine], no. 72 (1933): 72–74, esp. 72.

56. Roy Porter, *The Cambridge History of Medicine* (Cambridge: Cambridge University Press, 2006), 121–26.

57. Zhang, "Xu," 1.

58. Ibid.

59. Jia Chunhua 賈春華, "Gufangpai dui zhongguo jindai shanghanlun yanjiu de yingxiang" 古方派對中國近代傷寒論研究的影響 [Influence of the Ancient Formula School on shanghanlun studies in modern China], *Beijing zhongyiyao daxue xuebao* 北京中醫藥大學學報 [Journal of Beijing University of Traditional Chinese Medicine] 17, no. 4 (1994): 5–9.

60. Michael Worboys, *Spreading Germs: Disease Theories and Medical Practice in Britain, 1865–1900* (Cambridge: Cambridge University Press, 2000), 4.

61. Charles E. Rosenberg, "The Tyranny of Diagnosis: Specific Entities and Individual Experience," in *Our Present Complaint: American Medicine, Then and Now* (Baltimore, MD: Johns Hopkins University Press, 2007), 13.

62. Ibid., 13–14.

63. This is precisely the idea proposed by the hard-line conservative Zeng Juesou. He published an open letter to Zhang Taiyan to express his disagreement over Zhang's admission that Chinese medicine did not recognize disease causes. In light of Zeng's criticism and alternative proposal, the strategic nature of Zhang's idea becomes even more salient. See Zeng, "Zhi Zhang Taiyan shu," 72.

64. It is highly possible that Zhang Taiyan indeed borrowed from Watanabe the opposition between pattern and disease. When Watanabe's book was published in Japan in 1928, it was entitled *The True Character of Japanese-Chinese Medicine: The Therapeutics of Major Patterns*. As revealed by this title, two of Zhang's key emphases, therapeutics and pattern, had been used by Watanabe to highlight the special characteristics of Japanese-Chinese medicine.

65. It is noteworthy that Shi was praised later as the main architect for developing the theory of *bianzheng lunzhi* in the 1950s. Apparently, Shi appears to have changed his position around the time when he became the first professor in the Beijing College of Chinese medicine.

66. See Song Airen 宋愛人, "Kexue buzu cunfei guoyilun" 科學不足存廢國醫論 [Science is capable of deciding the fate of national medicine], *Yijie chunqiu* 醫界春秋 [Annals of the medical profession], no. 85 (1933): 17–19, esp. 17; Ye, "Chuanranbing zhi guoyi liaofa," 67.

67. Erwin H. Ackerknecht, *A Short History of Medicine* (Baltimore, MD: Johns Hopkins University Press, 1982), 232.

68. Ye, "Chuanranbing zhi guoyi liaofa," 68.

69. Ibid.

70. Paul Weindling, "The Immunological Tradition," in *Companion Ency-*

clopedia of the History of Medicine, ed. W. F. Bynum and Roy Porter (London: Cambridge University Press, 1995), 192–204, esp. 195.

71. Ye, "Chuanranbing zhi guoyi liaofa," 68.

72. Scheid, *Chinese Medicine in Contemporary China*, 203–4.

73. Watanabe, *Hehan yixue zhi zhensui*, 80–82.

74. Scheid, *Chinese Medicine in Contemporary China*, 200–37. See also Volker Scheid, "Convergent Lines of Descent: Symptoms, Patterns, Constellations and the Emergent Interface of System Biology and Chinese Medicine," *East Asian Science, Technology and Society: An International Journal*, forthcoming.

75. Eric I. Karchmer, "Chinese Medicine in Action: On the Postcoloniality of Medical Practice in China," *Medical Anthropology: Cross-Cultural Studies in Health and Illness*, 29, no. 3 (2010): 226–52, especially 246.

76. Watanabe, "Riben yixue boshi Dubianxi tiyi gedaxue tianshe hanyi jiangzuoshu," 79.

77. Ian Hacking, *Representing and Intervening* (Cambridge: Cambridge University Press, 1983), 23 and 262–75.

78. For the emergence and standardization of *bianzheng lunzhi* (辨證論治) in the 1950s, see Volker Scheid's excellent analysis in *Chinese Medicine in Contemporary China*, 200–37.

79. Most notably, Watanabe went so far as to claim that Japanese-style Chinese medicine was "a discipline without foundation."

CHAPTER 9

1. E. Leong Way, "Pharmacology," in *Sciences in Communist China*, ed. S. H. Gould (Washington, DC: American Association for the Advancement of Science, 1961), 363–82, esp. 364.

2. Fan Xingzhun 范行准, "Zazhongyi zhi poushi" 雜種醫之剖視 [An anatomy of mongrel medicine], *Zhongxi yiyao* 中西醫藥 [Journal of the Medical Research Society of China] 2, no. 11 (1936): 703–12, esp. 706.

3. Rao Yi 饒毅, Li Runhong 黎潤紅, Zhang Daqing 張大慶, "*Zhongyao de yanjiu fengbei* 中藥的研究豐碑 [New drugs from Ancient Chinese remedies: Unsung heroes in unusual times]," *Kexue wenhua pinglun* 科學文化評論 [Science and culture review], no. 4 (2011): 29–46.

4. I would like to thank Wenching Song for urging me to emphasize the important fact that the case of *changshan* was far from a normal or representative project of Scientific Research on Nationally Produced Drugs.

5. For a critical introduction to and reflection on the actor-network theory, see John Law and John Hassard, *Actor-Network Theory and After* (Oxford: Blackwell, 1999).

6. Bruno Latour, *The Pasteurization of France* (Cambridge, MA: Harvard University Press, 1988), 21.

7. *Shenbao* 申報 [Shen news], March 18, 1929.

8. Karl Gerth, *China Made: Consumer Culture and Creation of the Nation* (Cambridge, MA: Harvard University Asia Center, 2003).

9. Wang Qizhang, a practitioner of biomedicine and vocal critic of Chinese medicine, once admitted that he felt deeply distressed that he had no options but to prescribe foreign-made medications to his patients. See Wang Qizhang 汪企張, *Ershi nian lai zhongguo yishi chuyi* 二十年來中國醫事芻議 [Preliminary comments on the recent twenty years of medical affairs in China] (Shanghai: Zhengliao yibao she, 1935), 127.

10. In a series of historical studies on Chinese medicine, Fan Xingzhun 范行准, an early writer on Chinese medical history and a prolific critic of Chinese medicine, excavated the foreign origin of certain elements of Chinese medicine. Fan consciously appropriated historical studies to attack the cultural nationalists who had raised Chinese medicine to the status of a "national essence." See Fan Xingzhun 范行准, "Hu fang kao" 胡方考 [Research on foreign prescriptions], *Zhonghua yixue zazhi*中華醫學雜誌 [National medical journal of China] 22, no. 12 (1936): 1235–66.

11. Wang, *Ershi nian lai zhongguo yishi chuyi*, 164 and 278.

12. Practitioners of Western medicine obviously took the accusation of "salesman" very seriously; quite a few of them cited the statistics from the Chinese Maritime Customs Service to challenge the status of Chinese pharmaceuticals as domestic goods. See, for example, Fan Shouyuan 范守淵, "'Guoyao' yu guohuo"「國藥」與國貨 ["National drugs" and national goods], in *Fanshi yilunji* 范氏醫論集 [Mr. Fan's collected essays on medicine], vol. 2 (Shanghai: Jiu-jiu yixueshe, 1947), 171–75; Chen Fangzhi 陳方之, "Xiyao de louyi wenti" 西藥的漏溢問題 [Problem of trade deficit caused by Western pharmaceuticals], in *Yishi huikan* 醫事匯刊 [Collected essays on medical matters], no. 8 (1931): 3–4. Pang Jingzhou 龐京周, "Da jiouyi ji gao zhengfu zhugongwen" 答舊醫及告政府諸公文 [A response to old medical doctors and an announcement to the governmental and societal celebrities], in *Zhongyi jiaoyu taolunji* 中醫教育討論集 [Collected essays on Chinese medical education] (Shanghai: Zhongxi yiyao yanjiushe, 1939), 416–19.

13. Yan Ze 言者, "Cong mahuangjing lianxiang dao woguo de yiyaojie" 從麻黃精聯想到我國的醫藥界 [From ephedrine to the medical circle in China], *Yixue zhoukan* 醫學週刊 [Medical weekly], no. 36 (1930).

14. K. K. Chen, "Researches on Chinese Materia Medica," *Journal of the American Pharmaceutical Association* 20, no. 2 (1931): 110–13, esp. 111.

15. Zhang Changshao 張昌紹, "Sanshi nianlai zhongyao zhi kexue yanjiu" 三十年來中藥之科學研究 [Scientific research on Chinese herbs in the past 30 years], *Kexue* 科學 [Science] 31, no. 4 (1949): 99–116. Other secondary sources on the research on *changshan* include Zhang Minggao 張鳴皋, ed., *Yaoxue fazhan jianshi* 藥學發展簡史 [A brief history of the development of pharmacology] (Beijing: Zhongguo yiyao keji chubanshe, 1993), 148; Chen Xinqian 陳新謙 and Zhang Tianlu 張天祿, *Zhongguo jindai yiaoxueshi* 中國近代藥學史 [Modern history of Chinese pharmaceutics] (Beijing: Remin weisheng chubanshe, 1992), 128–31; Xue Yu 薛愚, *Zhongguo yaoxue shiliao* 中國藥學史料 [Historical documents of Chinese pharmaceutics] (Beijing: Remin weisheng chubanshe, 1984), 414–19.

16. Chen Guofu 陳果夫, "Changshan zhinüe" 常山治瘧 [Treating malaria

with changshan], in *Chen Guofu xiansheng quanji* 陳果夫先生全集第八集 [Complete works of Mr. Chen Guofu], 10 vols. (Hong Kong: Zhengzhong shuju, 1952), 8:263–68.

17. Ibid., 264.

18. In fact, biomedical pharmacologists used the same argument to urge the Nationalist government to establish its own modern pharmaceutical industry and research institutes. See Chen and Zhang, *Zhongguo jindai yiaoxueshi*, 114–16.

19. Francis G. Henderson, Charles L. Rose, Paul N. Harris, and K. K. Chen, "g-Dichroine, the Antimalaria Alkaloid of Chang Shan," *Journal of Pharmacology and Experimental Therapeutics* 95 (1948): 191–200.

20. Beginning in 1939, Liu Shaoguang and his colleagues at the Central Institute of Pharmacological Research pioneered the study of Chinese antimalaria drugs. See Liu Shaoguang 劉紹光, "Xi'nan kangnüe yaocai zhiyanjiu" 西南抗瘧藥材之研究 [Study of southwestern antimalaria medicinals], *Zhonghua yixue zazhi* 中華醫學雜誌 [Chinese medical journal] 27, no. 6 (1940): 327–42. For Zhang Changshao's subsequent experiments, which refuted Liu Shaoguang's conclusions, see Zhang Changshao 張昌紹 and Tingchong Zhou 周廷沖, "Guochan kangnüe yaocai zhi yanjiu" 國產抗瘧藥材之研究 [Study on nationally produced medicinals], *Zhonghua yixue zazhi* 中華醫學雜誌 [National medical journal of China] 29, no. 2 (1943): 137–42.

21. Ding Fubao 丁福保, *Zhongyao qianshuo* 中藥淺說 [Elementary introduction to Chinese drugs] (Shanghai: Shangwu yinshuguan, 1930), 2. For a brief introduction to Ding Fubao and his medical translations, see Ma Boying 馬伯英, Gao Xi 高晞, and Hong Zhongli 洪中立, *Zhongwai yixue wenhua jiaoliushi* 中外醫學文化交流史 [A history of intercultural communication between Chinese and Western medicine] (Shanghai: Wenhui chubanshe, 1993), 450–55.

22. With regard to the curative efficacy of *changshan*, F. Porter Smith simply rephrased what was recorded on *changshan* in the materia medica and stated that "all forms of the drug are used in fevers, especially those of malarial origin. There is no form of this disease for which it is not recommended." While acknowledging the traditionally claimed efficacy of *changshan*, Smith did not give his own evaluation. See F. Porter Smith, *Chinese Materia Medica: Vegetable Kingdom* (Taipei: Guting shuwu, 1969), 293; orig. pub. 1911.

23. Li Tao 李濤, "Woguo nüeji kao" 我國瘧疾考 [A study of malaria in China], *Zhonghua yixue zazhi* 中華醫學雜誌 [National medical journal of China] 18, no. 3 (1932): 415–19, esp. 419.

24. Xu Zhifang 許植方, "Guochan jienueyao zhi yanjiu" 國產截瘧藥之研究 [Research on the nationally produced antimalaria drug], *Yiyaoxue* 醫藥學 [Medicine and pharmaceutics] 1, no. 9 (1948): 31–34, esp. 34.

25. Chen, "Changshan zhinüe," 264.

26. Zhang Xichun 張錫純, *Yixue zhong zhong can xi lu* 醫學衷中參西錄 [The assimilation of Western medicine and Chinese medicine] (Shijiazhuan: Hebei kexue jishu chubanshe, 1995), *zhong ce* 中冊 [middle booklet], 125.

27. K. K. Chen, "Half a Century of Ephedrine," in *Chinese Medicine: New*

Medicine, ed. Frederick F. Kao and John J. Kao (New York: Institute for Advanced Research in Asian Science and Medicine, 1977), 21–27, esp. 22.

28. Carl F. Schmidt, "Pharmacology in a Changing World," *Annual Review of Physiology* 23 (1961): 1–15, esp. 7–8.

29. For a critical review of the problematic dichotomy between internal and external factors in science, see Steven Shapin, "Discipline and Bounding: The History and Sociology of Science as Seen Through the Externalism-Internalism Debate," *History of Science* 30 (1992): 333–69.

30. Chen, "Changshan zhinüe," 264.

31. The original prescription consists of the following ingredients: *changshan* 常山, areca nut (*binlang* 檳榔), tortoise shell (*biegia* 鱉甲), licorice (*gancao* 甘草), black plum (*wumei* 烏梅), red jujube (*hongzao* 紅棗), and raw ginger (*shengjiang* 生薑). Anon., *Changshan zhinüe chubu yanjiu baogao* 常山治瘧初步研究報告 [Preliminary research report on changshan as a malaria treatment] (Chongqing: Guoyao yanjiushi, 1944), 1.

32. Chen, "Changshan zhinüe," 265.

33. Anon., *Changshan zhinüe chubu yanjiu baogao*, 4.

34. Ibid., 9.

35. Ibid., 11.

36. Ibid., 9.

37. B. E. Read and J. C. Liu, "A Review of Scientific Work Done on Chinese Material Medica," *National Medical Journal of China* 14, no. 5 (1928): 326–27.

38. Actually, another pharmacognosist concluded in 1954 that more than one variety of *jigu changshan* were described in the Chinese materia medica. Moreover, the so-called *changshan* sold in the local markets around the nation was not the same herbal drug: see Lou Zhicen 樓之岑, "Changshan de shengyao jianding" 常山的生藥鑑定 [The pharmacognostic identification of *changshan*], *Zhonghua yixue zazhi* 中華醫學雜誌 [Chinese medical journal] 60 (1954): 869–70.

39. According to the scientific consensus at the time, when discrepancies were found between presently utilized Chinese medicinals and their traditional characterization, "one should simply get rid of the traditional name if no herbal drug could be found that matched the traditional characterization. On the other hand, if it was uncertain whether the herb in modern use was exactly identical with the traditional one, scientists should just record the properties of the modern herb instead." See Yan Zhe言者, "Yanjiu guochanyao yijian huilu" 研究國產藥意見匯錄 [A compilation of opinions on studying nationally produced drugs], *Dagongbao yixue zhoukan* 大公報醫學週刊 [Dagong Bao weekly medical supplement], no. 46 (1930): single-page issue.

40. Anon., *Changshan zhinüe chubu yanjiu baogao*, 11–14.

41. Ibid., 9.

42. See, for example, Smith, *Chinese Materia Medica: Vegetable Kingdom*, 292–93; and Bernard E. Read, *Chinese Medical Plants from the Pen Ts'ao Kang Mu (A. D. 1596) of a Botanical, Chemical and Pharmacological Reference List* (Taipei: Southern Material Center, 1982), 106; orig. pub. 1923.

43. Anon., *Changshan zhinüe chubu yanjiu baogao*, 12.

44. Saburô Miyashita, "Malaria in Chinese Medicine during the Chin and Yüan Periods," *Acta Asiatica*, no. 36 (1979): 10.

45. For example, in Chen Cunren's *Dictionary of Chinese Pharmacopoeia* (1935), *Orixa japonica* is simply listed as a "foreign name" in the entry for *changshan*. See Chen Cunren 陳存仁, *Zhongguo yaoxue dacidian* 中國藥學大辭典 [Dictionary of Chinese pharmacopoeia] (Shanghai: Shijie shuju, 1935), 1143.

46. Henderson et al., "g-Dichroine, the Antimalaria Alkaloid of *Chang Shan*"; Isabel M. Tonkin and T. S. Work, "A New Antimalarial Drug," *Nature* 156 (November 1945): 630.

47. J. W. Fairbairn and T. C. Lou, "A Pharmacognostical Study of *Dichroa febrifuga* Lour: A Chinese Antimalarial Plant," *Journal of Pharmacy Pharmacol* 2, no. 1 (1950): 162–77, esp. 164.

48. Zhang and Zhou, "Guochan kangnue yaocai zhi yanjiu," 137–42.

49. A recent study gives one plausible explanation why the plant extract of febrifugine has weaker side effects than synthetic febrifugine. See Anthony Butler and John Moffett, "The Anti-Malarial Action of *Changshan* (Febrifugine): A Review," *Asian Medicine: Tradition and Modernity* 1, no. 2 (2005): 423–31, esp. 427.

50. Chen, "Changshan zhinüe," 267.

51. Chen Guofu 陳果夫, 'Yi, er, san, si, wu haishi wu, si, san, er, yi' 一二三四五還是五四三二一 [12345 or 54321], in his *Yizheng mantan xubian* 醫政漫談續編 [Continued chat about medicine and politics]: 71–72.

52. The following elaboration on the two protocols is based on a paper by Yu Yan written in 1952. I have no direct evidence that Yu Yan's description was the one either Chen Guofu or his critic had in mind. Nevertheless, in describing the alternative procedure for conducting scientific research on Chinese drugs, Yu Yan used a metaphor, "acting in the reversed order" (*daoxingnishi* 倒行逆施), that is very close to the one used by Chen Guofu ("5–4–3–2–1"). I recognize that I am risking an anachronism, especially because Yu Yan's attitude toward Chinese medicine reportedly went through a great transformation after the Communist takeover in 1949. Nevertheless, my point is not to belabor Yu Yan's position but to illustrate the agendas of the two competing protocols. Besides, I provide secondary documentation to support Yu Yan's characterization of these two protocols. I have chosen to refer to Yu Yan's 1952 paper mainly because, in comparison to other writers, he articulated these two protocols in a very systematic fashion. Most importantly, he specifically cited Chen Guofu's *changshan* research as an example of "reverse-order research." See Yu Yan 余巖, "Xianzai gai yanjiu zhongyao le!" 現在該研究中藥了 [Now is the time to study Chinese drugs!], in *Zhongguo yaowu de kexue yanjiu* 中國藥物的科學研究 [Scientific research on Chinese medicinals], ed. Huang Lansun 黃蘭孫 (Shanghai: Qianqingtang chubanshe, 1952), 6–11.

53. Some biomedical scientists went even further and inserted pharmacognostic identification before the first step of chemical analysis. For example, one of the most important Chinese pharmacognosists, Zhao Yuhuang 趙燏黃, held

this view and insisted that it was meaningless to conduct a chemical analysis before pharmacognostic identification had been performed as the first step. See Zhao Yuhuang趙燏黃, "Zhongyao yanjiu de buzou" 中藥研究的步驟 [The procedure of studying Chinese drugs], *Xinyiyao* 新醫藥 [New medicine] 2, no. 4 (1934): 331–34.

54. See Zhang Minggao 張鳴皋, ed., *Yaoxue fazhan jianshi* 藥學發展簡史 [A brief history of the development of pharmacology] (Beijing: Zhongguo yiyao keji chubanshe, 1993), 151.

55. K. K. Chen and Carl F. Schmidt, "Ephedrine and Related Substances," *Medicine: Analytical Reviews of General Medicine, Neurology and Pediatrics* 9, no. 1 (1930): 1–131, esp. 1–7. For a brief summary of the history of *mahuang* research, see James Reardon-Anderson, *The Study of Change: Chemistry in China, 1840–1949* (Cambridge: Cambridge University Press, 1991), 149–51.

56. Chen and Schmidt, "Ephedrine and Related Substances," 4.

57. Tan Cizhong 譚次仲, "Zaicheng yanjiuyuan lun yaowu shiyan buyi hushi jingyan" 再呈研究院論藥物實驗不宜忽視經驗 [Resubmitting to the Academia Sinica: Why pharmaceutical experimentation should not ignore experience], in Tan Cizhong, *Yixue geming lunzhan* 醫學革命論戰 [Polemics of medical revolution] (Hong Kong: Qiushi chubanshe, 1952), 50; orig. pub. 1931.

58. K. K. Chen, "Research on Chinese Materia Medica," 112.

59. According to M. K. Eggleston, "It was Dr. Chen's own initiative which started the investigation of *mahuang* at a time when the whole group [of Schmidt] was losing interest in Chinese drugs, having a number of disappointing experiences." "Letter from M. K. Eggleston to Roger S. Greene, 12 December 1930," RF folder 873, box 120, IV 2 B9, 1930. Rockefeller Foundation Archive Center, Sleepy Willow, NY.

60. K. K. Chen, "Research on Chinese Materia Medica," 112.

61. Ibid.

62. Chen and Schmidt, "Ephedrine and Related Substances," 6.

63. Wang, *Ershi nian lai zhongguo yishi chuyi*, 61–2.

64. See Yu Yan 余巖, "Huanghan yixue piping" 皇漢醫學批評 [A critique of Japanese-style Chinese medicine], *Dagongbao yixue zhoukan* 大公報醫學週刊 [Dagong Bao weekly medical supplement], nos. 70–85 (1931), esp. no. 75; and his "Woguo yixue geming zhi jianshe yu pohuai" 我國醫學革命之建設與破壞 [The construction and destruction of our country's medical revolution], *Zhongxi yiyao* 中西醫藥 [Journal of the Medical Research Society of China] 2, no. 3 (1936): 164–78, esp. 169.

65. Anon., *Changshan zhinüe chubu yanjiu baogao*, 47.

66. Ibid.

67. Ibid.

68. In fact, scientists in later research publications consistently referred to *changshan* as a "new antimalarial drug." See for example Tonkin and Work, "A New Antimalarial Drug," 630; and David Hooper, "A New Anti-Malarial Drug," *Nature* 157 (January 1946): 106.

69. Chen Guofu 陳果夫, "Laobingren tan zhongyi xiyi" 老病人談中醫西醫 [An old patient on Chinese Medicine and Western medicine], in *Chen Guofu xiansheng quanji*, 6:10.

70. Tan, "Zaicheng yanjiuyuan lun yaowu shiyan buyi hushi jingyan," 50–55.

71. Anon., *Changshan zhinüe chubu yanjiu baogao*, 4.

72. Zhang Changshao 張昌紹, *Xiandai de zhongyao yanjiu* 現代的中藥研究 [Modern research on Chinese drugs] (Shanghai: Zhongguo kexue tushu yiqi gongsi, 1954), 141.

73. A graduate of Tokyo Imperial University, Zhao Yuhuang was a founding member of the Chinese Pharmacy Association, established in 1907. In terms of modern pharmacognostic research on Chinese materia medica, Zhao probably contributed more than anybody else. For a brief discussion of Zhao's background and work, see Fu Weikang 傅維康, *Zhongyaoxue shi* 中藥學史 [History of Chinese materia medica] (Chengdu: Bashu shuwu, 1993), 304–6. See also Zhao Yuhuang 趙燏黃, "Shuo zhongyao" 說中藥 [On Chinese herbs], *Yiyao pinglun* 醫藥評論 [Medical review], no. 39 (1930): 5–7; and Zhao, "Zhongyao yanjiu de buzou."

74. Zhao Yuhuang 趙燏黃, "Zhongyang yanjiuyuan nishe zhongyao yanjiusuo jihuashu" 中央研究院擬設中藥研究所計畫書 [A plan for establishing a research institute for Chinese herbs at the Academia Sinica], *Yiyao pinglun* 醫藥評論 [Medical review], no. 1 (1929): 44–47.

75. Li, "Woguo nüeji kao," 419.

76. Anon., "Jiuyi shangke gaizao ye?" 舊醫尚可改造耶? [Is there any way to reform old-style medicine?], *Zhonghhua yixue zhazhi* 中華醫學雜誌 [National medical journal of China] 19 (1933): 41.

77. Sean Hsiang-lin Lei and Gerard Bodeker, "*Changshan*—Ancient Febrifuge and Modern Antimalarial: Lessons for Research from a Forgotten Tale," in *Traditional Medicinal Plants and Malaria*, ed. Merlin Willcox, Gerard Bodeker, and Philippe Rasonanivo (London: CRC Press, 2004), 61–82, esp. 71. Also see Merlin Willcox et al., "Guideline for the Preclinical Evaluation of the Safety of Traditional Herbal Antimalarials" in the same volume, pp. 279–96.

78. This is actually one of the central themes in Latour's critique of modernity. To replace what he calls the "modern constitution"—an eternal nature and many relative cultures—Latour invents the notion of "natures-cultures," that is, the idea that each culture has an inseparable "nature" mobilized by its socio-technical network and practice. See Bruno Latour, *We Have Never Been Modern* (Cambridge, MA: Harvard University Press, 1993), 91–130.

79. Materially, *changshan* was transformed from one of the seven drugs in a prescription (Chen Guofu before 1940), to the raw extract (Dr. Cheng Peizhen in 1940), to an unidentified isolated alkaloid (chemist Jiang Juhuang, in the *Report*, 1944 [see n. 30 above]), to g-Dichroine (K. K. Chen in Eli Lilly & Company, 1948). Conceptually, it was transformed from a poisonous plant (in *Systematic Materia Medica*) to an antiseptic drug (1935), and then to a "Chinese drug with specific efficacy" for treating malaria (Dr. Cheng Peizhen in 1940).

Andrew Pickering has developed a systematic language to investigate this interstabilizing process between practice and scientific culture; see Andrew Pickering, *The Mangle of Practice: Time, Agency, and Science* (Chicago: University of Chicago Press, 1995), 68–112; and Joan H. Fujimura, "Crafting Science: Standardized Package, Boundary Objects, and 'Translation,'" in *Science as Practice and Culture*, ed. Andrew Pickering (Chicago: University of Chicago Press, 1992), 139–67.

80. I should emphasize that, unlike linguistic translation, the re-networking discussed here involved a translation between socio-technical networks. Therefore, in addition to "meaning," the central concern of this translation is material objects. For a systematic investigation of the problematic nature of translation between Chinese, Japanese, and Western languages, see Lydia H. Liu, *Translingual Practice: Literature, National Culture, and Translated Modernity—China, 1900–1937* (Stanford, CA: Stanford University Press, 1995). Moreover, for more information on the practice of translation that involves an "inequality of languages," see Talal Asad, "The Concept of Cultural Translation in British Social Anthropology," in his *Genealogies of Religion* (Baltimore, MD: Johns Hopkins University Press, 1993), 171–99.

81. For a discussion of this important phenomenon and its implication for contemporary research on Chinese drugs, see Lei and Bodeker, "*Changshan*—Ancient Febrifuge and Modern Antimalarial," 61–82, esp. 72–77.

82. Both the study of *changshan* and the world-famous discovery of artemisinin, derived from the Chinese drug *qinghao* 青蒿 in the 1960s, were motivated by the need to fight a war in areas where malaria was endemic. For more details on the story of artemisinin, see Elisabeth Hsu, "Reflections on the Discovery of the Antimalarial *Qinghao*," *British Journal of Clinical Pharmacology* 61, no. 6 (2006): 666–670.

83. Yu, "Xianzai gai yanjiu zhongyao le!" 6.

84. Tan, "Zaicheng yanjiuyuan lun yaowu shiyan buyi hushi jingyan," 50–55.

CHAPTER 10

1. Margaret Chan, "Keynote Speech at the International Seminar on Primary Health Care in Rural China," in *International Seminar on Primary Health Care in Rural China* (Beijing: World Health Organization, 2007), online at http://www.who.int/dg/speeches/2007/20071101_beijing/en/index.html (accessed July 17, 2013).

2. In his recent work, Xiaoping Fang challenges the conventional wisdom that the success of the barefoot doctor program lies in its use of acupuncture and traditional herbal medicine. Instead, he argues that "the key impact of the barefoot doctor program was facilitating the entry of Western medicine into villages hitherto dominated by Chinese medicine" (*Barefoot Doctors and Western Medicine in China* [Rochester, NY: University of Rochester Press, 2012], 3).

3. As Szeming Sze, the general secretary of the Chinese Medical Association succintly summarized, "For the 84% of the population who live in the rural areas and are incapable of paying for private medical care, it is generally agreed

that the only solution to the problem of their medical care is state medicine" (*China's Health Problems* [Washington, DC: Chinese Medical Association, 1944], 13).

4. Ibid.

5. According to Mary Bullock, the policy of State Medicine was the brainchild of an intensive debate between Grant and his colleagues at PUMC in the spring of 1928. See Mary Brown Bullock, *An American Transplant: The Rockefeller Foundation and Peking Union Medical College* (Berkeley and Los Angeles: University of California Press, 1980), 151.

6. Lan Ansheng 蘭安生 (John B. Grant), "Zhongguo jianshe xiandai yixue zhi fangzhen" 中國建設現代醫學之方針 [Guideline for constructing modern medicine in China], in *Yixue zhoukanji* 醫學週刊集 [Collected essays from Medical Weekly] (Beijing: Bingyin yixueshe, 1928), 296–302, esp. 302.

7. John B. Grant, "Provisional National Health Council," China Medical Board, folio 529, box 75, 1927. Rockefeller Foundation Archive Center, Sleepy Willow, NY.

8. Concerning the rise of British State Medicine, see Steve Sturdy, "Hippocrates and State Medicine: George Newman Outlines the Founding Policy of the Ministries of Health," in *Greater Than the Parts: Holism in Biomedicine 1920–1950*, ed. Christopher Lawrence and George Weisz (Oxford: Oxford University Press, 1998), 112–34.

9. John B. Grant, "State Medicine: A Logical Policy for China," *National Medical Journal of China* 14, no. 2 (1928): 65–80, esp. 65.

10. This was another salient contrast between the proposal to the British Boxer Indemnity Fund and Grant's proposal for State Medicine. The former had assumed that the development of public health in China would roughly follow the road that had been traveled by the Western countries: starting in the larger cities—often through private stimulus—developing into provincial Departments of Health, and finally moving on to a national health administration. In addition, rural public health was generally the last of the four types of health administration to become established. The author of that proposal therefore concluded, "To omit due consideration of these factors in formulating a national health program would be to invite failure" (Association for the Advancement of Public Health in China, *Memorandum on the Need of a Public Health Organization in China: Presented to the British Boxer Indemnity Commission by the Association for the Advancement of Public Health in China* [Beijing: Association for the Advancement of Public Health in China, 1926], 15).

11. "When the First Health Station was organized in Peking, this was a decided departure from the type of public health teaching in most medical schools in the United States and Europe and was the only thing of its kind in China" (Mary E. Ferguson, *China Medical Board and Peking Union Medical College: A Chronicle of Fruitful Collaboration 1914–1915* [New York: China Medical Board of New York, 1970], 58).

12. Anon., *The Rockefeller Foundation China Medical Board Twelfth Annual Report* (New York: Rockefeller Foundation China Medical Board, 1927), 12.

13. Grant, "State Medicine," 69.

14. This analysis and the numerical data of "six million excess deaths" were used in the following documents: Association for the Advancement of Public Health in China, *Memorandum*, 5; Chen Zhiqian 陳志潛, "Wuguo quanyi jianshe wenti" 吾國全醫建設問題 [The problem of constructing comphehensive medicine in our country], *Yixue zhoukan* 醫學週刊 [Medical weekly], nos. 89–92 (1928); and Huang Zifang 黃子方, "Zhongguo weisheng chuyi" 中國衛生芻議 [A humble discussion on China's hygiene], *Zhonghua yixue zazhi* 中華醫學雜誌 [National medical journal] 13, no. 5 (1927): 338–54, esp. 338–39.

15. Instead of emphasizing the balanced application of preventive and curative medicine, the proposal for the British Boxer Indemnity Committee completely downplayed the role of curative medicine. Accepting the objective of reducing the six million excess deaths in China, the proposal reached the radical conclusion that "in fact, *most of the gross excess mortality in China could be reduced without the existence of a curative medical profession* [original emphasis]" (Association for the Rockefeller Foundation Archive Center, Sleepy Willow, NY. of Public Health in China, *Memorandum*, 5).

16. Grant, "State Medicine," 69.

17. Ibid. A similar view was presented in the proposal for the British Boxer Indemnity Committee: "[China] has no tradition with which to break, and from the start her medical school could short-cut the evolution, possibly necessary in the West, to the predominant preventive outlook" (Association for the Advancement of Public Health in China, *Memorandum*, 32).

18. Grant, "Provisional National Health Council," 5.

19. Huang, "Zhongguo weisheng chuyi," 353–54.

20. "Regrettably, around the world medical enterprises had been controlled by private forces. People are so used to this fact that even intellectuals do not think that medicine should become state-owned public enterprise. This idea is viewed to be extremely strange" (Chen, "Wuguo quanyi jianshe wenti," 90–91).

21. Ibid., 92.

22. Lan, "Zhongguo jianshe xiandai yixue zhi fangzhen," 296.

23. Ruiheng Liu, "Chinese Ministry of Health," *National Medical Journal of China* 15 (1929): 135–48, esp. 147.

24. K. Chimin Wong and Lien-teh Wu, *History of Chinese Medicine* (Taipei: Southern Materials Center, 1985), 724; orig. pub. 1932.

25. Chen, "Wuguo quanyi jianshe wenti," 10.

26. Yu Huanwen 俞煥文, "Xiehe yiyuan yu dingxian pingjiaohui" 協和醫院與定縣平教會 [Peking Union Medical Hospital and Ding County Mass Education Movement], in *Liu Ruiheng boshi yu zhongguo yiyao ji weisheng shiye* 劉瑞恆博士與中國醫藥及衛生事業 [Dr. J. Heng Liu and medical and health development in China], ed. Liu Sijing 劉似錦 (Taipei: Taiwan shanhwu yinshuguan, 1989), 28.

27. Elizabeth Fee and Dorothy Porter, "Public Health, Preventive Medicine and Professionalization: England and America in the Nineteenth Century," *Medicine in Society: Historical Essays*, ed. Andrew Wear (Cambridge: Cambridge University Press, 1992), 249–76.

28. For more information on Alfred Grotjahn and his conception of social hygiene, see George Rosen, *From Medical Police to Social Medicine: Essays on the History of Health Care* (New York: Science History Publications, 1974), 60–119.

29. Hu Ding'an 胡定安, *Zhongguo weisheng xingzheng sheshi jihua* 中國衛生行政設施計畫 (Plan for public health administration in China) (Shanghai: Shangwu yinshuguan, 1928), 39–40.

30. Charles Hayford, "The Storm over the Peasant: Orientalism and Rhetoric in Constructing China," in *Contesting the Master Narrative: Essays in Social History*, ed. Shelton Stromquist and Jeffrey Cox (Iowa City: University of Iowa Press, 1998), 150–72, esp. 151.

31. I would like to thank Tong Lam and Peter Zarrow for the very helpful discussion about the rise of peasants as a political construction.

32. Charles W. Hayford, *To the People: James Yen and Village China* (New York: Columbia University Press, 1990), 112.

33. Ibid., 103.

34. Ibid., 126–27.

35. See for example C. C. Chen, *Medicine in Rural China: A Personal Account* (Berkeley and Los Angeles: University of California Press, 1989); Ka-che Yip, *Health and National Reconstruction in Nationalist China: Development of Modern Health Service, 1928–1937* (Ann Arbor, MI: Association for Asian Studies, 1996); AnElissa Lucas, *Chinese Medical Modernization: Comparative Policy Continuities, 1930s–1980s* (New York: Praeger, 1982); Yang Nianqun 楊念群, *Zhaizao bingren* 再造病人 [Remaking "patients": Politics of space in the conflicts between traditional Chinese medicine and Western medicine, 1832–1985] (Peking: Zhongguo renmin daxue chubanshe, 2006).

36. Chen, *Medicine in Rural China*, 72.

37. C. C. Chen, "The Rural Public Health Experiment in Tsinghsien," *Milbank Memorial Fund Quarterly* 14, no. 1 (1936): 66–80, esp. 66–67.

38. "In our country, the so-called governmental responsibility of protecting citizens' health is a newly imported concept. Not only do people not understand [this concept], even the authorities might not fully understand it. No wonder that the government does not pay much attention to people's health and that people do not support the health-care enterprises that the government endeavors to construct." See Chen Zhiqian (aslo known as C. C. Chen) 陳志潛, "Dingxian shehui gaizao shiye zhong de baojian zhidu" 定縣社會改造事業中的保健制度 [Health-care institution in the Dingxian social reform], in *Xiangcunjianshe shiyan di er ji* 鄉村建設實驗第二集 [The experiment on rural construction: Volume 2], ed. Zhang Yuanshan 章元善 and Xu Zhonglian 許仲廉 (Shanghai: Zhonghua shuju, 1935), 459–73, esp. 459.

39. Ibid., 463.

40. Paul Starr, *The Social Transformation of American Medicine* (New York: Basic Books, 1982).

41. Chen, "Dingxian shehui gaizao shiye zhong de baojian zhidu," 463.

42. Ibid.

43. Tong Lam, *A Passion for Facts: Social Surveys and the Construction of*

the Chinese Nation-State, 1900–1949 (Berkeley and Los Angeles: University of California Press, 2011), 154–55.

44. Ibid.

45. Theodore Porter and Dorothy Ross, *The Cambridge History of Science*, vol. 7, *The Modern Social Sciences* (Cambridge: Cambridge University Press, 2003), 505.

46. C. C. Chen, "Public Health in Rural Reconstruction at Tsinghsien," *Milbank Memorial Fund Quarterly* 12, no. 4 (1934): 370–78, esp. 370.

47. In response to the political campaign to make Chinese medicine a part of the national system of medical education, Chen published his most critical statements in *Minjian* 民間 (Among the people), the official journal of the Rural Reconstruction Movement.

48. C. C. Chen, H. W. Yu, and F. J. Li, "Seven Years of Jennerian Vaccination in Tinghsien," *Chinese Medical Journal* 51 (1937): 953–62, esp. 961.

49. Chen, "Dingxian shehui gaizao shiye zhong de baojian zhidu," 463. Interestingly, practitioners of traditional medicine had no problem adopting Jennerian vaccination to serve their communities in nineteenth-century Canton. See Angela Ki-che Leung, "The Business of Vaccination in Nineteenth-Century Canton," *Late Imperial China* 29, no. 1 (2008): 7–39.

50. Chen, "The Rural Public Health Experiment in Tsinghsien," 70.

51. Chen, *Medicine in Rural China*, 82.

52. Chen, "Public Health in Rural Reconstruction at Tsinghsien," 371.

53. Chen, Yu, and Li, "Seven Years of Jennerian Vaccination in Tinghsien," 961.

54. Ibid., 952.

55. Ibid., 961.

56. Chen, "The Rural Public Health Experiment in Tsinghsien," 77.

57. Chen, "Dingxian shehui gaizao shiye zhong de baojian zhidu," 464.

58. Chen, "The Rural Public Health Experiment in Tsinghsien," 79.

59. Liu Ruiheng, "Our Responsibilities in Public Health," *Chinese Medical Journal* 51 (1937): 1039–42, esp. 1040.

60. Ibid.

61. Ibid.

62. Without mentioning Chen's survey by name, Wu pointed out that "recent statistical investigations have made it clear that all the farmer is able to spend on medical relief amounts to an average of thirty cents annually." On the basis of this discovery, Wu calculated that supporting a qualified physician would cost the government six hundred dollars in minimum income and at least four or five hundred dollars per year in equipment for active work. As the result of this simple calculation, "10,000 people are necessary to support a single physician—a number of patients that no practitioner could possibly care for." See Wu Liande, "Fundamentals of State Medicine," *Chinese Medical Journal* 51 (1937): 777–78.

63. Ibid., 779.

64. Ibid.

65. Anon., "Chinese Medical Conference Proceedings: The Public Health Section," *Chinese Medical Journal* 51 (1937): 1065–71, esp. 1067.

66. A participant raised the question to the audience "whether the ten cents per capita suggested by Dr. C. C. Chen are sufficient to meet expenses for those elementary services the 'State Medicine' units are expected to offer to the population both in curative and preventive lines." Ibid.

67. C. C. Chen, "Some Problems of Medical Organization in Rural China," *Chinese Medical Journal* 51, no. 6 (1937): 803–14, esp. 813.

68. Ibid., 814.

69. F. Oldt, "State Medicine Problems," *Chinese Medical Journal* 51 (1937): 797–802, esp. 797.

70. Grant, "State Medicine," 65.

71. Oldt, "State Medicine Problems," 797.

72. Ibid., 798.

73. R. K. S. Lim and C. C. Chen, "State Medicine," *Chinese Medical Journal* 51 (1937): 781–95, esp. 784.

74. Ibid., 782.

75. Ibid., 793.

76. Michael Shiyung Liu, *Prescribing Colonization: The Role of Medical Practices and Policies in Japan-Ruled Taiwan, 1895–1945* (Ann Arbor, MI: Association of Asian Studies, 2009), esp. chap. 3.

77. Lim and Chen, "State Medicine," 785.

78. Yip, *Health and National Reconstruction in Nationalist China*, 95.

79. Du Xiaoxian 杜孝賢, "Yi gonggong weisheng zhuanjia Jin Baoshan" 憶公共衛生專家金寶善 [Remembrance of public health expert Jin Baoshan], in *Zhonghua wenshi zhiliao wenku* 中華文史資料文庫: 第十六卷 [Chinese collections of literary and historical documents, vol. 16] (Beijing: Zhongghuo wenshi chubanshe, 1996), 779–83, esp. 780.

80. Jin Baoshan 金寶善, "Jiu zhongguo de xiyi paibie yu weisheng shiye de yanbian" 舊中國的西醫派別與衛生事業的演變 [On the factions of Western medicine in Old China and the evolution of the public health enterprise], in *Zhonghua wenshi zhiliao wenku* 中華文史資料文庫: 第十六卷 [Chinese collections of literary and historical documents, vol.16] (Beijing: Zhongguo wenshi chubanshe, 1996), 844–50, esp. 848.

81. P. Z. King (also known as Jin Baoshan), "Thirty Years of Public Health Work in China," *Chinese Medical Journal* 64 (1946): 3–16, esp. 11.

82. Chen Zhiqian pointed out that "two types of individuals accounted for its [the Chinese Medical Association's] expansive influence on public health in the 1928–37 period: (1) medical scientists with a broad orientation, of the type represented by Lim, a British-educated physiologist; and (2) physicians specially trained in public health, such as Dr. P. Z. King (Jin Baoshan)." Chen, 62.

83. Jin Baoshan 金寶善, "Woguo weisheng xingzheng de huigu yu qianzhan" 我國衛生行政的回顧與前瞻 [The review and future of our country's medical administration], *Shehui weisheng* 社會衛生 [Social hygiene] 1, no. 3 (1944): 1–7, esp. 4.

84. Chen Zhiqian 陳志潛, "Neizhengbu xingzheng weisheng jishu huiyi" 內政部行政衛生技術會議 [Conference on medical administration and related technology, organized by the Ministry of Internal Affairs], *Minjian* 民間 [Among the people] 1, no. 3 (1934): 11–15, esp. 12.

85. Li Yushang 李玉尚, "The elimination of schistosomiasis in Jiangxi and Haining counties, 1948–58: Public health as political movement," in *Health and Hygiene in Chinese East Asia*, ed. Angela Ki Che Leung and Charlotte Furth (Durham, NC: Duke University Press, 2010), 204–21.

86. Jin Baoshan 金寶善, "Neizhengbu weishengshu xiangchun weisheng gongzhuo baogao" 內政部衛生署鄉村衛生工作報告 [Report on rural health by National Medical Administration, Ministry of Internal Affairs], in *Xiangcun jianshe shiyan diyiji* 鄉村建設實驗: 第一集 [Experiments on rural health, vol. 1], ed. Zhang Yuanshan 章元善 and Xu Shilian 許仕廉 (Shanghai: Zhonghua shuju, 1933), 117–25, esp. 123.

87. According to Liu Shiyong's important research, because Kitasato Shibasaburo's pupils had been engaged in a hostile academic rivalry against the faculty of Todai (Tokyo Imperial University) between 1899 and 1914, many alumni of the Kitasato Institute had been forced to leave Japan proper and had relocated to its colonies. As an unintended result of this academic rivalry, the Kitasato Institute had thus provided a network of physicians for Japanese colonial medicine extending from Taiwan, Shanghai, and Korea to Manchuria. See Shiyung Liu, "The Ripples of Rivalry: The Spread of Modern Medicine from Japan to Its Colonies," *East Asian Science, Technology and Society: An International Journal* 2, no. 1 (2008): 47–72.

88. Concerning the central role that hygienic habits played in the New Life Movement, especially their role in reviving the four cardinal virtues, see Sean Hsiang-lin Lei 雷祥麟, "Xiguan cheng siwei: Xinshenghuo yundong yu feijiehe fangzhi zhong de lunli jiating yu shenti" 習慣成四維：新生活運動與肺結核防治中的倫理、家庭與身體 [Habituating the four virtues: Ethics, family and the body in the anti-tuberculosis campaigns and the New Life Movement], *Zhongyang yanjiuyuan jindaishi yanjiusuo jikan* 中央研究院近代史研究所集刊 [Bulletin of the Institute of Modern History, Academia Sinica], no. 74 (2011): 133–77.

89. As Ka-che Yip pointed out, Chiang Kai-shek's interest in the programs in Ding County and similar efforts in rural areas in fact resulted from his attempts to find "ways of governing the areas cleared of [Communist] bandits." Yip, *Health and National Reconstruction in Nationalist China*, 181.

90. Jin Baoshan 金寶善, "Gongyi zhidu" 公醫制度 [Institution of the State Medicine], *Gonggong weishneng yukan* 公共衛生月刊 [Public health monthly] 2, no. 6 (1936): 255–59.

91. Ibid., 258.

92. Ibid.

93. Ibid., 259.

94. *Zhongguo guomingdang lici huiyi ji zhongyao jueyian huibian* 中國國民黨歷次會議及重要決議案彙編 [Collected volume of the manifestos and important resultions of the Chinese Nationlist Party] (N.p.: Zhongguo guomingdang xunlian weiyuanhui, 1941), 1187.

95. Ibid., 1189.

96. In the three-year plan drafted after this grand conference, public health implementation was listed as one aspect of the political endeavors that aimed at "establishing political organization at the local level and realizing the self-governance of local communities." Ibid., 1145.

97. Anon., *Diwujie bazhong qunahui jueyian xingzhengyuan banli qingxing baogaobiao* 第五屆八中全會決議案行政院辦理情形報告表 [With regard to the resolutioned passed in the Eighth Plenary Session of the Fifth Central Committee, a report of progress from the Executive Yuan] (1941), appendix 7. Library of the Institute of Modern History, Academia Sinica.

98. Since *weishengyuan* was designed to transcend the division between curative medicine and preventive medicine (and public health), its translation is anything but natural. The terms, Chinese and English, that Chen used to describe this three-level organization are the following: (1) county level: Health Center 保健院 (*baojianyuan*); (2) subcounty level: Health Station 保健所 (*baojiansuo*); (3) village level: Health Worker 保健員 (*baojianyuan*). The National Medical Administration used *weishengyuan* for both the county and subcounty levels.

99. Chen, "The Rural Public Health Experiment in Tsinghsien," 74.

100. Jiang Zhongzheng 蔣中正 (Chiang Kai-shek), *Zhongguo zhi mingyun* 中國之命運 [China's destiny] (Chongqing: Zhongyang xunlian tuan, 1951), 158; orig. pub. 1943.

101. Ibid., 154.

102. Perhaps because Yip's important research stopped in 1937 when the war broke out, he did not trace the wartime development of State Medicine as meticulously as he did for the period before the war. Nevertheless, he rightly emphasized the continuity across political regimes by pointing out that "organizationally, the Communist health structure built upon the Nationalist foundation of state medicine." Yip, *Health and National Reconstruction in Nationalist China*, 191. For medical development during the war time period, see Nicole Barnes, "Protecting the National Body: Medicine and Public Health in Wartime Chongqing, 1937–1945" (PhD diss., University of California, Irvine, 2012).

103. Jin Baoshan 金寶善, "'Gongyi' zhi shiming" 「公醫」之使命 [The mission of "State Medicine"], *Gongyi* 公醫 [State medicine] 1, no. 1 (1945): 1–2, esp. 1.

104. Investment in health care continued to grow even after the outbreak of the Sino-Japanese war in 1937. The central government's public health budget increased from $1.5 million dollars in 1936 to $5.5 million in 1939, more than tripling within three years. Nevertheless, Jin Baoshan admitted, "[This budget] only amounted to 0.5% of the total budget of our government; it is far behind that of more advanced countries." See Jin Baoshan 金寶善 and Xu Shijin 許世謹, "Woguo zhanshi weisheng sheshi zhi gaikuang" 我國戰時衛生設施之概況 [An outline of our country's public health infrastructure during the war], *Zhonghua yixue zazhi* 中華醫學雜誌 [National medical journal of China] 27, no. 3 (1940): 133–46, esp. 143.

105. Hayford, *To the People*, 201.

106. Anon., "Xian weishengjianshe zhuotanhui jilu" 縣衛生建設座談會紀錄 [A record of the group meetings on the construction of country public health], *Gongyi* 公醫 [State medicine] 1, no. 1 (1945): 53–60, esp. 59.

107. Chen, "Public Health in Rural Reconstruction at Tsinghsien," 371.

108. Chen Wanli 陳萬里, "Ruhe xunlian baoweishengyuan" 如何訓練保

衛生員 [How to train the village health worker], *Zhejiang zhengzhi* 浙江政治 [Political affairs of Zhejiang Province], no. 3 (1940): 30–37, esp. 31.

109. "[The village health workers] realize, and their neighbors also know, the limitation of their training; their membership in an organized group, the alumni association, subjects them to group opinion and censure; and they could easily be replaced if they should prove incompetent or should attempt too much" (Chen, "Public Health in Rural Reconstruction at Tsinghsien," 372).

110. Yip, *Health and National Reconstruction in Nationalist China*, 80.

111. Chen, *Medicine in Rural China*, 80.

112. Chen, "Ruhe xunlian baoweishengyuan," 31.

113. Federica Ferlanti, "The New Life Movement in Jiangxi Province, 1934–38," *Modern Asian Studies* 44, no. 5 (2010): 961–1000, esp. 968; Hans. J. van de Ven, "New States of War: Communist and Nationalist Warfare and State Building, 1928–34," in *Warfare in Chinese History*, ed. Hans. J. van de Ven (Leiden: Brill, 2000), 321–396, esp. 355–64.

114. Chen Wanli 陳萬里, "Dapo weishengyuan kunan de jumian" 打破衛生院困難的局面 [To overcome the difficulties facing the public health hospital], *Gongyi* 公醫 [State medicine] 1, no. 2 (1945): 3–5, esp. 4.

115. Ruan Buchan 阮步蟾, "Banli xianweisheng jige shiji wenti zhi shangque" 辦理縣衛生幾個實際問題之商榷 [A consideration about several practical problems in county-level public health construction], *Gongyi* 公醫 [State medicine] 1, no. 4 (1945): 40–42, esp. 42.

116. Ibid., 40.

117. Ibid., 41.

118. Deng Zongyu 鄧宗禹, "Ruhe shishi liudai quanhui guanyu yi1yao weisheng fangmian de zhishi" 如何實施六代全會關於醫藥衛生方面的指示 [How to realize the resolutions passed in the conference of the Sixth Central Committee concerning medicine and public health], *Gongyi* 公醫 [State medicine] 1, no. 6 (1945): 1–4, esp. 3.

119. Anon., "Xian weishengjianshe dierci zhuotanhui jilu" 縣衛生建設第二次座談會紀錄 [A record of the second group meetings on the construction of country public health], *Gongyi* 公醫 [State medicine] 1, nos. 6–7 (1945): 37–39, esp. 38.

120. Anon., "Changshashi guoyi gonghui feng kaiyou daidian" 長沙市國醫工會等快郵代電 [An express mail from the Association of National Medicine of Changsha City], *Yijie chunqiu* 醫界春秋 [Annals of the medical profession], no. 87 (1934): 40–41, esp. 40.

121. Ibid., 40–1.

122. Zhu Jingming 祝敬銘, "Guanyu wuquan dahui 'zhengfu duiyu zhongxiyi ying pingdeng daiyu yi hong xueshu er li minsheng an' zhi ganxiang yu xi wang" 關於五全大會「政府對於中西醫應平等待遇以宏學術而利民生案」之感想與希望 [Thoughts and hopes concerning the proposal "The Government Should Treat Chinese and Western Medicine Equally So as to Develop Scholarship and Benefit Peoples' Well-Being," passed in the fifth section conference of the Central Committee], *Yijie chunqiu* 醫界春秋 [Annals of the medical profession], no. 107 (1935): 1–4.

123. Zhu Dian 朱殿, *Jianshe sanqian ge nongcun yiyuan* 建設三千個農村

醫院 [Constructing three thousand village hospitals] (Shanghai: Nongcunyiyao gaijinshe, 1933).

124. Advertisement in *Gunghua yixue zazhi* 光華醫學雜誌 [Splendor: A medical journal] 1, no. 2 (1933).

125. Zhu, *Jianshe sanqian ge nongcun yiyuan*, 72–73.

126. Ibid., 107.

127. Ibid., 133.

128. Ibid., 138.

129. Zhu, *Jianshe sanqian ge nongcun yiyuan*, 132.

130. Chen, “Neizhengbu xingzheng weisheng jishu huiyi,” 14.

131. Zhu, *Jianshe sanqian ge nongcun yiyuan*, 134.

132. According to Chen Zhiqian’s recollections, perhaps because these practitioners of Chinese medicine had higher career ambitions than the laypeople he recruited in Ding County, the barefoot doctors were not willing to serve only as village health workers but proclaimed themselves to be “doctors,” even if barefooted. As Chen stated adamantly, “It was a cardinal principle in our system that the health worker should never act, or be called on to act, as a physician.” (83) From his point of view, this was a highly unfortunate transformation of his original concept of village health workers, because these barefoot doctors went beyond their limited training to practice medicine and treated patients with traditional or modern methods. See Chen, *Medicine in Rural China*, 130.

133. Anon., “Benhui di yibai wushi si ci zhenwen ti” 本會第一百五十四次徵文題 [The number 154 call for papers from our society], *Yixue zazhi* 醫學雜誌 [Journal of medicine], no. 87 (1936): 83.

134. Fan Guoyi 范國義, “Yiyao gongyou zhi zhi shishi jihua an” 醫藥公有制之實施計書案 [The plan for implementing state medicine], *Yixue zazhi* 醫學雜誌 [Journal of medicine], no. 87 (1936): 37–54, esp. 41; Shi Yiren 時逸人, “Yiyao gongyou zhi zhi shishi jihua an” 醫藥公有制之實施計書案 [The plan for implementing state medicine], *Yixue zazhi* 醫學雜誌 [Journal of medicine], no. 91 (1936): 1–11, esp. 2.

135. Quite a few practitioners of Chinese medicine supported Chen Guofu’s idea of building rural health clinics staffed by practitioners of Chinese medicine. See Ye Jinqiu 葉勁秋, “Chuangshe xiangchun yiyuan zhi jianyi” 創設鄉村醫院之建議 [Proposal for building rural health clinics], *Yijie chonqiu* 醫界春秋 [Annals of the medical profession], no. 105 (1935): 2–3, esp. 3.

136. Hu received his degree in Public Health from Berlin University, where he studied under Alfred Grotjahn (1869–1931), the great advocate of social hygiene and eugenics. See Hu Dingan 胡定安, *Zhongguo weisheng xingzheng sheshi jihua* 中國衛生行政設施計畫 [Plan for public health administration in China] (Shanghai: Shangwu yinshuguan, 1928).

137. Chen Guofu 陳果夫, “Jiangsu yizheng xueyuan de guoqu yu weilai” 江蘇醫政學院的過去與未來 [The past and future of the Jiangsu Provincial School of Medical Administration], in *Kukou tan yiyao* 苦口談醫藥 [Bitter chats on medicine], ed. Chen Guofu (Taipei: Zhengzhong shuju), 47–66, esp. 48.

138. Ibid.

139. Ibid., 51.

140. Chen Guofu 陳果夫, "Duiyu yixueyuan de qiwang" 對於醫學院的期望 [Expectations for medical schools], in *Kukou tan yiyao* 苦口談醫藥 [Bitter chats on medicine], ed. Chen Guofu 陳果夫 (Taipei: Zhengzhong shuju, 1949), 66–75, esp. 71.

141. Chen, "Jiangsu yizheng xueyuan de guoqu yu weilai," 65.

142. Ibid., 66.

143. Oldt, "State Medicine Problems," 800.

144. Anon., "*Jiaoyubu yixue weiyuanhui*" 教育部醫學委員會 [Committee on Medical Education, Ministry of Education], *Quanzonghao* 606, *Juananhao* 88, The Second Historical Archive of China, Nanjing.

145. Anon., "Jinggao quanguo zhongyishi tongrenshu" 敬告全國中醫師同仁書 [A letter to colleagues of Chinese medicine all over the country], *Jishi rbao* 濟世日報 [Jishi daily], October 31, 1947, 2.

146. In the proposal "Five-Year Plan for the Construction of Public Health" drafted in 1946, the first objective was "reducing the national mortality rate from thirty to fifteen per thousand of the population."

147. Chen Zhiqian 陳志潛, "Zhonguo gonggong weisheng yinggai zou yitiao xinlu" 中國公共衛生應該走一條新路 [We should find a new policy for China's public health], *Yichao* 醫潮 [Trend of medicine] 1, no. 5 (1947): 3–7, esp. 6.

CHAPTER 11

1. Zhang Zanchen 張贊臣, "Xuyan" 緒言 [Preface], *Yijie chunqiu* 醫界春秋 [Annals of the medical profession], no. 34 (1929): cover page.

2. Ibid.

3. Elisabeth Hsu, "Introduction for the Special Issue on the Globalization of Chinese Medicine and Meditation Practices," *East Asian Science, Technology and Society: An International Journal* 2, no. 4, (2009): 461–64; Joseph S. Alter, ed. *Asian Medicine and Globalization* (Philadelphia: University of Pennsylvania Press, 2005).

4. Michel Foucault, "Governmentality," in *The Foucault Effect: Studies in Governmentality*, ed. G. Burchell, C. Gordon, and P. Miller (Chicago: University of Chicago Press, 1991), 87–104.

5. Pierre Bourdieu, "Rethinking the State: Genesis and Structure of the Bureaucratic Field," *Sociological Theory* 12, no. 1 (1994): 1–18, esp. 16.

6. The 1935 Congress of the Nationalist Party passed a resolution entitled "Equal Treatment for Chinese and Western Medicine." In the following year, the government promulgated "Regulations for Chinese Medicine." In addition, the Ministry of Education promulgated "Temporary Outline for the Curriculum of Schools of Chinese Medicine" in 1939. Because of the beginning of the Sino-Japanese war in 1937, none of these regulations were ever put into practice.

7. John Fitzgerald, "The Misconceived Revolution: State and Society in China's Nationalist Revolution, 1923–26," *Journal of Asian Studies* 49, no. 2 (1990): 323–43, esp. 334–37.

8. In the case of India, for example, since the middle of the nineteenth

century "the possibility of incorporating indigenous practice into Western therapeutics . . . was ruled out." See Gyan Prakash, *Another Reason: Science and the Imagination of Modern India* (Princeton, NJ: Princeton University Press, 1999), 129.

9. Warwick Anderson, "The Possession of Kuru: Medical Science and Biological Exchange," *Comparative Studies in Society and History* 42, no. 4 (2000): 713–44, esp. 715.

10. Lu Yuanlei 陸淵雷, "Ni guoyiyao xueshu zhengli dagang caoan" 擬國醫藥學術整理大綱草案 [Proposal for putting in order Chinese medicine and pharmaceutics], *Shenzhou guoyi xuebao* 神州國醫學報 [Shenzhou bulletin of national medicine] 1, no. 1 (1932): 1–9, esp. 4.

11. Wang Shenxuan 王慎軒, "Qijing bamai zhi xinyi" 奇經八脈之新義 [A new meaning for the eight extraordinary vessels], in *Zhongyi xinlun huibian* 中醫新論彙編 [Collected essays on new discussions of Chinese medicine], vol. 1, ed. Wang Shenxuan 王慎軒 (Shanghai: Shanghai shudian, 1931), 39. See also Du Yaquan 杜亞泉, "Qi xue xinjie" 氣血新解 [New explanations for *qi* and *xue*], in *Zhongyi xunlun huibian*, 44–45.

12. Fu Yaocheng 傅嶢承, "Yingwei xinshi" 營衛新釋 [New explanations on ying and wei], in *Zhongyi xunlun huibian*, 40–42, esp. 41 (see n. 11 above).

13. Yang Zhiyi 楊志一, "Tiankui yu neifenmi" 「天奎」與「內分泌」 [Tiankui and the endocrine system], in *Zhongyi xunlun huibian*, 56–59 (see n. 11 above).

14. William C. Kirby, "Continuity and Change in Modern China: Economic Planning on the Mainland and on Taiwan, 1943–1958," *Australian Journal of Chinese Affairs*, no. 24 (1990): 121–41, esp. 121.

15. For example, in tracing the development of Chinese medicine since the Opium War (1839–42), Qian Xinzhong 錢信忠 (1911–2009), the minister of health of the People's Republic of China from 1965 to 1973, simply skips the Republican period and briefly characterizes Yu Yan's proposal for abolishing Chinese medicine as the policy of the Nationalist government. See Qian Xinzhong 錢信忠, *Zhongguo chuantong yiyaoxue fazhan yu xianzhuang* 中國傳統醫藥學發展與現狀 [Development and present situation of traditional Chinese medicine and pharmacy in China] (Taipei: Qingchun chubanshe, 1995), 42–43.

16. Kim Taylor, *Chinese Medicine in Early Communist China, 1945–63* (London: Routledge Curzon, 2005), 8.

17. Ibid.

18. Judith Farquhar, "Re-Writing Traditional Medicine in Post-Maoist China," in *Knowledge and the Scholarly Medical Traditions*, ed. D. Bates (Cambridge: Cambridge University Press, 1995), 251.

19. The reform-minded practitioners I have in mind include Lu Yuanlei 陸淵雷, Shi Jinmo 施今墨, Qin Bowei 秦伯未, and Cheng Menxue 程門雪. Scheid points out that "like most of his contemporaries, Cheng Menxue was prepared to surrender his autonomy in exchange for the state's commitment to protecting his tradition and establishing an institutional infrastructure that facilitated learning and research." See Volker Scheid, *Currents of Tradition in Chinese Medicine, 1626–2006* (Seattle, WA: Eastland Press, 2007), 326.

20. Taylor, *Chinese Medicine*, 151.

21. Prakash, *Another Reason*, 9.

22. Chen Zhiqian 陳志潛, "Dingxian shehui gaizao shiye zhong de baojian zhidu" 定縣社會改造事業中的保健制度 [Health care institutions in the Dingxian social reform], in *Xiangcunjianshe shiyan di er ji* 鄉村建設實驗第二集 [The experiment on rural construction: Volume 2], ed. Zhang Yuanshan 章元善 and Xu Zhonglian 許仲廉, (Shanghai: Zhonghua shuju, 1935), 459–73, esp. 463.

23. See John Grant, "State Medicine: A Logical Policy for China," *National Medical Journal of China* 14, no. 2 (1928): 65–80, esp. 75.

24. Taylor, *Chinese Medicine*, 33.

25. Although Mao Zedong did not put it in these words, his public speech entitled "The United Front in Cultural Work" in 1944 that touched upon the issue of traditional medicine was interpreted and popularized as a paired slogan: "scientization of Chinese medicine and popularization of Western medicine" (ibid., 17).

26. Farquhar, "Re-Writing Traditional Medicine," 261.

27. Volker Scheid, *Chinese Medicine in Contemporary China: Plurality and Synthesis* (Durham, NC: Duke University Press, 2002), 214.

28. Bruno Latour, *We Have Never Been Modern* (Cambridge, MA: Harvard University Press, 1993), 34.

29. Ibid., 99.

30. This is just one of many cases that reveal the interesting and potentially productive tension between the history of modern science in China and the STS approach. While STS scholarship has focused on making visible the hidden relationship between techno-science and its sociopolitical contexts, historical actors in the modern Chinese context sometimes explicitly promoted and valorized this relationship between science and politics. For an important effort to highlight this tension in the reevaluation of science in socialist China, see Sigrid Schmalzer, "On the Appropriate Use of Rose-Colored Glasses: Reflection on Science in Socialist China," *Isis* 98, no. 3 (2007): 571–83, esp. 581–82.

31. Latour, *We Have Never Been Modern*, 47.

32. Tan Cizhong 譚次仲, "Zaicheng yanjiuyuan lun yaowushiyan buyi hushi jingyan" 再呈研究院論藥物實驗不宜忽視經驗 [Resubmitting to the Academia Sinica: Pharmaceutical experiments should not ignore experience], in his *Yixue geming lunzheng* 醫學革命論爭 [Polemic of medical revolution] (Hong Kong: Qiushi chubanshe, 1952), 50–55; orig. pub. 1931.

33. For more information on the rise of experimental therapeutics, see Harry M. Marks, *The Progress of Experiment: Science and Therapeutic Reform in the United States, 1900–1990* (Cambridge: Cambridge University Press, 1997), 50; Miles Weatherall, "Drug Therapies," in *Companion Encyclopedia of the History of Medicine*, ed. W. F. Bynum and Roy Porter (London: Routledge, 1993), 915–38; K. K. Chen, ed. *The American Society for Pharmacology and Experimental Therapeutics, Incorporated: The First Sixty Years* (Washington, DC: Printed by Judd and Detweiler, 1969). Concerning Du Congming's remarkable and controversial proposal for the study of traditional medicine, see Sean Hsiang-lin Lei 雷祥麟, "Du Congming de hanyiyao yanjiu

zhi mi: Jianlun chuangzhao jiazhi de zhenghe yixue yanjiu" 杜聰明的漢醫藥研究之謎：兼論創造價值的整合醫學研究 [The enigma concerning Dr. Tsung-ming Tu's research of traditional East Asian Medicine: On the creation of value in integrative medicine], *Keji yiliao yu shehui* 科技、醫療與社會 [Techno-science, medicine and society] 11 (October 2010): 199–283.

34. Merlin L. Wilcox, Bertrand Graz, Jacques Falquet, Chiaka Diakite, Sergio Giani, and Drissa Diallo, "A 'Reverse Pharmacology' Approach for Developing an Anti-malarial Phytomedicine," Malaria Journal 10 (supp. 1), S 8 (2011): 1–10, esp. 1.

35. Latour, *We Have Never Been Modern*, 11.

36. Ibid., 113.

37. See Dipesh Chakrabarty, *Provincializing Europe: Postcolonial Thought and Historical Difference* (Princeton, NJ: Princeton University Press, 2000); Prakash, *Another Reason*.

Index

A page number followed by the letter f refers to an illustration.

Studies of the Weatherhead East Asian Institute
Columbia University

SELECTED TITLES

(Complete list at: http://www.columbia.edu/cu/weai/weatherhead-studies.html)

Beyond the Metropolis: Second Cities and Modern Life in Interwar Japan, by Louise Young. University of California Press, 2013.
Imperial Eclipse: Japan's Strategic Thinking about Continental Asia before August 1945, by Yukiko Koshiro. Cornell University Press, 2013.
The Nature of the Beasts: Empire and Exhibition at the Tokyo Imperial Zoo, by Ian J. Miller. University of California Press, 2013.
Reconstructing Bodies: Biomedicine, Health, and Nation-Building in South Korea Since 1945, by John P. DiMoia. Stanford University Press, 2013.
Tyranny of the Weak: North Korea and the World, 1950–1992, by Charles Armstrong. Cornell University Press, 2013
The Art of Censorship in Postwar Japan, by Kirsten Cather. University of Hawai'i Press, 2012.
Asia for the Asians: China in the Lives of Five Meiji Japanese, by Paula Harrell. MerwinAsia, 2012.
Lin Shu, Inc.: Translation and the Making of Modern Chinese Culture, by Michael Gibbs Hill. Oxford University Press, 2012.
Occupying Power: Sex Workers and Servicemen in Postwar Japan, by Sarah Kovner. Stanford University Press, 2012.

Redacted: The Archives of Censorship in Postwar Japan, by Jonathan E. Abel. University of California Press, 2012.

Empire of Dogs: Canines, Japan, and the Making of the Modern Imperial World, by Aaron Herald Skabelund. Cornell University Press, 2011.

Planning for Empire: Reform Bureaucrats and the Japanese Wartime State, by Janis Mimura. Cornell University Press, 2011.

Realms of Literacy: Early Japan and the History of Writing, by David Lurie. Harvard University Asia Center, 2011.

Russo-Japanese Relations, 1905–17: From Enemies to Allies, by Peter Berton. Routledge, 2011.

Behind the Gate: Inventing Students in Beijing, by Fabio Lanza. Columbia University Press, 2010.

Imperial Japan at Its Zenith: The Wartime Celebration of the Empire's 2,600th Anniversary, by Kenneth J. Ruoff. Cornell University Press, 2010.

Passage to Manhood: Youth Migration, Heroin, and AIDS in Southwest China, by Shao-hua Liu. Stanford University Press, 2010.

Postwar History Education in Japan and the Germanys: Guilty Lessons, by Julian Dierkes. Routledge, 2010.

The Aesthetics of Japanese Fascism, by Alan Tansman. University of California Press, 2009.

The Growth Idea: Purpose and Prosperity in Postwar Japan, by Scott O'Bryan. University of Hawai'i Press, 2009.

Leprosy in China: A History, by Angela Ki Che Leung. Columbia University Press, 2008.

National History and the World of Nations: Capital, State, and the Rhetoric of History in Japan, France, and the United States, by Christopher Hill. Duke University Press, 2008.